Smart Agriculture for Sustainable Practices

This book offers insights into how modern technologies are transforming agriculture with a focus on their effects in aiding sustainability. It explores the current applications of technologies such as AI, IoT, and machine learning, and their impact on improving crop yield, resource efficiency, and environmental impact. Through practical examples and case studies, it demonstrates how these technologies can address challenges in agriculture and enhance operational efficiency.

Key Features:

- Analysis of how IoT devices and sensor networks are used for real-time monitoring of farm operations, soil conditions, and weather patterns.
- Studies emerging technologies like machine vision, predictive modelling, and natural language processing for tasks such as crop monitoring and pest detection.
- Detailed discussion on machine learning algorithms, including supervised, unsupervised, and reinforcement learning, tailored for agricultural applications.
- Insights into how AI, IoT, and machine learning optimize resource use, improve crop yield and quality, and enhance sustainability.
- Case studies to highlight real-life examples from crop management to livestock and water conservation.

Smart Agriculture for Sustainable Practices is for researchers, professionals, and policymakers interested in leveraging technology for sustainable agricultural practices.

Computational Intelligence in Sustainable Agriculture

Series editors:
Souvik Pal, Sister Nivedita University, Kolkata and Basim Alhadidi, Al-Balqa Applied University, Jordan

This series explores the critical role of computational intelligence in advancing sustainable agriculture.

Smart Agriculture for Sustainable Practices
The Role of AI, IoT, and Machine Learning
Edited by S. N. Kumar, Sherin Zafar, Arun S, Sameena Naaz, M. M. Sufyan Beg, Eduard Babulak

For more information about this series, please visit https://www.routledge.com/Computational-Intelligence-in-Sustainable-Agriculture/book-series/CISA

Smart Agriculture for Sustainable Practices

The Role of AI, IoT, and Machine Learning

Edited by
S. N. Kumar
Sherin Zafar
Arun S
Sameena Naaz
M. M. Sufyan Beg
Eduard Babulak

CRC Press
Taylor & Francis Group
Boca Raton London New York

CRC Press is an imprint of the
Taylor & Francis Group, an **informa** business
A CHAPMAN & HALL BOOK

Front cover image: Mitthil Vectx/Shutterstock

First edition published 2026
by CRC Press
2385 NW Executive Center Drive, Suite 320, Boca Raton FL 33431

and by CRC Press
4 Park Square, Milton Park, Abingdon, Oxon, OX14 4RN

CRC Press is an imprint of Taylor & Francis Group, LLC

ISBN: 978-1-032-97501-6 (hbk)
ISBN: 978-1-032-97515-3 (pbk)
ISBN: 978-1-003-59409-3 (ebk)

DOI: 10.1201/9781003594093

Typeset in Times
by SPi Technologies India Pvt Ltd (Straive)

Contents

About the Editors

S. N. Kumar is an Associate Professor in the Department of Electrical and Electronics Engineering at Amal Jyothi College of Engineering. He holds a PhD in Electronics Engineering from Sathyabama Institute of Science and Technology, where his research focused on the development of novel medical image processing segmentation and compression algorithms for telemedicine applications. Dr Kumar has a Master's degree in Applied Electronics and is highly distinguished in his academic achievements, having received top marks and first-class honors throughout his education. With a rich academic career, he has contributed significantly to the field of medical image processing, including segmentation and compression techniques for CT/MRI images. Dr Kumar has also been actively involved in funded projects, including a Co-Principal Investigator role for a DST-sanctioned project. Additionally, he has published numerous high-impact research papers and holds several patents in areas like green energy harvesting, machine learning, and IoT applications in healthcare. His work continues to shape innovations in telemedicine and AI-based medical image analysis.

Sherin Zafar holds a PhD in Computer Science from MRIU Faridabad, specializing in secure protocols for MANET and blockchain applications. She also earned her MTech in Computer Engineering and BE in Computers. Dr Zafar has a significant research profile with over 125 citations on Google Scholar and an h-index of 7. She has actively supervised PhD students and contributed to several research projects, including a DST-funded project. Additionally, Dr Zafar is a joint supervisor for several PhD students working on topics such as trust-based optimized routing approaches for MANET and blockchain integration. As an Assistant Professor at Jamia Hamdard, she has a strong academic and research background, actively participating in conferences, reviewing papers for multiple journals, and being a member of professional bodies such as CSI and IACSIT. Her work includes multiple published patents, demonstrating her contributions to the field of secure communication protocols and innovative routing solutions.

Arun S is an Associate Professor in the Department of Electrical & Electronics Engineering at Amal Jyothi College of Engineering, Kottayam, Kerala, with 19 years of academic experience. He earned his BE degree in Electronics and Communication Engineering from Madurai Kamaraj University in 2003, followed by an ME degree in Power Electronics and Drives from Anna University in 2005. He completed his PhD in Power Electronics from APJ Abdul Kalam Technological University in 2023. Dr Arun also holds an advanced certificate in Artificial Intelligence & Machine Learning from the IBM Innovation Centre for Education. His research interests are centered around power converters, solar PV systems, embedded systems, and SoPC design. He has contributed to various peer-reviewed journals, book chapters, and conference proceedings. In addition, he has conducted peer reviews for several journal papers, further demonstrating his involvement in the academic and research community.

Sameena Naaz is a distinguished academic with over 22 years of experience in teaching, research, and administration. She earned her PhD in Computer Science from Jamia Hamdard University, focusing on fuzzy-based algorithms for load balancing in distributed environments. In addition to her academic achievements, Dr Naaz has a robust publication record, contributing to well-regarded journals and conferences in the fields of Artificial Intelligence, medical image processing, and telemedicine. She is also involved in several innovative projects related to IoT and AI applications in health and education. Currently, she is a Senior Lecturer at the University of Roehampton, London, and has served as a professor at Jamia Hamdard. Dr Naaz holds multiple patents in the areas of robotics, IoT, and health metrics, underlining her role as an innovator in her field. Her passion for mentoring students and contributing to administrative responsibilities further strengthens her profile in academia.

M. M. Sufyan Beg is an accomplished academic with extensive experience in the field of Computer Engineering. He holds a PhD in Computer Technology from IIT Delhi and has completed postdoctoral research at UC Berkeley. Currently serving as the Principal of Zakir Husain College of Engineering and Technology at Aligarh Muslim University (AMU), he has held several leadership roles, including Chair of the Computer Engineering Department. With a career spanning over two decades, he has written numerous publications and contributed to various conferences, specializing in such areas as web search quality, fuzzy logic, machine learning, and data mining. He has also been involved in advanced research in cloud computing, IoT, and electric vehicles.

Eduard Babulak is a is a globally recognized scholar, researcher, consultant, and professional engineer with over 30 years of international experience. A polyglot fluent in 16 languages, he has served in esteemed academic, editorial, and leadership roles across renowned institutions such as MIT, University of Cambridge, Tokyo, Oxford, and Purdue. Prof. Babulak is a Fellow of the Royal Society of Arts (UK), a Chartered IT Fellow and Elite Group Member of the British Computer Society, a Senior Member of IEEE and ACM, and an expert advisor for the European Commission and the US National Science Foundation. He holds advanced degrees from institutions in the UK, Czech Republic, and Slovakia, and has contributed significantly to engineering, computer science, and global educational initiatives.

Contributors

Abubakar Kabir Adam
Department of Food Science and Technology, School of Agriculture, SR University
Warangal, India

Naved Alam
Department of Computer Science, Jamia Hamdard
New Delhi, India

Ghadah Naif Alwakid
Department of Computer Science, College of Computer and Information Sciences, Jouf University
AI Jouf, Saudi Arabia

Ankit
Department of Horticulture (Fruit Science), Lovely Professional University
Jalandhar, India

A. Anushya
College of Computer Science and Engineering, University of Hail
Saudi Arabia

Roshini Bakur
Department of Computer Science, Christ Academy Institute for Advanced Studies
Bangalore, India

Isha Batra
School of Computer Science and Engineering, Lovely Professional University
Punjab, India

Poonam Chaudhary
CSE Department, The NorthCap University
Gurgaon, India

P. K. A. Chitra
SRM Institute of Science land Technology
Tiruchirappalli, India

Monika Dandotiya
Poornima University
Jaipur, India

Dhiviya Bharathi Duraimanickam
School of Computing, SASTRA University
Thanjavur, India

Nikhil Kumar Goyal
Poornima University
Jaipur, India

Ambreen Hamadani
Sher-e-Kashmir University of Agricultural Sciences and Technology of Kashmir
Kashmir, India

Henna Hamadani
Sher-e-Kashmir University of Agricultural Sciences and Technology of Kashmir
Kashmir, India

Himanshu
Department of Horticulture (Fruit Science), Lovely Professional University
Jalandhar, India

S. Jagir Hussain
B S Abdur Rahman Crescent Institute of Science and Technology, Vandalur
Chennai, India

Khan Jabroot
Department of Horticulture (Fruit Science), Lovely Professional University
Jalandhar, India

Dhayalini K
K. Ramakrishnan College of Engineering, Tiruchirappalli
Tamil Nadu, India

Sneha Kandacharam
CSE Department, The NorthCap University
Gurgaon, India

Abirami Karthikeyan
School of Electrical and Electronics Engineering, SASTRA Deemed to be University
Thanjavur, Tamil Nadu, India

Rahmatullah Kazimi
Department of Horticulture, Daikundi of Higher Education Institution
Daikundi- 2402, Afghanistan

Joseph Abraham Sundar Koilraj
School of Computing, SASTRA University
Thanjavur, India

B. Santhana Krishnan
Mississippi State University
Mississippi, USA

S. N. Kumar
Amal Jyothi College of Engineering
Kanjirappally, India

Singh Shailesh Kumar
Department of Horticulture, School of Agriculture, Lovely Professional University
Jalandhar, India

Monika Kumari
Poornima University
Jaipur, India

Rakshith M
Department of MCA, BMS Institute of Technology & Management
Bangalore, India

Arun Malik
School of Computer Science and Engineering, Lovely Professional University
Punjab, India

Midhun P Mathew
Dept. of Computer Science, Amal Jyothi College of Engineering (Autonomous)
Kottayam, India

Aquib Mustafa
Sensia (Rockwell Automation + Schlumberger)
United States of America

Rajshekhar M Patil
Department of ISE, Don Bosco Institute of Technology
Bangalore, India

Shantakumar B Patil
Department of CSE, ISE, Sai Vidya Institute of Technology
Bangalore, India

Ananda R
Department of CSE, BMS Institute of Technology & Management
Bangalore, India

Muthaiah Rajappa
School of Computing, SASTRA University
Thanjavur, India

A. Rajesh
School of Electrical and Electronics Engineering, SASTRA Deemed to be University
Thanjavur, Tamil Nadu, India

Jerina Rugji
University of Wisconsin-Madison
Wisconsin, USA

Abhinand S
Dept. of Computer Science, Amal Jyothi College of Engineering (Autonomous)
Kottayam, India

Muneshwara M S
Department of CSE, BMS Institute of Technology & Management
Bangalore, India

Elakkiya Rajasekar
Department of Computer Science, Birla Institute of Technology & Science
Pilani, Dubai Campus, Dubai, UAE

Tithli Sadhu
Department of Food Science and Technology, School of Agriculture, SR University
Warangal, India

Abuzar Saeed
Department of Computer Science, Jamia Hamdard
New Delhi, India

Tanushree Sanwal
KIET School of Management, KIET Group of Institutions
Delhi, India

Shikha Sharma
Poornima University
Jaipur, India

Amal Sasikumar
Dept. of Computer Science, Amal Jyothi College of Engineering (Autonomous)
Kottayam, India

Deepika Saxena
Department of Horticulture, School of Agriculture, Lovely Professional University
Jalandhar, India

Alan Sebastian
Dept. of Computer Science, Amal Jyothi College of Engineering (Autonomous)
Kottayam, India

Seval Selimoglu
Anadolu University
Eskisehir, Turkey

Prabu Selvam
SRM Institute of Science and Technology
Tiruchirappalli, India

Shuprajhaa T
ICAR – National Research Centre for Banana, Tiruchirappalli
Tamil Nadu, India

Shrishti Singh
KIET School of Management, KIET Group of Institutions
Delhi, India

Sofia Singla
University Institute of Computing, Chandigarh University
Mohali, India

C. Sivananth
ICAR – National Research Centre for Banana, Tiruchirappalli
Tamil Nadu, India

Navdeep Singh Sodhi
University Institute of Computing, Chandigarh University
Mohali, India

Suresh Kumar P
ICAR – National Research Centre for Banana, Tiruchirappalli
Tamil Nadu, India

Swastika
The NorthCap University
Gurgaon, India

Shivakumara T
Department of MCA, BMS Institute of Technology & Management
Bangalore, India

Abhishek Tanwer
Lincoln Institute for Agri-Food Technology, School of Agri-Food Technology and Manufacturing, University of Lincoln
United Kingdom

Adheena Thomas
Dept. of Computer Science, Amal Jyothi College of Engineering (Autonomous)
Kottayam, India

Priya Thomas
Department of MCA, New Horizon College of Engineering
Bangalore, India

Meenakshi Tyagi
KIET School of Management, KIET Group of Institutions
Delhi, India

S. Vijayanand
Sri Venkateswara College of Engineering, Sriperumbudur
Chennai, India

Ajantha Vijayakumar
School of Computing, SASTRA University
Thanjavur, India

Kaebeh Yaeghoobi
Faculty of Computer Engineering, K. N. Toosi University of Technology
Tehran, Iran

Resül Yazici
Bilecik Seyh Edebali University
Turkey

Idris Kaida Zubairu
Department of Food Science and Technology, Federal University Dutsin-ma
Katsina, Nigeria

1 Optimizing Agricultural Supply Chains through AI and IoT
A Pathway to Sustainable Practices

Shrishti Singh, Meenakshi Tyagi, and Tanushree Sanwal

1.1 INTRODUCTION

The consumer is diverse. A very crucial point about agriculture nowadays is that it has to undergo many ups and downs to feed the growing food demands and, on the other hand, save environmental sustainability. The Food and Agriculture Organization (FAO) projects that our world will have a population of 9.7 billion people by the year 2050, which will increase food needs, and calls for agriculture to be more efficient, sustainable, and resilient. At the same time, many agricultural systems around the world are already facing immediate challenges on multiple fronts – resource management, climate change, and food security – all of which require attention and action to meet this increased demand for food production [1].

1.1.1 Agricultural Challenges

Resource management is one of the main challenges for modern agriculture. The United Nations has indicated that the top restrictions to agricultural productivity are water shortages and land use practices. Across the globe, water resources are strained with over-extraction for irrigation in some regions and a lack of water access for crops in others. In addition, soils are being degraded by bad agricultural practices and deforestation, which puts agricultural land at risk of being unable to sustain food production in the future. What might be an even more formidable obstacle to agriculture is climate change. Extreme weather events such as droughts and floods are hitting the globe regularly, upending farming systems. The FAO said some immediate effects of these shifts in weather patterns are already having an effect on crops and food production. Food security could be further aggravated, as the World Bank estimates that by the year 2050, climate change may lead to a decline of up to 30% in agricultural productivity across numerous developing countries [2].

DOI: 10.1201/9781003594093-1

1.1.2 AI and IoT Solutions

In order to tackle these challenges, modern technologies such as artificial intelligence (AI) and the Internet of Things (IoT) are now helping in the modernization of agriculture. These technologies are creating new ways to be more productive, minimize waste, and enhance sustainability. Through predictive analytics and machine learning, AI helps farmers mitigate risk and make more informed decisions in real time. For instance, AI-based solutions can estimate the production of crops, predict climatic changes, and enhance resource utilization. In the United States, the National Agricultural Library (NAL) puts out a lot of reports on how AI is transforming agriculture. According to these reports, AI can improve precision agriculture by enabling farmers to fine-tune planting times, irrigation and fertilization practices, and harvesting schedules for optimum yields with minimal wastage of resources [3].

At the same time, IoT is revolutionizing agriculture by allowing farmers to monitor their crops with sophisticated monitoring systems. Farmers can identify and analyze many environmental variables through IoT devices, including sensors, drones, and connected machines for soil moisture and temperature, together with pest activity surveys. Real-time insight enables a more precise management approach to farming, reducing associated costs and improving efficiency, as noted by the U.S. Department of Agriculture (USDA). Thanks to IoT applications, farmers can also check on their fields from virtually anywhere in the world, so they will be alerted early to issues such as pest infestations or water shortages. The combination of AI with IoT can strengthen the resilience of agriculture against climate change and resource constraints, allowing farmers to adapt to uncertainty while also continuing their farming activities. Together, these technologies pave a way forward for agriculture to go into relatively more sustainable practices, associating food security with the global climate and environmental character.

1.2 CURRENT AGRICULTURAL SUPPLY CHAINS

Agricultural supply chains link farmers to consumers, markets, and suppliers and play a key role in the global food system. But these supply chains also face huge challenges from production and distribution inefficiencies to global trade and market accessibility. Against the backdrop of an ever-increasing demand for food worldwide, it is becoming increasingly urgent to deal with these inefficiencies. There has been a plethora of research from the likes of the World Bank and Organisation for Economic Co-operation and Development (OECD) as they found that conventional agricultural supply chains are sub-optimal and tech integration is increasingly urgent [4].

1.2.1 Traditional Supply Chain Issues

The most common problem with the typical agricultural supply chain is that it is inefficient. The supply chains are very long with multiple intermediaries between the farmer and the consumer, which can result in higher costs, lower margins for producers, and increased food waste. As the World Bank has pointed out, in many developing countries, inefficiencies in storage, transport, and distribution are as much a barrier to food security as are production constraints. For example, the lack of proper infrastructure, like refrigeration and transportation networks, leads to huge amounts of harvested crops rotting before they get to market. The FAO states that around 14% of the food produced worldwide is lost from harvest to retail, mostly attributed to inadequate infrastructure and inefficient supply chains. The fragmentation of agricultural markets is another crucial challenge [2]. High logistics and economic costs preclude large-scale markets for many farmers, particularly small-scale producers from developing countries. While they could be highly competitive in a globalizing agricultural market, they will remain mired in trade restrictions because they tend to fail quality standards or struggle to reach consumers even when regulations permit trade [4, 5]. Consequently, world agrarian trade continues to be imbalanced, with justified pricing for producers in certain regions becoming a real challenge (Table 1.1).

Climate change further complicates supply chain stability. The unpredictability of weather events and changing seasonal patterns directly affect crop yields, adding another layer of uncertainty to the traditional supply chain. Droughts, floods, and other extreme weather events can damage crops, disrupt transportation routes, and cause supply shortages, which, in turn, raise prices and reduce food availability. As reported by the OECD, climate change has the potential to exacerbate these supply chain disruptions, further straining global food systems.

1.2.2 Need for Technological Integration

With such inefficiencies and pain points, experts are increasingly agreeing that integrating technology remains critical to agriculture supply chain modernization. The FAO, along with the USDA, has highlighted the importance of high-end technology to boost productivity, gain better access to markets, and curtail waste in agricultural systems. There are a lot of developing needs in the industry, and it is exploring new technologies

TABLE 1.1
Agricultural Supply Chain Flow

Stage	Description	Challenges
Farm Production	Initial stage of growing crops or raising livestock	Limited access to resources, weather impact
Post-Harvest Handling	Sorting, cleaning, and packing	Storage facilities, waste management
Transportation	Movement from farm to processing facility or market	Infrastructure issues, transport costs
Processing	Turning raw materials into finished products	Energy consumption, processing inefficiency
Distribution	Wholesale and retail delivery of products	Distribution inefficiencies, lack of cold chain
Consumer	End consumption of agricultural products	Pricing, availability, food security

Adapted from FAO's "Agricultural Supply Chains: Global Overview" [1, 6–8].

TABLE 1.2
Agricultural Trade and Market Distribution (World Bank Agricultural Trade Reports) [3, 9–11]

Region/Country	Agricultural Exports (in Billion USD)	Agricultural Imports (in Billion USD)	Trade Balance (Billion USD)
North America (USA, Canada)	182.4	98.1	+84.3
European Union	145.6	106.3	+39.3
China	102.7	131.4	−28.7
India	40.2	18.7	+21.5
Brazil	85.9	17.2	+68.7

in order to optimize the agricultural work with a variety of operations to develop agricultural processes that also eliminate wastes, improve efficiency of resources, and strengthen relationships between farmers and consumers through digital links with the help of the Internet such as integrating AI, IoT, and blockchain technologies [3, 9]. For example, AI-based platforms will help farmers with an efficient prediction of crop yields for a balanced harvest and minimizing wastage. IoT devices such as sensors and tracking technologies can keep track of the movement of goods along the supply chain, thereby enhancing transparency and traceability. In addition, as the USDA notes, this is in part thanks to technologies that allow real-time information about transportation conditions of products, minimizing the risk of contamination or spoilage [10].

Another emerging solution to improve supply chain efficiency is blockchain technology, which provides secure and transparent record-keeping. With the potential to provide improved traceability of products from farm to market, blockchain technology has been tipped as a way to help combat incidences of food fraud, improve trust between stakeholders, and enable consumers to make more informed choices about where their food comes from and how it is produced. According to a number of reports by FAO, this blockchain decentralized system, in which everyone can have their own market based on transparency and efficiency, is a potential game changer for agricultural supply chains (Table 1.2).

It shows that, especially in emerging nations, traditional agricultural supply chains suffer from inefficiency and a lack of integration. However, the possibility of technological solutions – from blockchain to AI and IoT – offers a chance to update these systems, enhancing market accessibility, productivity, and the agricultural supply chain's overall resilience.

1.3 AI AND IOT SOLUTIONS

1.3.1 Predictive Analytics and Real-Time Monitoring

AI and IoT applications – where unthinkably smart machines learn from millions of observations, integrating the ability to act autonomously with data and experience – are now a part of our daily lives to help us grow more food using fewer natural resources. This technology is making a significant impact in predictive analytics and real-time monitoring, helping farmers to make educated decisions and respond to challenges sooner.

Agricultural predictive analytics is the use of big data, algorithms, and machine learning models to predict what will happen in the future. Predictive analytics allows you to predict crop yields, pest outbreaks, weather patterns, and market trends by using historical data and observing the connections among the observations. The USDA, for example, employs predictive models to forecast crop yields and identify threats to food security [1]. This way, farmers can modify their practices when the advice comes before drought or disease hits hard enough to reduce yields. In addition to these, predictive analytics can be used for making sure that resources are allocated optimally and in a planned manner. Farmers can use weather forecasts, soil moisture level predictions, and growth stage data of crops to create simple or advanced irrigation schedules, plan fertilizer use, and reduce pesticide application. The National Institute of Food and Agriculture (NIFA) has funded several activities to strategically apply AI and machine learning for precision agriculture [2]. The goal of these initiatives is to drive efficiency and lower environmental impacts through data-driven predictions that inform agricultural operations. The data sensor communicates with cloud-based platforms and AI algorithms and downloads the information in real time to implement actionable insights.

Real-time monitoring – While predictive analytics deals with the future, real-time monitoring complements it and offers current information on where crops, soil, and weather really are. The IoT devices, including sensors and drones, determine temperature, humidity, soil moisture, and light intensity on a constant basis. The sensors collect the data, and it is sent to cloud-based platforms, which use AI algorithms to analyze the information coming from various endpoints in real-time, providing actionable insights. The U.S. Department of Energy's Office of Energy Efficiency & Renewable Energy (EERE) talks about IoT on its website – in agriculture, for example, where it monitors conditions on the farm and allows farmers to adjust accordingly [10].

For instance, soil-based IoT sensors can measure moisture levels and notify you when it is time to irrigate. With drones, farmers can get an aerial view of their fields to detect areas where there are pest infestations or nutrient deficiencies. Such quick feedback enables timely and better targeted interventions, ensuring a healthier crop while reducing the resource inputs.Farmers can take proactive rather than reactive approaches to problems with the implementation of AI-driven predictive models and IoT-based monitoring systems. Especially in the age of climate change that can impact crop yield, such proactivity is a top priority. This also increases the resilience of agricultural systems by allowing farmers to see changes in weather patterns and adapt accordingly, often within a near-real-time window.

1.3.2 Enhancing Efficiency and Reducing Waste

AI and IoT technologies contribute significantly to faster operation and also help to reduce wastage in the agricultural supply chain. Food waste represents one of the most severe problems in food production, occurring at various stages all throughout the supply chain – including production, distribution, and consumption. The FAO estimates that about one-third of all food produced globally is wasted [3]. This report finds that AI and IoT technologies can mitigate this problem by making supply chains more efficient, preventing overstocking, and assisting food to end up on people's tables rather than in the landfill.

The supply chain management in agriculture can be greatly discussed under the implementation of AI and IoT. Such real-time monitoring of goods moving through the supply chain allows for greater transparency and helps eliminate inefficiencies. For instance, smart sensors can be used to monitor the state of perishable items like fruits and vegetables while in transit. Such sensors can monitor temperature, humidity, and vibration levels to ensure that food is stored under optimal conditions, reducing spoilage and waste.

Alternatively, demand forecasting and logistics optimization are some of the key areas in which one can use AI. Based on historical sales data, weather predictions, and consumer behavior, AI models can make future demand forecasting much more accurate. Such forecasts assist farmers and suppliers to plan the production and distribution schedules such that there is minimal overproduction, or underproduction – one of the root causes of waste. As we covered earlier, the USDA's Economic Research Service (ERS) had studies that demonstrated the value of technology in further reducing wastage to better market access and forecasting that AI these days validates and optimizes supply chain process outputs [6].

Not only do they minimize the waste during transportation, but even at the farm level through AI and IoT. Using data from IoT sensors, precision farming tools allow farmers to better apply inputs like water, fertilizer, and pesticides. However, AI can significantly help farmers analyze and map this data so that they might apply the right inputs exactly in the right places. It also avoids waste where it is least needed, thus saving cost and minimizing environmental impact. The FAO has been promoting such technologies because they are capable of efficiently improving resource management, especially in areas where issues related to water scarcity and soil degradation seem to occur most commonly [4].

This process of waste reduction leads to the saving of not only food but also inputs used in farming. Accurate tracking of crop growth, soil conditions, and environmental factors using AI and IoT prevents unwanted use of water and fertilizers. This eliminates runoff that can easily pollute water systems and lowers the carbon footprint of the agricultural practice. With such AI-driven insights, the farmer can then make smarter decisions on applying inputs only when necessary and in the best way possible. Similarly, the same IoT systems can help in monitoring and optimizing the efficiency of irrigation systems. The smart irrigation system applies water based on real-time measurement of soil moisture content and local weather conditions to avoid unnecessary losses of water because of overuse in areas that have a perpetual drought condition or even short supplies of water. These devices may also be equipped with weather forecasts for the timing of rain showers or events, which can adjust the schedule so that overwatering will not be experienced because of precipitation. Extended to consumer behavior, AI, and IoT can play a part in cutting down the quantity of food wasted. Analysis of consumer trends as well as preferences will enable producers to better adjust their supply to meet demand. Data analytics can also identify those areas in the supply chain where waste is most probable and, therefore, allows for more targeted interventions to prevent it.

1.4 BENEFITS OF AI AND IOT IN AGRICULTURE

Agriculture is one of the most important sectors that maintains life around the globe, yet the sector still faces more challenges than ever, accompanied by the spread of climate change and population growth, along with the urgent need for more sustainable agriculture. With this in mind, agriculture has immense opportunities to integrate AI and the IoT at different phases in improving productivity, sustainability, and resource efficiency. This section unveils the latest developments in how AI and IoT technologies upgrade agricultural practice, especially within the context of improved resource management and through case studies of real-world applications.

1.4.1 Improved Resource Management

Among the main drivers in agricultural resource management for more resource-efficient means is the implementation of AI and IoT. Resource management in agriculture refers to the preservation of both water and energy, as well as proper care of soil health. These are still essential factors for proper farming. When gauged against the inefficient and wasteful traditional agricultural practices, it's within these technologies that are transforming them through more efficient usage of resources and less environmental impact.

1.4.1.1 Water Management

Perhaps the largest challenge in contemporary agriculture is that of water scarcity. Water is the basic ingredient for irrigation purposes, and for irrigation alone, 70% of the world's freshwater supply is used. However, most of this water is wasted because of improper irrigation techniques. The USDA Natural Resources Conservation Service guidelines on using modern irrigation methods include automated irrigation through IoT sensors [12]. These sensors capture the real-time rate of soil moisture and feed this back into AI systems to analyze the data. The AI model then optimizes the irrigation schedule so that the right amount of water reaches the crops, thus saving water and making irrigation much more efficient. IoT-enabled smart irrigation systems can also be linked to weather forecasting data. The predictability of these systems regarding rainfall enables them to automatically alter their schedules so that the water does not get applied in cases where it is probably going to rain. This technology helps prevent the waste of precious water resources, which may be a huge problem in dry or water-scarce areas.

1.4.1.2 Soil Health and Fertility

The health and fertility of soil are essential for long-term agricultural sustainability. Soils are degraded through traditional farming due to poor fertility, resulting in a low yield. AI and IoT allow farmers to monitor soil conditions in real time and adjust to better health. For example, sensors placed into the soil measure variables such as pH, moisture, temperature, and nutrient levels. Through the analysis of data collected through these sensors, AI systems identify the peak levels of nutrients so that fertilizers are applied precisely instead of too much, thus not having a high impact on running into nearby water bodies and enhancing the health of the soil over time.

The FAO report on sustainable soil management emphasized the role of technology in addressing organic matter depletion and erosion [13]. Tools powered by AI allow farmers to monitor the status of soils in real time and then apply practices having long-term benefits for soil fertility, such as crop rotation and precision fertilization.

1.4.1.3 Energy Efficiency

Another area for improvement through AI and IoT is in the use of energy in agriculture. Traditional farming operations, including machinery operation, irrigation, and harvesting, often depend on fossil fuels and inefficient energy systems. Through smart technologies, farmers can trace and monitor their consumption of energy across various operations and find loopholes in efficiency. For example, the IoT sensors installed on farm machines may track data on fuel utilization and the performance of these machines. The AI networks analyze this data and identify alterations in the functioning of the machines, which help to reduce energy wastage, thus conserving money on fuel consumption and minimizing the carbon impact. The USDA has sponsored research initiatives that have incorporated the use of AI for energy-efficient agriculture, promoting the use of renewable sources of energy and smart grid technologies for efficient provision of power to their farm infrastructure [7] (Table 1.3).

With these technologies, the farm will be able to minimize dependence on external resources, cutting costs as well as maximizing production. This will go hand in hand in ensuring sustainable use of energy and contributing to a greener model of farming with low carbon dioxide emissions [1].

1.4.2 Case Studies: Real-World Applications

AI and IoT technology in farming is extremely practical and already exemplified by many real-world applications in bringing about the value it proves. The following case studies reflect how these technologies are turning farming practices upside down, improving yields, saving waste, and contributing to sustainability.

TABLE 1.3
Examples of Resource Optimization through AI and IoT in Agriculture [10, 13, 14]

Resource	Technology/Method	Reduction in Resource Use	Source
Water	IoT soil moisture sensors	20–30% reduction in water use	USDA Natural Resources Report
Fertilizer	AI-driven soil nutrient analysis	15–25% reduced fertilizer use	FAO "Soil Health" Publication
Energy	AI in greenhouse climate control	10–15% energy savings	National Institute of Food and Agriculture (NIFA)
Pesticides	AI-powered pest detection systems	Up to 40% reduction	European Commission Agriculture

1.4.2.1 Case Study 1: Smart Irrigation in California

In farming operations where the state bases part of its operation, smart irrigation technologies have been of major importance because it's a place of periodic drought. A good example is the deployment of IoT-based irrigation systems in vineyards that rely for their irrigation on soil moisture sensors, weather data, and AI models to optimize the use of the water source. The amount of water needed can be determined based on real-time data, with a minimal amount of waste. The California Department of Water Resources is collaborating with other local agricultural companies to test and implement the technologies on thousands of acres. Major outcomes include saving up to 20% in water usage while maintaining a higher quality grape product [5]. This case study indicates the possibility of AI and IoT for the solution of water scarcity problems, making farming operational and economically viable in regions that face challenges with regard to water.

1.4.2.2 Case Study 2: Precision Farming in India

India, among the biggest producers of agriculture in the world, uses AI and IoT to execute its mission through high productivity, efficient resource usage, and food security. The implementation of AI and IoT in international agriculture has been led by the FAO. Through flagship projects such as 1000 Digital Villages, FAO is trying to harness the digital divide by bringing smart technology into rural villages. The Water Scarcity Initiative and the e-Agriculture Strategy Guide are two projects that focus on leveraging digital solutions to enhance resource use efficiency and climate resilience. In addition, the FAO Geospatial Platform supports data-driven decision-making for land and resource planning, and platforms such as AMIS (Agricultural Market Information System) apply AI to increase market transparency. Precision farming is being actively fostered in India by the Indian Agricultural Research Institute (IARI) in collaboration with the FAO [2]. Intelligent irrigation systems and AI-based nutrient management technologies have been implemented in rural areas as part of the Sustainable Agriculture Program. Smallholder farmers have gained financially and environmentally due to these innovations, which have caused fertilizer use efficiency to rise by 25% and water use to fall by 30% [11]. Such programs are an indication of how dedicated FAO is towards improving sustainable agriculture through partnerships and innovation. This case study shows how technology can be the driver of sustainable farming practices to enhance the livelihoods of smallholder farmers.

1.4.2.3 Case Study 3: AI for Crop Prediction in Sub-Saharan Africa

In Sub-Saharan Africa, food insecurity has been a major emphasis, and AI/IoT were applied in improving agricultural productivity and access to market. FAO has supported several projects based on AI-driven tools that produce crop yield and weather pattern forecasts. Countries such as Kenya and Uganda use AI systems that process the satellite data, weather reports, and historical crop data to forecast production and let farmers plan ahead. This technology allows better decisions on planting, harvesting, and selling crops to avoid crop failure and post-harvest losses. It also helps improve market access by sharing accurate crop forecasts with local markets

TABLE 1.4
AI and IoT Case Studies in Agriculture by Region [11, 12, 16, 17]

Region	Application	Technology	Outcome
India	Precision farming in rice fields	IoT moisture and temperature sensors	Increased yield by 20%
Brazil	Coffee crop monitoring	Drones and climate data	15% reduction in crop loss
United States	Vineyard health tracking	AI-based imaging for pest detection	Reduced pesticide use by 30%
Kenya	Real-time crop health monitoring	Mobile platforms with IoT sensors	Improved farmer income by 25%

and buyers, thereby reducing price volatility and improving food security. AI-driven solutions have proved very important in helping farmers adapt to climate variability and mitigate effects of changing weather patterns [15].

1.4.2.4 Case Study 4: AI and IoT for Livestock Management in Europe

The European Commission has funded several projects applying AI and IoT technologies to livestock farming, such as AI-driven health monitoring systems for cattle. In these systems, sensors attached to the animals capture health indicators such as temperature, heart rate, and activity levels. In real time, this data is sent to the AI systems that identify early signs of a disease or distress. Farmers can intervene immediately to treat any animals and greatly reduce the probability of illness spreading throughout the herd while reducing the use of antibiotics. The feeding schedule and grazing patterns are also optimized, and this improves general health and productivity within livestock and reduces environmental impact in farming operations. This case study illustrates how AI and IoT can help improve animal welfare and efficiencies in farming practices with respect to sustainable livestock farming [7] (Table 1.4).

1.5 CHALLENGES TO IMPLEMENTATION

While integrating AI and IoT in agriculture does promise large benefits in terms of productivity, resource use, and sustainability, many hurdles need to be overcome before these technologies are fully and widely exploited effectively. This section describes two major hurdles on which the prosperous use of these technologies hinges: data privacy and cost barriers, and the digital divide. These two hurdles must be overcome because both are vital in supporting the inclusive development of technology in agriculture [1].

1.5.1 Data Privacy and Cost Barriers

The amount of data collected from AI and IoT in agriculture is tremendous. The data is utilized to make on-the-spot decisions, predict best practices, and optimize farm operations. However, one of the concerns relating to data usage relates to data privacy and security, which may act as a barrier to the adoption of these technologies [10].

1.5.1.1 Data Privacy Concerns

This all becomes critical in the collection, storage, and later analysis of the data in AI and IoT solutions. Sensors, drones, and other smart devices collect data about soil conditions, crop health, weather patterns, and even water usage, among many other things. Much of this data is sensitive, however, and concerns about who owns it and how it is used will be a major barrier to adoption [7]. Government bodies such as the European Commission and the USDA have raised arguments on privacy concerns in the process as personal as well as operational data is collected. Farmers may be resentful to have their data collected as the information may be used in some adverse way or even stolen, especially if it's collected and processed by private firms or third-party vendors [6]. This presents a special challenge where the data collected may be put to use in making significant business decisions, such as changing market prices, competition, or even insurance.

The National Agricultural Statistics Service is a service of the USDA that surveys and collects information on a regular basis on agricultural producers in the United States. Although these surveys are crucial for policy and promotional activities of agricultural projects, some farmers oppose these surveys owing to the fear that they may have to reveal sensitive information that might violate issues of data privacy [18]. Besides, since precision agriculture involves linking various sources of data that cut across different stakeholders like farm equipment manufacturers and input providers, it is in this context that one begins to lose control over privacy in data.

1.5.1.2 Cost Barriers

Introducing AI and IoT into agriculture is an expensive activity for small-scale farmers because they often lack the financial capabilities to invest in advanced technologies. Installation, maintenance, and training on sensors, drones, and software systems are costly and act as a huge barrier preventing a small-scale farmer from involving such projects, especially where profit margins are very minimal and credit access is being hampered [12]. The European Commission and FAO also added that the other very significant obstacle for the uptake of these technologies in agriculture by smallholder farmers is that these modern technologies are expensive, especially for small-scale farmers [7]. The OECD report concluded that the highest costs of implementing AI and IoT solutions are unaffordable to small farms since that holding size has more financial barriers. That creates an enormous barrier to the entry of technology, especially in developing countries where access to capital and credit is limited [4].

To address the problem, governments and international institutions attempt to lighten the financial burden on farmers. For example, the FAO has introduced programs that would allow farmers to obtain subsidization or low-interest loans to purchase precision agriculture technologies [19]. These initiatives are aimed at supporting the incurred investment cost when starting a venture, but they alone cannot compensate for the long-term cost of maintenance and implementation. The other challenge is the high cost of IoT infrastructure, network connectivity, and data storage, especially in remote areas where internet access is limited. It is further worsened by frequent and continuous software updates and system maintenance, which can drive the total cost of ownership upward. [12].

1.5.2 Addressing the Digital Divide

The term "digital divide" defines the gap between those who have access to modern digital technologies and those who don't have such access. In agriculture, this divide would primarily be made evident in rural areas with reduced access to high-speed internet, advanced technologies, and digital literacy. Therefore, it poses significant challenges to the wide application of AI and IoT technologies in agriculture [20].

1.5.2.1 Lack of Broadband Access

In fact, the proper adoption of IoT and AI systems into agriculture would require a prior ability to access broadband internet. These technologies rely on real-time collection and transmission of large volumes of data. A reliable internet connection is needed for this kind of large-scale data transmission. While infrastructure has been developed for broadband connectivity, it is still underrepresented or even completely lacking in many rural areas, especially in developing countries [15]. For the United States, the Federal Communications Commission (FCC) noted rural-urban broadband gap as one of the major barriers to the adoption of technology in agriculture. According to the 2019 Broadband Deployment Report by the FCC, 39% of rural Americans lack broadband access, whereas only 4% of urban residents do [20]. The lack of connectivity, hence, significantly limits the rural farming production system from reaping the maximum benefits that come with the IoT systems, as they hinge on the flow of data. The internet access is thus perceived as being low or unreliable, and thus, agricultural farmers cannot utilize the smart sensors, cloud-based software, and real-time monitoring systems in their activities. This thus results in losses in efficiency made by various firms that embrace such technologies. The *USDA Rural Development* program provides funding for rural broadband infrastructure projects to bridge the gap, but the progress is strenuous, and most areas are still considered rural and lack the right foundation for AI and IoT solutions [21].

1.5.2.1.1 Digital Literacy and Training

Besides the physical infrastructure challenges, the other challenge rural communities face is digital literacy. Most farmers lack the skills to be able to use advanced technologies, such as AI and IoT, to their fullest capacities. In instances where tools are available, there could still be a lack of capability to apply such tools in their entirety [18]. The *USDA* has realized the need for digital literacy and has instituted programs for farmers to develop the necessary skills in the operation and maintenance of precision farming equipment. In this regard, the *NIFA* has provided funding for several initiatives aimed at farmer training and education, especially in rural areas [13]. Such a program would enhance digital literacy, teaching farmers how they can efficiently run their farms while maximizing production, whether through data analysis or sensors and other applied technologies. However, these digital literacy efforts will also require substantial investments in education and training that may not readily be available to most farmers. In this regard, existing younger farmers will face a challenge about funding or community-level support meant to encourage them to embrace digital technologies [21].

TABLE 1.5
Digital Divide in Agriculture by Region [5, 11, 15, 16, 22]

Region	Internet Penetration (%)	Main Barriers	Initiatives for Improvement
Sub-Saharan Africa	35%	Lack of infrastructure	USAID's Digital Access Program
South Asia	50%	Affordability and skills gap	Indian Government's Digital India Initiative
Latin America	65%	Cost and availability	FAO Regional Initiative for Family Farming
North America	90%	Mainly rural access issues	USDA Rural Broadband Program

1.5.2.2 Global Challenges

The digital divide is even more apparent in the developing world, where agriculture plays a more central role in the economy. According to recent reports by the FAO, many developing countries have embraced mobile technologies, but a lack of reliable access to the internet and digital literacy continues to be a huge barrier to progress. Smallholder farmers face the double burden of limited access to modern agricultural technologies on their plate, as well as a lack of knowledge to operate them in developing regions [15]. To bridge the digital gap in agriculture, governments, international organizations, and private sector players are needed. The World Bank and the United Nations Development Programme (UNDP) started some activities to extend more digital access and literacy initiatives in the rural regions, especially focusing on the development of connectivity and capacity-building events for farmers [6]. The efforts will give farmers all the support they need to capitalize on AI and IoT for effective exploitation of these technologies, hence bridging the digital divide and promoting more-inclusive agricultural development (Table 1.5).

1.6 FUTURE DIRECTIONS

1.6.1 Blockchain for Transparency

Blockchain has become a very strong technology in agricultural supply chains because of its inherent ability to use a transparent and immutable ledger in tracking goods' movements, quality products, and building trust among chain stakeholders. Now, the USDA, as well as the European Union (EU), looks into blockchain applications in agriculture as an effort to increase traceability and reduce inefficiencies throughout the supply chain. One of the most significant benefits of blockchain for agriculture is the capability it gives to ensure full transparency about the origin and journey of agricultural products. In a report by *USDA* about blockchain and agriculture, the department shows how blockchain helps farmers trace produce from farm to market and ensures consumers know where their food comes from and how it was produced. This traceability would limit fraud, prove organic claims, and enhance food safety assurance regarding the source, age, or health condition with its quick identification and cure for contamination or fraud. The USDA's *Agricultural Marketing Service (AMS)* has already experimented with blockchain-based traceability systems on how to enhance accountability and transparency in the agricultural supply chain, which are a pre-requisite for ensuring food safety and fair trade.

Similarly, the *EU* has been the pioneer of using blockchain for ensuring supply chain transparency. The EU has also implemented a succession of projects to test and implement blockchain systems in agriculture. Central to the improvement of traceability of food products, improved traceability has always been at the heart of such projects. For example, *the Horizon 2020* initiative by the EU has funded several blockchain-based projects geared toward helping make agricultural supply chains more efficient. Such projects would not only ensure a better follow-through for food products but also secure access to any party in the supply chain, from the consumer all the way to the retailer and also to regulatory bodies. Blockchain also helps in reducing food waste by being transparent and minimizing fraud. Moreover, blockchain provides producers with an opportunity to rapidly identify inefficiencies in distribution. Therefore, as technology matures, blockchain is going to play a far more significant role as agriculture supply chains become increasingly efficient and consumer-centric in the future (Table 1.6).

1.6.2 Advancements in AI Models

Modern developments like AI and machine learning are going ahead with unprecedented speed, and their implications for agriculture are growing deeper with new models especially designed to handle agricultural situations. Agencies at the governmental level, as well as organizations such as NIFA and FAO, in turn, research applying such breakthroughs to improvements in productivity, sustainability, and resilience in farming [2, 13]. The important innovation in AI for Agri-activities is the advanced machine learning model that can process very large volumes of data in real time. These models enable farmers to make more informed decisions on irrigation, pest management, crop conditions, and yield predictions. For instance, NIFA has funded some research on machine learning in agriculture, particularly its application in precision agriculture and data-driven decision-making. They train to predict weather, soil health, and crop growth so farmers can make better use of resources, reduce waste, and predict where problems might arise in advance [13].

TABLE 1.6
Blockchain Applications in Agricultural Supply Chains by Use Case [4, 6, 10, 19]

Use Case	Function	Benefit
Supply Chain Transparency	Track products from farm to consumer	Enhanced consumer trust, traceability
Smart Contracts	Automate payments upon delivery	Faster transactions, reduced fraud
Quality Assurance	Verify organic or fair-trade certification	Ensures authenticity, attracts premium prices

TABLE 1.7
Emerging AI Models for Agriculture and Their Applications [1, 18, 20, 21, 23]

AI Model	Application	Advancement over Traditional Models
Deep Learning (DL)	Disease detection through image analysis	Higher accuracy in identifying diseases
Reinforcement Learning (RL)	Optimal irrigation scheduling	Adapts to environmental changes in real time
Transfer Learning	Yield prediction in data-poor regions	Utilizes existing data to improve predictions
GANs (Generative Adversarial Networks)	Synthetic data generation for model training	Reduces need for extensive real-world data

The FAO has also taken a leading role in heralding AI-based solutions to global agricultural challenges that are currently foremost regarding climate change and scarcity of resources. These include the application of AI models in agricultural practice to enhance crop monitoring by drones, soil health assessment through sensors, and predictive analytics that could predict crop failures. The AI applications are particularly relevant in climate-vulnerable regions, as they enable farmers to adapt to changing environmental circumstances and make better water and land use decisions [2]. There will then be further developments in the coming years so as to make use of AI in such areas with increasingly powerful models, which will make very accurate predictions and recommendations. These discoveries are likely to catapult the future of precision agriculture toward still more efficiency and sustainability for farming operations (Table 1.7).

1.7 CONCLUSION

The better use of AI, IoT, and blockchain technologies in agriculture is transformative in nature and would revolutionize farming. These technologies are already starting to improve resource management, reduce waste, and drive farm productivity as discussed in this chapter [6, 10]. AI and IoT solutions, as evinced by potent solutions, are still subject to a series of barriers. Data privacy issues, high implementation costs, and the digital divide are among the major barriers more visible in underdeveloped rural areas with poor infrastructural facilities. Governments, international organizations, and private sector players must work in tandem to develop ways to provide equitable access to the benefits of such technologies [1, 2, 7]. The future of agricultural technology holds tremendous promise, with advancements in AI models and increasing applications of blockchain technology in supply chains. Blockchain, holds significant potential for increasing transparency, reducing fraud, and enhancing traceability. Meanwhile, AI models on weather pattern predictions, optimization of irrigation levels, and crop health monitoring will continue to advance and will help the farmers make more objective decisions [2, 24]. Therefore, agricultural technology has a great future ahead, but more investment, innovation, and collaboration are still required to deal with future challenges with the fullest possible support to ensure food security and sustainability for generations to come [6, 21].

REFERENCES

[1] International Telecommunication Union (ITU). (2023). *Bridging the digital divide for sustainable agriculture*. ITU Publications.
[2] Food and Agriculture Organization of the United Nations (FAO). (2023). *The state of food and agriculture 2023*. FAO.
[3] National Agricultural Library (NAL). (2023). *Research on AI and IoT in agriculture*. USDA.
[4] Organisation for Economic Co-operation and Development (OECD). (2023). *Agricultural policy monitoring and evaluation 2023*. OECD.
[5] California Department of Food and Agriculture (CDFA). (2023). *Smart agriculture initiatives in California*. CDFA.
[6] World Bank. (2023). *Digital agriculture and innovation for rural development*. World Bank Publications.
[7] European Commission. (2023). *Digital transformation in EU agriculture and rural development*. European Commission Reports.
[8] UN Economic Commission for Africa (UNECA), Agricultural and Rural Development Division. (2023). *Digital agriculture and AI in African farming*. UNECA Reports.
[9] Australian Bureau of Agricultural and Resource Economics and Sciences (ABARES). (2023). *Advances in agricultural technology: AI, IoT, and data-driven solutions*. ABARES.
[10] United States Department of Agriculture (USDA). (2023). *Artificial intelligence in agriculture: Emerging trends and applications*. USDA.
[11] Indian Council of Agricultural Research (ICAR). (2023). *Precision and sustainable agriculture in India*. ICAR Reports.
[12] U.S. Department of Energy – Office of Energy Efficiency and Renewable Energy (EERE). (2023). *Smart agriculture and IoT applications*. EERE Publications.
[13] National Institute of Food and Agriculture (NIFA). (2023). *Precision agriculture research and AI advancements*. USDA NIFA Reports.
[14] Canadian Ministry of Agriculture and Agri-Food. (2023). *Precision agriculture and resource management*. Government of Canada Publications.
[15] World Bank. (2023). *Technology adoption in African agriculture: The role of IoT*. World Bank Publications.
[16] Embrapa (Brazilian Agricultural Research Corporation). (2023). *Drones and IoT for Brazilian agriculture*. Embrapa Reports.
[17] USDA Natural Resources Conservation Service (NRCS). (2023). *Resource conservation in agriculture: Water, soil, and energy*. USDA NRCS Publications.
[18] National Agricultural Statistics Service (NASS). (2023). *Data privacy and technology in agriculture*. USDA NASS Reports.
[19] Food and Agriculture Organization (FAO), & International Fund for Agricultural Development (IFAD). (2023). *Future directions for AI and IoT in global agriculture*. FAO/IFAD Reports.
[20] Federal Communications Commission (FCC). (2023). *Broadband access and the digital divide in agriculture*. FCC Reports.
[21] Indian Ministry of Electronics and Information Technology. (2023). *Digital inclusion in rural India: Addressing the digital divide*. Government of India Publications.
[22] Twiga Foods. (2023). *IoT case study: Smallholder farms in Kenya*. Twiga Foods Reports.
[23] European Parliament. (2023). *Data protection and digital innovation in agriculture*. European Parliament Reports.
[24] National Institute of Standards and Technology (NIST). (2023). *Standards for blockchain in agricultural supply chains*. NIST Publications.

2 Smart Irrigation

Harnessing AI for Sustainable Agriculture

Ankit, Himanshu, Deepika Saxena, Singh Shailesh Kumar, Rahmatullah Kazimi, and Abhishek Tanwer

2.1 INTRODUCTION

The increasingly important decision-making process of irrigation scheduling is needed to achieve the best use, and most importantly, efficient use, of water [1]. Crop quality and yield are highly affected by frequency and watering levels. The purpose of irrigation scheduling is to apply enough water to a given crop at the proper time. Inappropriate irrigation scheduling techniques can lead to over- or under-irrigation. Therefore, using groundwater for irrigation causes loss of water in over-irrigated areas, leaching of nutrients, and high levels of soil salinity, while under-irrigated fields suffer a financial loss. Science and technology-based techniques of irrigation scheduling can help increase crop profits as well as reduce interactions with the environment through controlling agricultural water stress [2]. The modern artificial intelligence (AI) revolutionized the process of irrigation by offering several ways to save water by increasing efficiency. With the use of AI systems, irrigation decisions will be made faster and efficiently, especially in the face of climate change. The usage of AI in the field of agriculture is becoming a very important area for development and research, because an increasingly significant field of application is the agro use of AI that aims to increase the irrigation water usage efficiency [3–5]. The digital revolution happening in the water management industry through the integration of real-time data analytics and insights to make smart choices has already started using AI. AI-based predictions are used in apps to track crop health, to trace the spread of disease, and to optimize resource use. These data-driven methods help users make better and faster decisions about when to grow plants, and how to irrigate, fertilize, or apply pesticides to increase yield and economic efficiency while reducing environmental impact [6–8].

Also, smart irrigation systems could automate components automatically as they can precisely adjust the amount of water delivered with multiple factors, including crop status, soil moisture, vegetation index, and irrigation strategy. Their advantages are described in Figure 2.1. Under- or over-watering removal helps to achieve accuracy in irrigation programming, allowing plants to receive the correct amount of

DOI: 10.1201/9781003594093-2

water at the right time. AI algorithms allow prediction for future patterns of climate change and water needs. They can assist farmers to avoid watering their crops during rain and provide water during dry conditions. Using AI algorithms to identify leaks and supply inefficiencies in irrigation equipment can avoid water loss to a great extent. Scheduling irrigation to avoid periods of high electricity demand will reduce the energy consumption of AI systems. AI is changing the way we practice irrigation by finally helping growers transition from their current, static conventional scheduling to modern, data-driven management. This transition is of great water-saving, high-yield crops, and agricultural sustainability. The AI algorithms eliminate human intervention as much as possible, consume the largest amount of input possible, and run the most automated industrial processes. However, AI applications have the opportunity to greatly contribute to managing water resources, conserving the fertility of the soil, and thereby preserving biodiversity and supporting management of water in an agronomically sustainable way. AI systems can optimally utilize soil and water resources, manpower through the recording and forecasting of the demand for water, infrastructure installation, and the monitoring and prediction of inevitable disasters [9–11].

Smart sensor and actuator systems are being increasingly used in smart irrigation systems to monitor and manage crop health, weather, etc., in real time. This data will be used to optimize watering schedules, thereby reducing water waste.

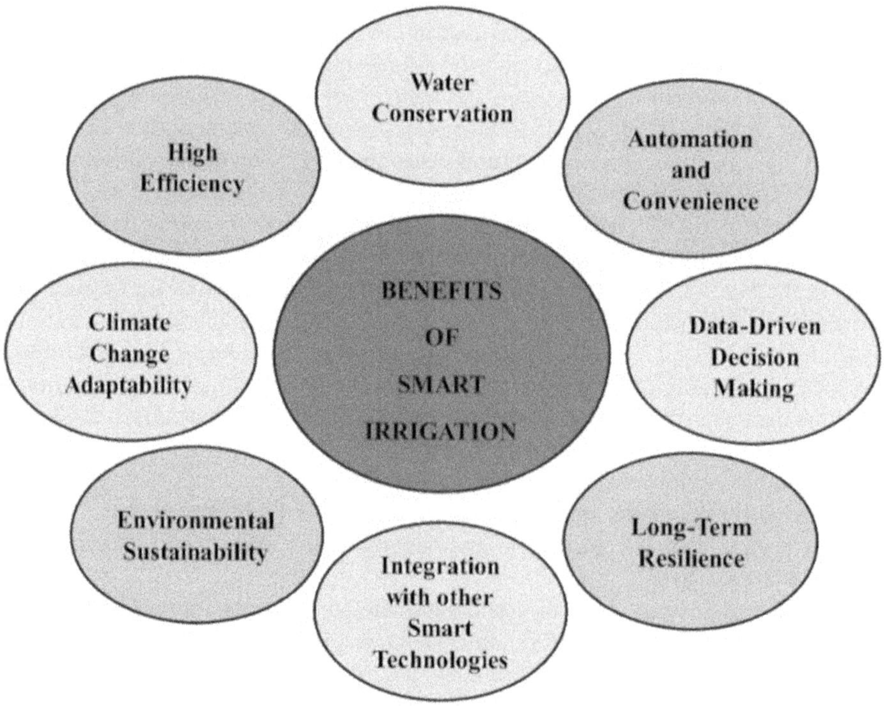

FIGURE 2.1 Key advantages of adopting smart irrigation systems.

In addition to monitoring vital moisture in the soil, these sensors would also be able to analyze nutrient levels, infestations of pests, and even plant health. This real-time data will allow systems to adjust irrigation according to water requirements and the specific needs of different plants over different growth stages. Additionally, AI-powered decision-making algorithms can elevate predictive analytics by allowing systems to predict weather, change temperature forecasts, and operate schedules for irrigation according to their changes. With these types of algorithms, users can gain more flexibility and efficiency to remotely monitor and control irrigation over computers or cell phones by pattern-finding parameter adjustments concerning real-time and previous data inputs. They can also improve interoperability across different farm management systems so that users may have a more complete picture of how agricultural operations work. Sensors and actuators collect enormous volumes of data, and big data analytics can be leveraged to tap into this large volume of data. Farmers will be able to pick up trends, predict what they are going to need in terms of water in the future, and derive data-driven choices in better techniques of managing water [12].

2.2 AI TECHNOLOGIES IN SMART IRRIGATION

Smart irrigation includes various AI technologies that aid in sustainable irrigation management, which are illustrated in Figure 2.2 and described as follows.

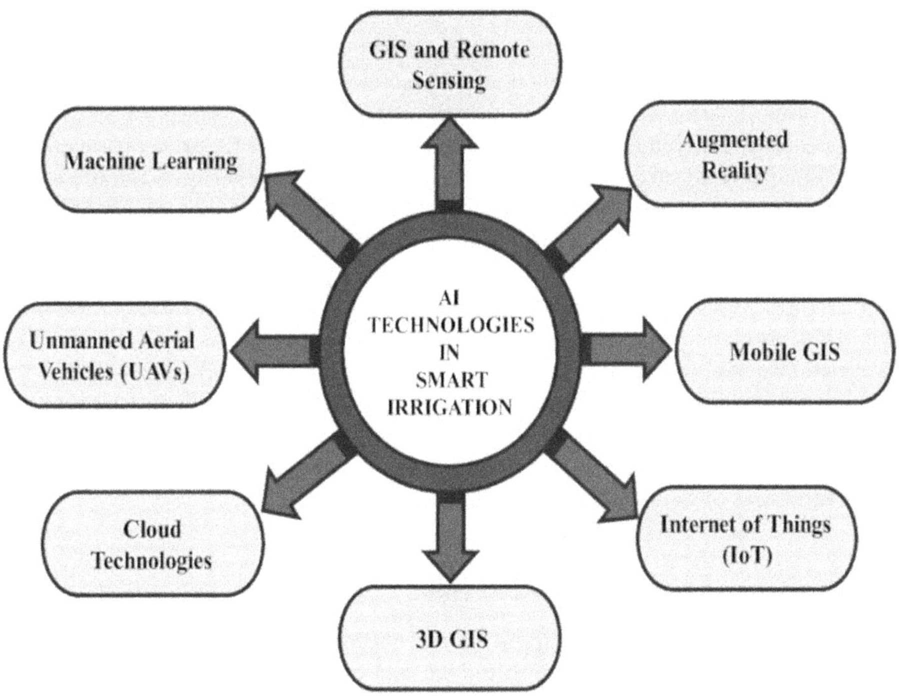

FIGURE 2.2 Innovative AI applications transforming smart irrigation.

2.2.1 GIS and Remote Sensing

Geographic Information System (GIS) can capture, store, manipulate, manage, and analyze geographically referenced datasets and can also produce output. It is therefore a crucial tool for decision-making, modeling, resource allocation, and optimization. GIS and Remote Sensing (RS) are strong and efficient tools for identifying land cover and land use changes. Counting insects in crops, estimating crop area and crop yield, crop identification, checking soil conditions and plant health and status, estimating overall crop health and potential yield, agricultural water management, etc., have been studied under GIS combination with other techniques [13]. GIS is a tool of data input, data storage, data retrieval, data manipulation, data analysis, and data presentation. GIS is critically important to geographic decision-making. By its nature, GIS allows integration of different forms of data based on locational and geographic characteristics of the data. GIS can help people solve complex problems, build arguments, integrate data, visualize situations, make maps, and find workable solutions. Moreover, GIS and associated technology are being recognized as having value for natural resource management and inventory studies. At heart, GIS is used to help define quantitative research and decision-making. Additionally, if the location of the object or phenomenon is critical to the research, GIS is used in the collection, archiving, and assessment of the object or phenomenon. GIS is superior as an information system because of spatial scanning and overlay features. GIS is an advanced and efficient planning tool for decision-makers and resource managers to quickly calculate and analyze enormous datasets. These features have made them crucial for widespread use in agriculture, business, natural resource management, urban and rural planning, as well as in military applications [14].

Remote sensing is the art, science, and technology of gathering data about an object, often determined with satellites or airplanes. Remote sensors had detected earthly energy, which they had recorded as data. Light, force fields, acoustic energy, or any other type of electromagnetic radiation is a form of energy. A remote sensor can be a standard camera, which detects light from the object that travels back to a light-sensitive film. If the light hits the film, then the picture has been captured, and a latent image is formed. The photo lab develops the photo for the analysis. Medical, architecture, archaeology, industrial quality control, robots and extraterrestrial mapping [15], and other sectors use remote sensing with an image as input. RS and GIS in agriculture have enabled the spatiotemporal visualization of field data to track crop and animal conditions in real time. Among the many uses of GIS in irrigation are watershed management, irrigation delivery monitoring, hydrology, groundwater assessment, water productivity estimation, water demand calculation, and spatiotemporal variations of irrigation needs [16].

2.2.2 Machine Learning and Big Data

To develop sustainable water applications, irrigation managers must work with techniques like machine learning, big data, and blockchain. Data-driven precision irrigation is the most promising strategy to deal with today's and future problems with these emerging technologies [17–19]. Using a large volume of data from an irrigated area allows farmers to make informed decisions about irrigation patterns. Data-driven agriculture can increase agricultural productivity, reduce costs, and ensure sustainability. Big data, remote sensing, GIS, and machine learning, when combined, promote the automation of collecting, analyzing, and decision-making of agricultural data. Automated field management tools, irrigation scheduling, mapping, machine learning tools, big data, and GIS-based data management systems are compiled to help producers with certain irrigation-related tasks, such as weather forecasting. Furthermore, the development of algorithms from field data collected from satellites, Unmanned Aerial Vehicles (UAVs), and sensors is possible as larger computing resources are available [20].

With data-driven machine learning models for irrigation, the physical/mechanistic models have been replaced to predict soil moisture [21–23] since they require less data for calibration than mechanistic models. In addition, machine learning models can outperform physical models in prediction and can quickly process spatial and temporal data. A machine learning model-based irrigation system closed-loop controller uses feedback from comparing pre-processed data and real-time measured data to learn from the calibration [24].

2.2.3 Cloud Computing and GIS in Smart Irrigation

To guarantee variable rate application, precision irrigation technology will be dependent on cloud computing and geographic information systems. By using cutting-edge algorithms and software in the cloud, the prediction of geographically distributed crop water requirements based on field circumstances becomes reliable and efficient [25]. Information is directed to an irrigation controller system for actuation. The farmer can regulate the system with a single click anywhere by understanding the conditions of soil moisture with his smartphone or a web interface, reducing operating expenses and conserving irrigation water. A study proposed that the Irriman Platform, a cloud computing-based low-cost wireless data logging system, gathered data from a variety of sensors and an innovative software architecture, allowing secure storage and processing of such data for irrigation management and crop monitoring [26]. This platform facilitated the optimization of irrigation operations by agronomists through a web-based tool that allowed users to create intricate irrigation strategies for evaluating each crop [27].

2.2.4 Integration of Internet of Things in Irrigation Systems

Farmers have historically scheduled irrigation using heuristic methods, which frequently leads to either too much or too little watering [19]. Closed-loop control technique assures the supply of as much water as necessary according to the soil moisture and the current crop environment. IoT provides a platform to perform precise irrigation by connecting many soil sensors wirelessly with other farm equipment. To automate irrigation systems, GIS and IoT must be integrated, bringing dynamic changes to the irrigation industry and other related sectors. In IoT technology, the data is passed wirelessly to the server through the standard Internet Protocol. With weather and soil moisture sensors, solenoids mounted over the sprinklers, Lora modulation, and a 4G connection, irrigation needs can be supplied in real time [28]. Data was transmitted via a wireless sensor network (WSN), and the plant environment and water stress were tracked with weather sensors and plant sensors. The system was used to make real-time irrigation scheduling possible. Real-time detection of irrigation water pollution [29], calculation of leaching need [30], and scheduling of irrigation for water management [31] are other uses of IoT integrated with Geographic Information System/RS [32].

2.2.5 Mobile GIS

The combination of mobile devices, GPS, GIS, and wireless connections that provide access to the internet-based geographic information systems offers an opportunity for smart irrigation. Mobile GIS updates and modified real-time data provide faster and more accurate results. With major advances in mobile technology, many mobile GIS applications have been created for specific uses and are much more than just simple mapping. Several of these applications were created for use by companies in the private sector. Examples of mobile GIS applications created specifically for use in irrigated agricultural systems include diagnosing plant diseases, predicting the weather, and getting market pricing.

2.2.6 Unmanned Aerial Vehicles

Armed forces have used mostly drones or UAVs since the debut of drone technology. UAVs have another name in technological society, referred to as dynamically remotely operated navigation equipment (DRONE) [33]. Drones are any aircraft or apparatus capable of being flown from the ground while under the control of a human. For hovering, takeoff, and flying, rotors or propellers are used. The drone is raised on the principle hover and movement in the air, by the rotor. These flying robots operate off-site, and through their applied systems, they act on software battle plans and are operated by a team of humans, whose actions are based on output from sensors. A UAV is a system with three main components: the ground control system, sensor payloads or cameras, and the actual unmanned aircraft. Crop progress assessment is based upon a wide range of physical measurements, including thermal, hyperspectral, and multispectral cameras, and Light Detection and Ranging (LiDAR) systems [34].

2.2.7 3D GIS

3D GIS provides objects with an additional dimension (z) to increase visibility and detail. While 2D maps have elevations, 3D GIS gives objects elevations. Using 3D technology in GIS maps, explanatory images are generated, showing the scale of the real-world items. An excellent appearance is created by three-dimensional models, and they help in surveying several sectors. 3D technologies can aid in better visualizing, assessing, and analyzing the effects that change in irrigation layouts and their sustainability [35]. A standard 3D model would include data, satellite imagery, and irrigation infrastructure information and is used to aid irrigation designers in selecting the most cost-effective paths of irrigation conveyance systems. This 2D and 3D GIS can be integrated with ArcGIS for better visualization and simulations of the irrigation system [36]. The integration of holographic computation in GIS spatial analysis aids in the assessment of the impacts of floods on irrigated agriculture using the three-dimensional space. A basic, useful, and inexpensive approach based on RS coupled with 3D GIS was offered. Texture data was used to build both models and is linked to a program called ArcGIS. A colorful and lifelike 3D virtual image of an irrigation field was produced, allowing field managers to gain the ease and comfort of seeing their irrigation field, with simple GIS technologies for data collection and analysis [37].

2.3 DATA COLLECTION AND MANAGEMENT

The performance and reliability of AI systems directly depend on data collection and management, especially in agriculture, where the decisions are based on recently collected data. For example, AI-based smart irrigation needs specific weather data, temperature and soil sensors, and other IoT device data to give good and accurate recommendations. But accommodating a large amount of data means dealing with all the information being generated every day and ensuring that this information is of high quality and accurate [38].

2.3.1 Importance of High-Quality Data for AI Systems

Data is consumed during model development as well as when making a decision or when trying to optimize an existing model. Data is the raw material of AI in smart irrigation and the quality of the data in crop production forecast, soil, and water suggestion. The poor data quality results in inaccurate judgments, which are often expensive and can be detrimental to crops and the environment. There are several things they have to adhere to while working on quality data that is accurate, relevant, timely, and consistent. For example, if an AI model that estimates the amount of water needed for crops is useless or uses old data describing soil moisture levels, the model might be under- or over-irrigating a crop, releasing unnecessary water, and potentially affecting crops. In agricultural environments, high-quality data has a direct effect on production, efficiency, resource use, and sustainability. The quality of data could be influenced by several components, such as the quality of the sensors, the quality of the conditions under which the data is collected, and the ways in which the data is acquired [39]. The sensors are to be checked and/or calibrated over time to ensure that they accurately measure readings. Heat or rain can alter the conditions of survey and questioning, and thereby influence statistics as produced. Unless the quality input is regulated, the AI systems are capable of learning incorrect or incomplete quality input. The results of the model can look distorted, which in turn leads to the wrong conclusions. Data preparation techniques like normalization, filtering, and data curating are mainly the core values in improving the quality of data. A redundancy check and the establishment that the data source is periodically can be used to improve data recovery. Keeping high-quality data is a continuous process that needs supervision and repeated checking over a period of time so that the data is accurate and current [40].

2.3.2 IoT Devices for Data Collection: Soil and Temperature Sensors, Weather Stations

The new trend of introducing IoT devices, especially in the farming industry, has revolutionized data collection for building applications for AI. IoT-connected equipment is very effective at collecting real-time data for efficient irrigation and farming. Table 2.1 lists some of these devices, such as weather stations, temperature sensors, and moisture sensors in the soil. In terms of accumulating data points in instances in the field area, each device has its function [16].

2.3.2.1 Soil and Temperature Sensors

The artificial intelligence (AI) systems are able to calculate the amount of water to apply to irrigation shaping, depending on the sensor information of the moisture level on the ground and sophisticated processing tools that will not over-diagnose water and under-irrigated crops. Air and soil temperature are also to be monitored, which in turn affects the crop health and growth rates. For example, fluctuating soil temperature indicates that the problems associated with certain stress factors are to be solved immediately. But IoT sensors are not free of challenges in terms of data

TABLE 2.1
Different Smart Sensors in AI-Based Irrigation Management

Sensors	Type	Function	References
Soil Moisture Sensors	Time-Domain Reflectometry (TDR)	Uses the dielectric constant of the soil to determine the moisture content by sending an electrical pulse through it	[41]
	Gypsum Blocks	Utilizing a substance that changes electrical resistance when it absorbs water, one can determine the moisture content of soil	[42]
Soil Temperature Sensors	Thermocouples	Make a voltage differential between two metals to determine the temperature	[43]
	Digital Soil Thermometers	Give current information on soil temperature, frequently incorporated into automated systems	[44]
EC (Electrical Conductivity) Sensors	Two-Electrode Sensors	Measure the conductivity of the soil using two electrodes	[45]
	Four-Electrode Sensors	For more precise readings, lessen the impact of soil contact resistance	[46]
pH Sensors	Glass pH Sensors	Standard for measuring pH, however, can be fragile and needs to be handled with care	[47]
	Solid-State pH Sensors	More robust, employing substances such as soil-application ion-selective electrodes	[48]
Soil Nutrient Sensors	Ion-Selective Electrodes	Determine the amounts of particular ions, such as ammonium and nitrate	[49]
	Optical Sensors	Examine the chemistry of the soil using reflectance or light absorption	[50]

collection. Finally, such device performance can be compromised by a deterioration associated with continuous contact with water and exposure to extremely hot or cold temperatures. Those faulty, damaged, or ill-functioning sensors must be repaired, recalibrated, and serviced frequently to ensure data accuracy. Regrettably, most sensors have three major problems: (1) reliance on batteries as a power source, requiring regular recharging or maintenance and hard to manage in the remote farming regions; (2) high cost of installation and upkeep, which restrict their use on the scale of large-scale farming; and (3) their sensitivity to environmental factors, including dust, heat, or humidity, which can impair the accuracy and duration [51].

This smart irrigation system is manufactured employing AI technology in an effort to triage the temperature soil sensors in order to obtain field data. The system uses predictive modeling to get precise, automatic control of irrigation with cloud tools, thereby improving its usage and crop yield.

2.3.2.2 Weather Stations

Data-gathering instruments for macro-environmental information, such as rainfall, humidity, wind, and amount of light, are weather stations. Weather variables are critical for AI irrigation systems that use weather in scheduling the application of water. An AI system may optimize the irrigation period if a weather station over the next few hours predicts a rainy climate, to avoid over-watering. Another problem of data integration is synchronization from different weather stations. It is proposed that a variation in the measurement is allowed due to the intrinsically different location of each station. Also, data obtained from weather stations cannot be accurate because of the micro-climate; hence, there is a need to place weather stations in different locations to collect a lot of reliable data. Performance is always an inevitable issue of not just hardware but also connectivity while using IoT devices in places with bad internet connection. Some devices encourage delays in the transmission of data that affect real-time decision-making, and data inputs are continuous and require a stable connection [52].

The list of weather stations used in various crops is shown in Table 2.2.

2.3.3 Managing Large Data Volumes and Ensuring Accuracy

As IoT devices are being used in the field of agriculture, the quantity of data generated is growing. One of the major challenges of using AI-based smart irrigation is to handle large volumes of data generated and to improve its accuracy. The individual sensors or weather stations, in which large amounts of data are stored, need to be managed and analyzed as the data is collected and transmitted regularly [59]. Storing and consuming such big data needs platform structures to handle it, when data has to be stored and used for a long time for model training and history analysis. In smart agriculture, cloud storage can be used to achieve this quite often. There are however problems with cost, delay, and security if the data are held in the cloud. In case farmers are not even connected to the Internet, saving their data in the cloud is a challenge [60].

The second challenge is to keep the data valid over a large coverage. While collecting data from multiple IoT devices, problems associated with the environment or errors on the devices are visible. The predictions and the decisions can all be

TABLE 2.2
Various Weather Stations Utilized in AI-Based Irrigation Management for Crops

Weather Station	Crop	Model	Outcome	References
Davis Vantage Pro2	Tomato	Stochastic Segmentation of input Energy (SSiE)	For effectively obtaining crop data from standard greenhouse measurements to accomplish more accurate irrigation control.	[53]
WeatherHawk 232 Direct Connect	*Cinnamomum camphora*	MixSIR model	To better comprehend the soil-plant-atmosphere continuum's water cycle in the East Asian monsoon region.	[54]
Onset HOBO	Wheat	–	Grain-filling time was reduced by 3.33 days, kernel weight was reduced by 1.25% per degree Celsius, grain number was reduced by 2.36% per degree Celsius, and yield was reduced by 3.58% per degree Celsius.	[55]
Campbell Scientific (CR1000X)	"Hass" avocado	–	Changes in the hourly time courses of fruit volume, stem water potential, and pedicel sap flow under drought imply that the stomatal response inhibited more significant increases in fruit outflow.	[56]
Spectrum WatchDog	Corn	Linear mixed model	Enhance conventional tillage, corn's ability to suppress weeds in the early season.	[57]
	Maize	Linear mixed model	The utilization of exotic open-pollinated populations in US maize breeding, taking into account phenotypic stability and unfavorable alleles.	[58]

influenced by the heterogeneous data at the model level. A discrepancy in data relayed from one soil moisture sensor versus the others might lead to incorrect decisions about irrigation. Techniques of data verification, such as cross-validation or redundancy, can help us manage such problems. When multiple sensors are used to register a single parameter, comparison between the data shows the error [61]. Intelligent irrigation systems should respond to instantaneous water application change, which, however, requires irregular data requirements. To filter and apply those extracted conclusions within a short time, highly computational systems and algorithms are used to analyze large amounts of data in real time. This requires technical and software prerequisites such as the ability to acquire such high-frequency data and also to process these data streams without suffering issues in latency. However, the reported procedure may be further optimized by the incorporation of machine learning in conjunction with data management systems aimed at high-scale applications. Hence, the need to optimize AI for real-time data streams and make sound/quicker decisions in agriculture is still a tech issue [62].

2.4 ENVIRONMENTAL IMPACT AND SUSTAINABILITY

Due to climate change and water stress, sustainable agriculture is still a call. Traditional ways of water distribution for plant purposes are deteriorating; however, the AI advancements bring along the ways through which the water resources can be sustained in smart irrigation. The smart irrigation systems make water application effective, without applying chemical fertilizers, while getting the maximum output [63].

2.4.1 Reducing Water Consumption with AI

One of the most important environmental benefits of AI in smart irrigation is reducing water consumption. AI system is another example of an advanced irrigation system that would calculate the needed amount of water based on the forecasted weather conditions, stage of crop plant, and soil moisture content. The system learns from past and present data to generate appropriate irrigation advice for farmers. For example, patterns can be modified by using machine learning algorithms to find trends of weather prediction and adjust the irrigation rate to add water to those dry seasons and remove water from the wet ones. Satellite data and soil sensors do help; however, they cannot measure specific levels of soil moisture. Water use is cut down greatly by this strategy; some estimates indicate that by using AI-enhanced smart irrigation, water usage is reduced by one-third to one-half compared to conventional methods. Advanced irrigation using AI helps the agriculture sector ease the burden on specific ecosystems and supplies of the precious commodity, water. In the arid or drought region, these techniques are becoming especially effective to save water sources and help the agricultural department [64].

2.4.2 Minimizing Chemical Fertilizer Use through Targeted Irrigation

AI is a benefit of smart irrigation because it can minimize the application of chemical fertilizers through proper watering methods. Many chemicals are often applied on, for example, huge amounts of land using techniques like the spraying of fertilizers, which washes the nutrients into the water sources till they are contaminated and ruined. Instead, AI-driven mechanized irrigation controls the dispersing of water mixed with fertilizer, thus reducing chemical input. Moreover, it can evaluate the fertilizer needs of the crops based on the soil nutrient content and automatically provide water, only if the nutrient deficiency is recognized. This technology could be used together with drip irrigation, which delivers water directly to the root areas, allowing optimum use of both water and nutrients. In addition to minimizing fertilizer consumption, it also decreases the possibility of creating a dangerous high runoff, a major cause of eutrophication and water source contamination, including rivers and lakes. Precision irrigation optimizes the ground and environmental conditions and reduces the harm to the environment caused by farming. Using intelligent irrigation, the topsoil sustainability can be maintained with little or no abuse of nutrients. Future improvement of agricultural sustainability can be made known by the fact that farmers should abandon the use of synthetic fertilizer [65].

2.5 FUTURE PERSPECTIVES ON AI IN SMART IRRIGATION

In the future, AI will play a role in influencing the smart irrigation that will be unimaginable, in terms of the growing role that it will play in having higher precision and efficiency in how to manage irrigation water in agriculture. The promise of AI-based smart irrigation system applications, such as machine learning, big data, and IoT sensors, aids in improving the rate of crop cultivation and conserves water in a more sustainable approach. Using information including crop demand, weather information, and the amount of moisture in the soil, crop water control systems can regulate the quantity and timing of the water supply to crops, reducing wastage and, at the same time, increasing production. In addition, the AI systems of the present time may not only solve the problem in irrigation and foresee the drought but also suggest the type of crop needed based on the probable conditions of the weather. An autonomous system supported by a drone and a satellite may provide site-specific and on-the-spot effective irrigation management plans, requiring less involvement of humans. Additionally, AI will cause long-term devices to remember their experience and function according to new environmental conditions.

Cost and technical skills that are beyond the reach of smallholder farmers are major challenges in using AI for smart irrigation. However, as costs drop and technology improves, smart irrigation systems using AI will probably see the greatest reduction in price. The change is captivating the whole world of agriculturists to rational water usage, boosting agricultural yields, and addressing the effects of climate change.

2.6 CONCLUSION

The current economic environment has led technology to become a management tool for companies across industries to grow the right way! From this perspective, the existing irrigation structures and their application will be optimized to achieve the right performance and operation efficiency. While IoT is utilized in increasing the yields and mitigating their environmental impacts on the environment, conserving resources, the same sensory systems are being used by farmers. Although these technologies have been with organizations for a medium amount of time, few have successfully implemented, executed, and captured the value in them. On one hand, the water crisis is a situation characterized by water scarcity, lack of water, and problems of providing a water supply. Currently, there is the notion of water management of sorts, and organizations are interested in finding out whether water conservation is possible and still increases productivity. Today's irrigation systems in business environments work with their capability to support organizational usage of technology to accomplish performance goals. The costs of technology are brought down by IoT, and therefore, the probability of controlling the irrigation process monitoring system becomes low. Additionally, WSN for real-time supported irrigation, as well as for precision farming monitoring, is available.

REFERENCES

1. Saggi, M. K., & Jain, S. (2022). A survey towards decision support system on smart irrigation scheduling using machine learning approaches. *Archives of Computational Methods in Engineering*, 29(6), 4455–4478.
2. Zhang, J., Guan, K., Peng, B., Jiang, C., Zhou, W., Yang, Y., Cai, Y. (2021). Challenges and opportunities in precision irrigation decision-support systems for center pivots. *Environmental Research Letters*, 16(5), 053003.
3. Ahmed, Z., Gui, D., Murtaza, G., Yunfei, L., & Ali, S. (2023). An overview of smart irrigation management for improving water productivity under climate change in drylands. *Agronomy*, 13(8), 2113.
4. Disasa, K. N., Yan, H., Wang, G., Zhang, J., Zhang, C., & Zhu, X. (2024). Projection of future precipitation, air temperature, and solar radiation changes in southeastern China. *Theoretical and Applied Climatology*, 155(6), 4481–4506.
5. Wei, H., Xu, W., Kang, B., Eisner, R., Muleke, A., Rodriguez, D. Harrison, M. T. (2024). Irrigation with artificial intelligence: Problems, premises, promises. *Human-Centric Intelligent Systems*, 4(2), 187–205.
6. Abioye, E. A., Hensel, O., Esau, T. J., Elijah, O., Abidin, M. S. Z., Ayobami, A. S., … Nasirahmadi, A. (2022). Precision irrigation management using machine learning and digital farming solutions. *AgriEngineering*, 4(1), 70–103.
7. Bondad, J., Harrison, M. T., Whish, J., Sprague, S., & Barry, K. (2023). Integrated crop-disease models: New frontiers in systems thinking. *Farming System*, 1(1), 100004.
8. Drogkoula, M., Kokkinos, K., & Samaras, N. (2023). A comprehensive survey of machine learning methodologies with emphasis in water resources management. *Applied Sciences*, 13(22), 12147.
9. Oliveira, R. C. D., & Silva, R. D. D. S. E. (2023). Artificial intelligence in agriculture: Benefits, challenges, and trends. *Applied Sciences*, 13(13), 7405.
10. Tzachor, A., Devare, M., King, B., Avin, S., & Ó hÉigeartaigh, S. (2022). Responsible artificial intelligence in agriculture requires systemic understanding of risks and externalities. *Nature Machine Intelligence*, 4(2), 104–109.
11. Vallejo-Gomez, D., Osorio, M., & Hincapie, C. A. (2023). Smart irrigation systems in agriculture: A systematic review. *Agronomy*, 13(2), 342.
12. Lakhiar, I. A., Yan, H., Zhang, C., Wang, G., He, B., Hao, B., … Rakibuzzaman, M. (2024). A review of precision irrigation water-saving technology under changing climate for enhancing water use efficiency, crop yield, and environmental footprints. *Agriculture*, 14(7), 1141.

13. Zhao, Q., Haseeb, M., Wang, X., Zheng, X., Tahir, Z., Ghafoor, S., Almutairi, K. F. (2024). Evaluation of land use land cover changes in response to land surface temperature with satellite indices and remote sensing data. *Rangeland Ecology & Management*, 96, 183–196.
14. Obi Reddy, G. P., Dwivedi, B. S., & Ravindra Chary, G. (2023). Applications of geospatial and big data technologies in smart farming. In: Pakeerathan, K. (eds) *Smart Agriculture for Developing Nations: Status, Perspectives and Challenges* (pp. 15–31). Singapore: Springer Nature Singapore.
15. Nina, A., Radovanović, M., & Popović, L. (2021). Extraterrestrial influences on remote sensing in the earth's atmosphere. *Remote Sensing*, 13(5), 890.
16. Morchid, A., Alblushi, I. G. M., Khalid, H. M., El Alami, R., Sitaramanan, S. R., & Muyeen, S. M. (2024). High-technology agriculture system to enhance food security: A concept of smart irrigation system using Internet of Things and cloud computing. *Journal of the Saudi Society of Agricultural Sciences*. https://doi.org/10.1016/j.jssas.2024.02.001
17. Alfred, R., Obit, J. H., Chin, C. P. Y., Haviluddin, H., & Lim, Y. (2021). Towards paddy rice smart farming: A review on big data, machine learning, and rice production tasks. *IEEE Access*, 9, 50358–50380.
18. Bhat, S. A., & Huang, N. F. (2021). Big data and ai revolution in precision agriculture: Survey and challenges. *IEEE Access*, 9, 110209–110222.
19. Kashyap, P. K., Kumar, S., Jaiswal, A., Prasad, M., & Gandomi, A. H. (2021). Towards precision agriculture: IoT-enabled intelligent irrigation systems using deep learning neural network. *IEEE Sensors Journal*, 21(16), 17479–17491.
20. Kasimati, A., Psiroukis, V., Panoutsopoulos, H., Mouseti, S., Mylonas, N., & Fountas, S. (2023). Management information systems and emerging technologies. In: Vougioukas, S.G., Zhang, Q. (eds) *Advanced Automation for Tree Fruit Orchards and Vineyards*, 195–218.
21. Abioye, E. A., Abidin, M. S. Z., Aman, M. N., Mahmud, M. S. A., & Buyamin, S. (2021). A model predictive controller for precision irrigation using discrete Lagurre networks. *Computers and Electronics in Agriculture*, 181, 105953.
22. Bwambale, E., Abagale, F. K., & Anornu, G. K. (2022). Smart irrigation monitoring and control strategies for improving water use efficiency in precision agriculture: A review. *Agricultural Water Management*, 260, 107324.
23. Bwambale, E., Abagale, F. K., & Anornu, G. K. (2023). Data-driven model predictive control for precision irrigation management. *Smart Agricultural Technology*, 3, 100074.
24. Abioye, E. A., Abidin, M. S. Z., Mahmud, M. S. A., Buyamin, S., Ijike, O. D., Otuoze, A. O., Olajide, O. M. (2023). A data-driven Kalman filter-PID controller for fibrous capillary irrigation. *Smart Agricultural Technology*, 3, 100085.
25. Kumar, P., & Choudhury, D. (2024). Innovative technologies for effective water resources management. In: Izah, S.C., Ogwu, M.C., Loukas, A., Hamidifar, H. (eds) *Water Crises and Sustainable Management in the Global South*, 555–594.
26. Forcén-Muñoz, M., Pavón-Pulido, N., López-Riquelme, J. A., Temnani-Rajjaf, A., Berríos, P., Morais, R., & Pérez-Pastor, A. (2021). Irriman platform: Enhancing farming sustainability through cloud computing techniques for irrigation management. *Sensors*, 22(1), 228.
27. Fuentes-Peñailillo, F., Gutter, K., Vega, R., & Silva, G. C. (2024). Transformative technologies in digital agriculture: Leveraging Internet of Things, remote sensing, and artificial intelligence for smart crop management. *Journal of Sensor and Actuator Networks*, 13(4), 39.

28. Rehman, A., Saba, T., Kashif, M., Fati, S. M., Bahaj, S. A., & Chaudhry, H. (2022). A revisit of internet of things technologies for monitoring and control strategies in smart agriculture. *Agronomy*, 12(1), 127.
29. Lin, Y. P., Mukhtar, H., Huang, K. T., Petway, J. R., Lin, C. M., Chou, C. F., & Liao, S. W. (2020). Real-time identification of irrigation water pollution sources and pathways with a wireless sensor network and blockchain framework. *Sensors*, 20(13), 3634.
30. Bashir, R. N., Bajwa, I. S., & Shahid, M. M. A. (2019). Internet of things and machine-learning-based leaching requirements estimation for saline soils. *IEEE Internet of Things Journal*, 7(5), 4464–4472.
31. Roy, S. K., Misra, S., Raghuwanshi, N. S., & Das, S. K. (2020). AgriSens: IoT-based dynamic irrigation scheduling system for water management of irrigated crops. *IEEE Internet of Things Journal*, 8(6), 5023–5030.
32. Bwambale, E., Naangmenyele, Z., Iradukunda, P., Agboka, K. M., Houessou-Dossou, E. A., Akansake, D. A., Chikabvumbwa, S. R. (2022). Towards precision irrigation management: A review of GIS, remote sensing and emerging technologies. *Cogent Engineering*, 9(1), 2100573.
33. Ortega-Farias, S., Esteban-Condori, W., Riveros-Burgos, C., Fuentes-Peñailillo, F., & Bardeen, M. (2021). Evaluation of a two-source patch model to estimate vineyard energy balance using high-resolution thermal images acquired by an unmanned aerial vehicle (UAV). *Agricultural and Forest Meteorology*, 304, 108433.
34. Zeyliger, A. M., Muzalevskiy, K. V., Zinchenko, E. V., & Ermolaeva, O. S. (2022). Field test of the surface soil moisture mapping using Sentinel-1 radar data. *Science of the Total Environment*, 807, 151121.
35. He, L., Han, B., Ji, H., Mao, G., & Chen, J. (2024). A method to integrate hydraulic structure models into 3D terrain models for irrigation infrastructure visualization. *Scientific Reports*, 14(1), 21255.
36. Putri, P. I. D., Dharmayasa, I. G. N. P., Anggreni, N. L. J., Suryanti, I., & Nuraga, I. K. (2024, July). Application of remote sensing and geographic information systems to transform sustainable irrigation water management in Bali. In *2024 10th International Conference on Smart Computing and Communication (ICSCC)*, Bali, Indonesia (pp. 294–299). IEEE.
37. Gupta, A., Ruan, J., & Jaiswal, S. (2024). The potential of geographic information systems as a tool to achieve sustainable development goals. In Sharma, C., Shukla, A.K., Pathak, S., Singh, V.P. (eds) *Sustainable Development and Geospatial Technology: Volume 1: Foundations and Innovations* (pp. 185–196). Cham: Springer Nature Switzerland.
38. Júnior, M. R. B., de Almeida Moreira, B. R., dos Santos Carreira, V., de Brito Filho, A. L., Trentin, C., de Souza, F. L. P., Shiratsuchi, L. S. (2024). Precision agriculture in the United States: A comprehensive meta-review inspiring further research, innovation, and adoption. *Computers and Electronics in Agriculture*, 221, 108993.
39. Yao, W., & Sun, Z. (2023). The impact of the digital economy on high-quality development of agriculture: A China case study. *Sustainability*, 15(7), 5745.
40. Buelvas, J., Múnera, D., Tobón V, D. P., Aguirre, J., & Gaviria, N. (2023). Data quality in IoT-based air quality monitoring systems: A systematic mapping study. *Water, Air, & Soil Pollution*, 234(4), 248.
41. Abdulraheem, M. I., Chen, H., Li, L., Moshood, A. Y., Zhang, W., Xiong, Y., Zhang, Y., Taiwo, L. B., Farooque, A. A., & Hu, J. (2024). Recent advances in dielectric properties-based soil water content measurements. *Remote Sensing*, 16(8), 1328. https://doi.org/10.3390/rs16081328
42. Maiti, S., Jain, N., & Malik, J. (2023). A comprehensive review of flue gas desulphurized gypsum: Production, properties, and applications. *Construction and Building Materials*, 393, 131918.

43. Subramanian, S. K., Choudhary, P., Singh, V. R., Meenu, M., & Kalra, S. (2024). Microwave temperature measurement methods in food processing industry. In Pratap Singh, A., Erdogdu, F., Wang, S., Ramaswamy, H.S. (eds) *Microwave Processing of Foods: Challenges, Advances and Prospects: Microwaves and Food* (pp. 151–170). Cham: Springer International Publishing.
44. Cahyani, D. E., Gumilar, L., Afandi, A. N., Abd Samat, A. A. B., Safi'i, A., & Prasetyo, M. W. (2023, September). Connectivity of agricultural soil fertility online monitoring devices with smartphones. In *2023 8th International Conference on Electrical, Electronics and Information Engineering (ICEEIE)*, Malang City, Indonesia (pp. 1–6). IEEE.
45. Toselli, M., Baldi, E., Ferro, F., Rossi, S., & Cillis, D. (2023). Smart farming tool for monitoring nutrients in soil and plants for precise fertilization. *Horticulturae*, 9(9), 1011.
46. Horváth, J., Kátai, L., Szabó, I., & Korzenszky, P. (2024). An electrical conductivity sensor for the selective determination of soil salinity. *Sensors*, 24(11), 3296.
47. Faqir, Y., Qayoom, A., Erasmus, E., Schutte-Smith, M., & Visser, H. G. (2024). A review on the application of advanced soil and plant sensors in the agriculture sector. *Computers and Electronics in Agriculture*, 226, 109385.
48. Mudila, H., Kumar, A., Prasher, P., Talniya, N. C., & Verma, S. (2024). Electrochemical methods for determination of fertilizers in soil. In Kamel A. Abd-Elsalam (eds) *Nanofertilizer Delivery, Effects and Application Methods* (pp. 111–126). Elsevier.
49. Kumari, A., Mishra, P., Chaulya, S. K., Prasad, G. M., Nadeem, M., Kisku, V., Chowdhury, A. (2024). Estimation of soil nutrients and fertilizer dosage using ion-selective electrodes for efficient soil management. *Communications in Soil Science and Plant Analysis*, 55(13), 1920–1941.
50. Ameer, S., Ibrahim, H., Kulsoom, F. N. U., Ameer, G., & Sher, M. (2024). Real-time detection and measurements of nitrogen, phosphorous & potassium from soil samples: A comprehensive review. *Journal of Soils and Sediments*, 24(7), 2565–2583.
51. Veeramanju, K. (2024). Predictive models for optimal irrigation scheduling and water management: A review of AI and ML approaches. *International Journal of Management, Technology and Social Sciences (IJMTS)*, 9(2), 94–110.
52. Roy, T., Kalambukattu, J. G., Biswas, S. S., & Kumar, S. (2023). Agro-climatic variability in climate change scenario: Adaptive approach and sustainability. In: Chatterjee, U., Akanwa, A.O., Kumar, S., Singh, S.K., Dutta Roy, A. (eds) *Ecological Footprints of Climate Change: Adaptive Approaches and Sustainability* (pp. 313–348). Cham: Springer International Publishing.
53. Florakis, K., Trevezas, S., & Letort, V. (2023). Predicting tomato water consumption in a hydroponic greenhouse: Contribution of light interception models. *Frontiers in Plant Science*, 14, 1264915.
54. Dai, J., Zhang, X., Luo, Z., Wang, R., Liu, Z., He, X., Guan, H. (2020). Variation of the stable isotopes of water in the soil-plant-atmosphere continuum of a Cinnamomum camphora woodland in the East Asian monsoon region. *Journal of Hydrology*, 589, 125199.
55. Hein, N. T., Bheemanahalli, R., Wagner, D., Vennapusa, A. R., Bustamante, C., Ostmeyer, T., Jagadish, S. K. (2020). Improved cyber-physical system captured post-flowering high night temperature impact on yield and quality of field grown wheat. *Scientific Reports*, 10(1), 22213.
56. Kaneko, T., Gould, N., Campbell, D., & Clearwater, M. J. (2024). Isohydric stomatal behaviour alters fruit vascular flows and minimizes fruit size reductions in drought-stressed 'Hass' avocado (Persea americana Mill.). *Annals of Botany*, 133(7), 969–982.

57. Silva, T. S., Arneson, N. J., DeWerff, R. P., Smith, D. H., Silva, D. V., & Werle, R. (2023). Preemergence herbicide premixes reduce the risk of soil residual weed control failure in corn. *Weed Technology*, 37(4), 410–421.
58. Perkins, A., Lima, D. C., Kaeppler, S. M., & de Leon, N. (2024). Performance and phenotypic stability of maize hybrids containing exotic introgressions in multi-environment trials. *Crop Science*, 64(2), 756–771.
59. Mowla, M. N., Mowla, N., Shah, A. S., Rabie, K. M., & Shongwe, T. (2023). Internet of things and wireless sensor networks for smart agriculture applications: A survey. *IEEE Access*, 11, 145813–145852.
60. Morchid, A., Jebabra, R., Khalid, H. M., El Alami, R., Qjidaa, H., & Jamil, M. O. (2024). IoT-based smart irrigation management system to enhance agricultural water security using embedded systems, telemetry data, and cloud computing. *Results in Engineering*, 23, 102829.
61. Wang, W. H., & Hsu, W. S. (2023). Integrating artificial intelligence and wearable IoT system in long-term care environments. *Sensors*, 23(13), 5913.
62. Panduman, Y. Y. F., Funabiki, N., Fajrianti, E. D., Fang, S., & Sukaridhoto, S. (2024). A survey of AI techniques in IoT applications with use case investigations in the smart environmental monitoring and analytics in real-time IoT platform. *Information*, 15(3), 153.
63. Amede, T., Konde, A. A., Muhinda, J. J., & Bigirwa, G. (2023). Sustainable farming in practice: Building resilient and profitable smallholder agricultural systems in Sub-Saharan Africa. *Sustainability*, 15(7), 5731.
64. Bhavsar, D., Limbasia, B., Mori, Y., Aglodiya, M. I., & Shah, M. (2023). A comprehensive and systematic study in smart drip and sprinkler irrigation systems. *Smart Agricultural Technology*, 5, 100303.
65. Taseer, A., & Han, X. (2024). Advancements in variable rate spraying for precise spray requirements in precision agriculture using Unmanned aerial spraying Systems: A review. *Computers and Electronics in Agriculture*, 219, 108841.

3 Resource Optimization in Agriculture

Priya Thomas and Roshini Bakur

3.1 INTRODUCTION

Smart agriculture refers to the practice of adopting advanced technologies into traditional farming with an aim to optimize farming practices. Adopting advanced technologies equips farmers to make more informed, efficient, and sustainable decisions in managing agricultural resources. By integrating digital tools with traditional farming methods, smart agriculture offers a path toward achieving sustainable agricultural practices and addressing food security challenges, environmental degradation, and climate change [1]. Sustainable agriculture should be practiced as an important extension of smart farming, which involves farming practices that satisfy the current food requirements without limiting the capacity of future generations. This commonly includes conserving existing resources and minimizing environmental impacts while fostering economic sustainability and social equity. Smart agriculture contributes to sustainability by improving efficiency, minimizing waste, and reducing the agricultural environmental impact.

Smart agriculture should efficiently use the different categories of resources, such as natural, human, capital, biological, and energy resources. Sustainable agriculture focuses on practices that are environmentally responsible, economically viable, and socially equitable [2]. The main types of resources in agriculture, along with sustainable practices designed to ensure their efficient and long-term use, are detailed in this chapter. The major modules of resource management and optimization are modeled in Figure 3.1.

- Water Usage Optimization: Water scarcity is a major global concern, and efficient water usage is a critical component of sustainable agriculture. Smart agriculture uses precision irrigation systems that incorporate sensors, weather data, and artificial intelligence (AI) models to optimize water usage. This ensures that crops receive the exact quantity of water they need based on available real-time monitoring data. This will greatly contribute to reduced water waste, thereby conserving a vital resource.
- Fertilizer and Pesticide Usage Optimization: Through data analytics, precision farming technologies can determine the required amount of fertilizers and pesticides at each stage of crop growth. Smart agriculture uses IoT sensors and AI algorithms to monitor soil health and nutrient levels, decreasing the over-application of fertilizers and chemicals, which in turn reduces water pollution and soil degradation.

 DOI: 10.1201/9781003594093-3

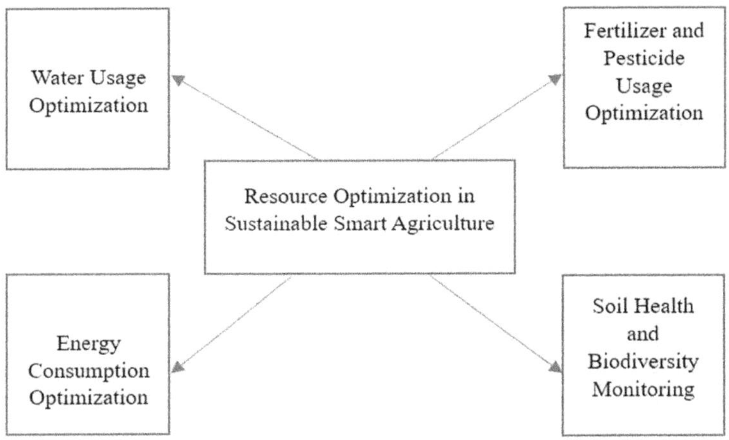

FIGURE 3.1 Types of resource optimization.

- Energy Consumption Optimization: Energy consumption in farming, particularly through mechanized equipment and irrigation systems, can be minimized by using autonomous systems (e.g., drones, automated tractors) that operate efficiently and with minimal waste [3]. AI-based optimization algorithms can also help reduce energy use by controlling machinery usage based on weather conditions and crop requirements.
- Soil Health and Biodiversity Monitoring: Soil health is one of the central pillars of sustainable agriculture, playing key roles in maintaining ecosystem services, improving productivity, and fostering resilience to climate change. By nurturing both soil health and biodiversity, farmers can improve yields by minimizing the need for synthetic inputs like fertilizers and pesticides.

There are a lot of technologies driving smart agriculture for sustainability. The Internet of Things (IoT) is one of the promising technologies driving smart agriculture, where IoT sensors and enabled devices can collect data from the field in real time [4]. The collected data will be forwarded to cloud systems to be analyzed to generate informed decisions. AI models can be used to analyze large datasets from multiple sources to make predictions that help to optimize farming practices. AI-driven decision support systems help farmers manage resources more efficiently and adapt to environmental changes. Machine learning (ML) algorithms help to analyze historical and real-time data to detect patterns and make predictions about crop yields, pest outbreaks, soil health, and weather patterns. These predictions will help farmers to take proactive measures that reduce resource use and increase efficiency.

Drones integrated with multispectral cameras and thermal sensors provide farmers with detailed, real-time data about crop health, soil conditions, and water stress. Autonomous machinery, such as robots for planting, weeding, and harvesting, can operate efficiently and with minimal resource waste. These technologies reduce labor costs and increase the precision of tasks like harvesting, which can significantly reduce crop loss. The resources available should be properly utilized, and wastage has to be reduced. Resource optimization strategies play a major role in adopting sustainable agriculture.

3.2 RESOURCE OPTIMIZATION

Resource optimization has to be performed at different levels to effectively leverage the benefits offered by various technologies. Resource optimization is central to the concept of sustainable agriculture, which aims to produce food, fiber, and other agricultural products while ensuring that the environmental, social, and economic systems that support agriculture remain viable for future generations [5]. Resource optimization in sustainable agricultural practices focuses on making the most efficient and effective use of the resources available, such as land, water, energy, and nutrients, while cutting off waste, reducing environmental harm, and promoting long-term ecological balance.

Given the growing global population, increasing demand for food, and the environmental pressures caused by climate change and resource depletion, optimizing agricultural resources is more important than ever. This process not only supports food security and livelihoods but also helps in mitigating the environmental footprint of agriculture.

3.2.1 Challenges for Resource Optimization

The need for resource optimization in agriculture stems from several key challenges facing the global food system. The main challenges include the following:

3.2.1.1 Resource Scarcity

- Water Scarcity: Agriculture is one of the largest consumers of freshwater globally, with over 70% of freshwater withdrawals used for irrigation. In many regions, access to water is limited, and inefficient irrigation practices can deplete water supplies, lead to aquifer depletion, and damage ecosystems.
- Soil Degradation: Unsustainable farming practices, such as excessive tillage, overuse of chemicals, and monocropping, degrade soil health and fertility. Depleted soils result in lower agricultural productivity, requiring more inputs to maintain yields.
- Fossil Fuel Dependence: The agricultural sector is a significant consumer of energy, particularly fossil fuels for machinery, transportation, and input production (e.g., fertilizers, pesticides). Reducing level of energy consumption and converting it to renewable energy sources is essential for mitigating agriculture's carbon footprint.

3.2.1.2 Climate Change

- Changing Weather Patterns: Unreliable weather patterns such as droughts, floods, and heatwaves stress the agricultural systems. Optimizing resources, particularly water and soil, can help crops withstand these changes and reduce vulnerability to climate extremes.

- Greenhouse Gas Emissions: Agriculture is considered a major source of greenhouse gas emissions. Efficient resource management strategies like reduced fertilizer application and optimized livestock feed can help reduce emissions while maintaining productivity.

3.2.1.3 Economic Pressures

- Cost Efficiency: Farmers face increasing input costs (e.g., seeds, water, fertilizers, fuel) due to inflation and supply chain disruptions. Resource optimization allows farmers to reduce costs by using fewer inputs while maintaining or improving yields.
- Market Demand: Consumers and businesses are increasingly demanding sustainably produced food. Efficient and resource-conserving practices are becoming essential for meeting these market preferences.

3.3 RESOURCE OPTIMIZATION STRATEGIES WITH SUSTAINABLE PRACTICES

Resource optimization in agriculture focuses mainly on four primary resources: land, water, soil, and energy. Section 3.3.1 details how to optimize each resource with related strategies. The application of IoT, ML, and AI for each resource is considered separately, and collective conclusions are generated. The advantages offered are also briefed. The approach followed for modelling resource optimization strategies is modeled in Figure 3.2.

3.3.1 Water Usage Optimization

Water is one of the most critical resources in agriculture, and optimizing its use is essential for sustainable farming. Resource optimization can be adopted in water management by integrating IoT, big data, and ML techniques. Smart Irrigation

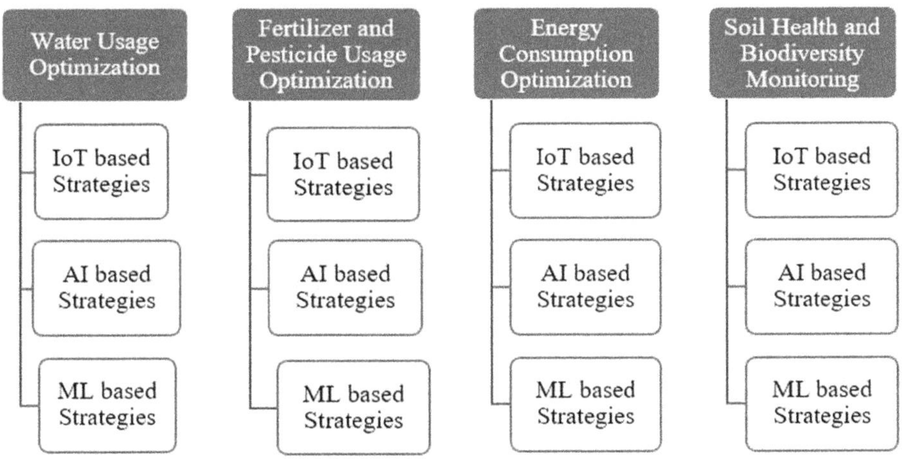

FIGURE 3.2 Strategies for different types of resource optimization.

Systems are equipped with IoT sensors to measure soil hydration levels and monitor weather conditions [6]. This real-time data can be used by farmers to decide when and how much water to apply. Automated irrigation systems can be designed to adjust water delivery based on this data, preventing over-watering or under-watering [7]. Rainwater harvesting policies using IoT can be used to track rainfall patterns, allowing for optimal water storage and usage. When rainwater is collected, IoT sensors can even manage how it is distributed to crops, reducing reliance on groundwater.

3.3.1.1 Key Strategies

3.3.1.1.1 IoT-Based Water Usage Optimization Systems

IoT-enabled systems can be used for continuous and real-time monitoring of soil hydration levels, weather patterns, and crop water usage, which is essential for making data-driven decisions about irrigation.

- Soil Moisture Sensors: IoT-enabled soil moisture sensors embedded in the soil can measure water retention at different depths. These sensors are capable of generating real-time data on soil conditions, equipping farmers to know exactly when crops need water.
- Weather Stations: IoT-based weather stations can track environmental conditions, such as precipitation, thermal conditions, humidity, wind velocity, and solar power, which can all impact irrigation needs.
- Water Flow Sensors: These sensors measure water consumption in irrigation systems, enabling farmers to detect any leaks or inefficiencies in the system. Real-time flow data helps in optimizing water distribution.

Benefits

- Real-Time Data: Continuous monitoring of soil and atmospheric conditions enables precise irrigation decisions.
- Early Detection of Issues: IoT sensors help identify under- or over-watering issues, equipment malfunctions, or water leakages, preventing water wastage.
- Resource Efficiency: By ensuring that water is applied only when necessary, IoT systems help conserve water and minimize the dependency on external water sources.

3.3.1.1.2 AI-Driven Water Usage Optimization

AI helps to analyze and interpret enormous volumes of data generated from IoT sensors, weather forecasts, and crop models to optimize irrigation schedules and water distribution [8]. AI systems can make autonomous decisions based on input data and provide actionable insights for farmers.

How AI Optimizes Water Management

- Smart Irrigation Systems: AI-powered irrigation systems use real-time data from IoT devices to precisely determine the amount of water required for crops. The system automatically adjusts water flow based on factors like soil moisture, temperature, and weather forecasts.

- Predictive Analytics: AI uses historical weather patterns, crop types, and environmental factors to predict future water requirements, allowing farmers to plan irrigation schedules more accurately.
- Optimization Algorithms: AI can optimize water use by running simulations and analyzing data for different scenarios. AI can be used to plan irrigation timelines based on the crop's growth stages, soil conditions, and weather forecasts, ensuring minimal water use for maximum crop productivity.

Benefits

- Reduced Water Waste: AI ensures that water is applied only when and where it is required, cutting down on excess watering.
- Cost Savings: By optimizing irrigation schedules and minimizing water waste, AI helps to reduce related energy and costs associated.
- Increased Crop Yields: AI can enhance crop health and yields by ensuring optimal watering at every growth stage.

3.3.1.1.3 Machine Learning for Water Usage Optimization

ML can be used to build predictive models that forecast water demand based on historical data, environmental factors, and crop growth stages. These models provide farmers with valuable insights into when water demand will peak, allowing them to plan irrigation more efficiently.

How ML Optimizes Water Management

- Water Demand Forecasting: ML models can be trained on historical irrigation data, climate data (e.g., temperature, humidity, rainfall), soil moisture levels, and crop type to predict future water usage. These predictions help in optimizing irrigation schedules, reducing over-watering or under-watering.
- Crop Growth Modeling: ML can analyze the growth stages of crops (e.g., germination, flowering, fruiting) and calculate the water required at each stage. This information helps farmers plan irrigation based on the specific needs of the crops.
- Weather Prediction Models: ML can be applied to weather prediction models that estimate rainfall, temperature, and other factors influencing irrigation [9]. By anticipating weather patterns, farmers can adjust irrigation schedules to take advantage of rainfall or avoid over-irrigating during a wet period.

Benefits

- Improved Efficiency: ML-driven irrigation systems can predict water demand with high accuracy, ensuring that water is applied in the right amounts at the right times.
- Adaptive Learning: ML systems tend to constantly learn and adapt based on incoming new data, enhancing the accuracy of predictions and irrigation schedules over time.
- Sustainability: By predicting water requirements based on real-time and historical data, ML minimizes unnecessary water use, reducing environmental impact and conserving valuable water resources.

Combining IoT, AI, and ML enables adaptive water management systems that can generate timely decisions based on a variety of dynamic factors, such as weather conditions, soil moisture, crop type, and more. Adaptive water management systems continuously adjust irrigation schedules based on real-time data inputs from IoT sensors and weather forecasts. If rain is expected or if the soil is sufficiently moist, the system may automatically delay or reduce irrigation, optimizing water use. Farmers can use mobile apps or web platforms to monitor water usage and adjust irrigation systems remotely.

AI and ML algorithms analyze the data, provide recommendations, and allow farmers to make adjustments in real time. Such adaptive systems offer numerous benefits such as flexibility as these systems can automatically adjust to varying environmental conditions, such as unexpected rainfall, sudden temperature shifts, or unexpected changes in crop growth, improved sustainability by dynamically optimizing water usage, adaptive systems ensure that water is used as efficiently as possible, helping to reduce waste and conserve water [10]. Data-driven decisions will make it easier for farmers to make insightful decisions based on accurate data rather than relying on intuition or outdated practices.

3.3.2 Fertilizer and Pesticide Usage Optimization

In smart sustainable agriculture, optimizing the use of fertilizers and pesticides is critical for improving crop yields, reducing environmental hazards, and minimizing costs. Traditional agricultural practices often lead to over-usage of fertilizers and pesticides, resulting in soil quality reduction, environmental pollution, and increased greenhouse gas emissions [11]. However, emerging technologies, including IoT, AI, and ML, offer powerful solutions to optimize these inputs, ensuring they are used more efficiently and effectively.

3.3.2.1 Key Strategies

3.3.2.1.1 IoT in Fertilizer and Pesticide Usage Optimization

The IoT plays a foundational role in collecting real-time data from the field, enabling specific and tailored application of fertilizers and pesticides. IoT devices, such as sensors, drones, and connected equipment, provide a detailed picture of the farm's conditions, which is critical for optimizing input usage.

- Soil and Crop Monitoring
  IoT sensors placed in the soil can measure key factors like moisture levels, nutrient content (e.g., nitrogen, phosphorus, potassium), pH, and temperature. This data helps farmers assess whether the soil requires additional fertilizers or if existing nutrient levels are adequate [12]. Similarly, crop health can be monitored using IoT sensors or drones equipped with imaging technologies to report early warnings of nutrient imbalances or pest invasions.
- Precision Application
  IoT-enabled precision agriculture allows for site-specific applications of fertilizers and pesticides. For example, variable rate technology (VRT) can be used to apply varying quantities of fertilizer or pesticides to different sections

of a field based on real-time sensor data. This guarantees that inputs are only used where needed, reducing wastage and environmental contamination.

Benefits

- Precision Application and Resource Efficiency: IoT-enabled sensors and devices allow for site-specific application of fertilizers and pesticides, ensuring they are applied only when required.
- Real-Time Monitoring and Early Detection: IoT provides timely data on soil health, crop conditions, and pest intrusions, enabling farmers to detect warning signs of nutrient deficiencies, pests, or diseases, and intervene at the right time with targeted applications of fertilizers and pesticides
- Data-Driven Decision-Making and Predictive Insights: IoT integrates with AI and ML tools to interpret huge amounts of data from sensors, weather patterns, and historical trends.

3.3.2.1.2 AI-Driven Fertilizer and Pesticide Usage Optimization

AI can analyze data from IoT sensors, weather forecasts, and soil health assessments to make decisions about the optimal timing, type, and quantity of fertilizer application. AI systems learn from past data, such as previous crop cycles, weather conditions, and soil performance, to suggest or automatically control fertilizer inputs for the best crop growth.

- Fertilizer Scheduling: AI-powered systems can schedule fertilizer applications based on multiple variables like soil conditions, crop growth stage, weather forecasts, and nutrient deficiencies. These systems can help optimize fertilizer applications, reducing overuse, improving nutrient uptake by plants, and preventing nutrient loss to the environment.
- Smart Fertilizer Recommendations: AI algorithms can interpret soil health data (e.g., from IoT sensors) and recommend the type and amount of fertilizer needed, considering the specific crop's nutrient requirements at different growth stages. This ensures that the soil's nutrient profile is maintained while preventing excessive fertilizer use.
- Predictive Analytics for Fertilizer Needs: AI-based models predict future nutrient requirements based on historical data, including previous harvest yields, crop rotation practices, and soil health trends. These predictions help in planning fertilizer application in advance and reduce reliance on reactive measures.

Benefits

- Enhanced Nutrient Efficiency: AI ensures that nutrients are applied in a way that maximizes crop uptake while minimizing wastage.
- Optimized Yield: AI-driven recommendations enable farmers to achieve maximum crop yields by ensuring that plants get the right nutrients at the right time.
- Sustainable Practices: By preventing over-fertilization and reducing nutrient runoff, AI helps protect the environment, particularly near water bodies, from nutrient pollution.

3.3.2.1.3 Machine Learning for Fertilizer and Pesticide Usage Optimization

ML involves training algorithms to recognize patterns in large datasets [13]. ML models are particularly useful for predicting nutrient requirements and optimizing fertilizer use by investigating vast amounts of historical data on soil types, weather patterns, crop types, and previous fertilizer applications.

- Soil Fertility Prediction Models: ML algorithms can predict changes in soil fertility over time by analyzing historical data on nutrient levels, crop rotation patterns, and fertilizer applications. These models can identify nutrient trends and predict the future nutrient needs of the soil.
- Variable Rate Fertilization: ML can be used to implement variable rate fertilization, where fertilizers are applied at different rates depending on soil nutrient levels in different parts of the field. By using sensor data (e.g., from IoT soil nutrient sensors) and processing it through ML models, the system can determine the most efficient amount of fertilizer to apply in each zone.
- Nutrient Deficiency Detection: ML-powered image recognition models can be trained to analyze field images (captured by drones or satellites) to detect visual symptoms of nutrient deficiencies, such as chlorosis (yellowing of leaves). These systems can identify specific nutrient deficiencies early, helping farmers take corrective actions promptly.

Benefits

- Data-Driven Fertilizer Decisions: ML models process large amounts of data to provide accurate, evidence-based fertilizer recommendations tailored to specific crops, fields, and growing conditions.
- Improved Yield and Quality: By optimizing fertilizer use and ensuring nutrients are available when crops need them, ML enhances crop growth, yield, and quality.
- Reduced Waste and Environmental Impact: ML systems help avoid over-fertilization and excessive nutrient runoff by applying precise amounts of fertilizer.

3.3.3 Energy Optimization

Energy efficiency in agriculture is critical for reducing environmental impacts, lowering production costs, and enhancing sustainability. Agricultural operations, such as irrigation, heating greenhouses, processing, and transportation, are energy-intensive, often relying on fossil fuels [14]. The integration of IoT, AI, and ML into agricultural systems provides significant opportunities for optimizing energy usage, enhancing productivity, and minimizing carbon emissions. These advanced technologies enable farmers to monitor energy consumption, optimize energy use, and transition toward more sustainable energy practices.

3.3.3.1 Key Strategies

In this context, energy resource optimization focuses on improving energy use through smart systems, reducing reliance on non-renewable energy sources, and incorporating renewable energy into agricultural operations.

3.3.3.1.1 IoT for Energy Optimization in Agriculture

The IoT uses interconnected sensors and devices to collect real-time data, providing insights into energy utilization patterns, thereby helping farmers to make informed decisions. Concerning energy optimization, IoT plays a critical role in monitoring energy use, detecting inefficiencies, and automating energy-intensive processes [15].

- Energy Consumption Monitoring: IoT-based smart meters and sensors can monitor energy usage at different points in agricultural operations, such as irrigation systems, lighting, heating systems for greenhouses, and machinery. These devices can track real-time energy consumption and provide insights into areas where energy is being wasted.
- Smart Irrigation Systems: IoT sensors, such as soil moisture sensors, weather stations, and temperature sensors, can be integrated with irrigation systems to optimize water and energy use. For example, using IoT, farmers can automate irrigation based on soil hydration levels and evaporation rates, which ensures water is applied only when needed, and irrigation pumps are not running unnecessarily.
- Precision Greenhouse Management: IoT sensors in greenhouses can monitor light intensity, temperature, humidity, and CO_2 levels. Based on this data, AI algorithms can adjust climate control systems, such as heating, cooling, and lighting, optimizing energy consumption while maintaining ideal growing conditions.

Benefits

- Real-Time Monitoring: IoT enables real-time tracking of energy consumption, allowing farmers to pinpoint inefficiencies and optimize usage.
- Reduced Energy Waste: By controlling irrigation, heating, and other energy-intensive processes based on actual data, IoT minimizes unnecessary energy use.
- Cost Savings: Monitoring energy usage allows farmers to track specific areas where energy cost reduction has to be incorporated, leading to significant savings.

3.3.3.1.2 AI-Powered Energy Optimization

AI leverages large datasets to optimize decision-making in agricultural processes. By analyzing data collected from IoT sensors, weather forecasts, and historical trends, AI can suggest energy-saving strategies, forecast energy demands, and automate energy-intensive activities, making energy management more efficient [16].

- Predictive Energy Management: AI models use historical data and real-time sensor data to predict future energy needs. For example, AI can forecast energy requirements for irrigation based on weather patterns and soil moisture levels, enabling farmers to plan energy use in advance, avoiding peak demand periods, and reducing reliance on non-renewable energy sources.

- Energy Demand Forecasting: AI can be used to predict periods of high energy demand, such as during peak irrigation seasons or extreme weather events. By understanding these patterns, farmers can adjust energy usage or switch to energy-efficient or renewable sources (e.g., solar or wind) when energy demand spikes.
- Optimal Operation of Agricultural Machinery: AI algorithms can optimize the scheduling and operation of energy-intensive farm machinery, like tractors, harvesters, and pumps, based on factors like soil moisture, crop growth stages, and weather conditions. This ensures machinery operates only when needed and uses energy most efficiently.

Benefits

- Enhanced Energy Efficiency: AI helps reduce unnecessary energy consumption by predicting and optimizing energy use across different agricultural activities.
- Load Shifting: AI-based systems can shift energy-intensive tasks to off-peak hours when energy costs are less or renewable energy is more readily available.
- Renewable Energy Utilization: AI can optimize the incorporation of renewable energy sources into agricultural operations, ensuring that energy from solar, wind, or biogas is utilized most effectively.

3.3.3.1.3 Machine Learning for Energy Optimization

ML models use data-driven models to identify patterns and make predictions. ML algorithms can be used to investigate large datasets from sensors, weather forecasts, and energy consumption records to make more accurate decisions about energy usage and optimize operations for efficiency.

- Energy Consumption Pattern Recognition: ML algorithms can analyze energy consumption patterns in agricultural processes and identify inefficiencies or trends. By learning from historical data, ML models can recommend adjustments to operations, such as when to operate irrigation pumps or when to power down machinery.
- Real-Time Optimization of Energy Systems: ML can be applied to continuously monitor and adjust energy systems in real time. For example, ML models can determine the optimal time for charging batteries or when to switch between grid power and renewable sources to minimize energy costs.
- Energy-Efficient Crop Management: ML algorithms can optimize crop management practices by adjusting irrigation schedules, light levels, and heating systems based on factors such as weather forecasts, plant growth stages, and soil conditions. This helps optimize energy usage and minimize waste.
- Autonomous Energy Systems: ML can drive autonomous systems that make real-time decisions about energy use. For example, a greenhouse could use ML to regulate temperature, lighting, and humidity without human intervention, based on dynamic data inputs.

Benefits

- Increased Energy Efficiency: ML helps reduce energy consumption by learning from historical data and making optimized decisions for energy use.
- Cost Reduction: By improving the efficiency of energy-intensive operations like irrigation and machinery usage, ML can significantly reduce energy costs.
- Scalable Solutions: ML-based systems can be applied to farms of all sizes, from small-scale operations to large commercial farms, making energy optimization scalable and accessible to diverse farming systems.

The utilization of renewable energy sources such as solar power, wind, and biogas in agricultural operations is a key strategy for reducing dependency on fossil fuels and improving the sustainability of farming. IoT, AI, and ML can help optimize the use of these renewable sources to meet energy demands more efficiently [17]. The ultimate goal of energy optimization in agriculture is to create smart farm energy systems that integrate renewable energy, energy-efficient machinery, and real-time energy management.

3.3.4 Soil Health and Biodiversity

Soil health is a cornerstone of sustainable agriculture, as it directly affects crop productivity, resilience, and the environmental impact of farming practices. Healthy soil will be rich in organic matter, nutrient content, and beneficial microorganisms, which contribute to the fertility, structure, and biodiversity of the soil ecosystem. In smart sustainable agriculture, the goal is to optimize soil health and biodiversity while maximizing agricultural productivity and minimizing environmental impact [18]. Leveraging advanced technologies provides farmers with tools to monitor, analyze, and optimize these critical aspects. Below, we explore specific soil health and biodiversity optimization strategies using these technologies and highlight the benefits of each.

3.3.4.1 Key Strategies

IoT technologies play a critical role in tracking and enhancing soil health and biodiversity in farming practices by monitoring ecosystems, wildlife, and plant diversity.

3.3.4.1.1 IoT in Soil Health and Biodiversity Optimization

IoT technologies involve sensor-enabled devices that collect data on soil conditions, enabling precision farming practices that optimize soil health.

- Soil Moisture Monitoring: IoT sensors track moisture levels in soil, enabling farmers to adjust irrigation systems to apply water only when needed. This helps maintain the right balance of water, prevents over-irrigation leading to nutrient leaching, and ensures optimal growing conditions.
- Soil Nutrient Monitoring: IoT sensors can continuously monitor the levels of essential nutrients present in the soil, informing precision fertilization. This reduces overdose of fertilizers and improves plant health by maintaining soil nutrient balance.

- Soil Temperature and pH Monitoring: Sensors can track soil temperature and pH, both of which are critical for microbial activity and nutrient availability [19]. Data from IoT devices helps farmers adjust practices like crop selection, fertilization, and planting schedules to maintain favorable conditions for soil organisms.
- Pollinator Monitoring: IoT-enabled sensors and cameras can track pollinator activity (e.g., bees, butterflies), which is essential for crop productivity. By monitoring the presence of pollinators, farmers can assess the health of their farm's ecosystem and make adjustments to support pollinator populations.
- Invasive Species Detection: IoT sensors can detect invasive species (plants or pests) by monitoring environmental conditions and tracking pest populations. Early detection of invasive species enables farmers to take timely action to control or eliminate them, protecting local biodiversity.
- Habitat Monitoring: IoT devices monitor habitats within the farm, including wetlands, forests, and grasslands, to ensure these areas are conducive to biodiversity. For example, temperature, humidity, and soil conditions can be monitored to support specific habitats for beneficial organisms like birds, earthworms, and beneficial insects.

Benefits

- Improved Resource Efficiency: By monitoring key soil parameters, farmers can reduce input waste (e.g., water, fertilizers) and apply resources only when and where they are needed.
- Minimized Environmental Impact: IoT helps reduce runoff, leaching, and other forms of soil and water pollution by ensuring more specific and focused application of fertilizers and water.
- Enhanced Soil Structure: By maintaining optimal moisture, temperature, and pH levels, IoT can help preserve soil structure, improve microbial health, and promote long-term soil fertility.
- Increased Pollinator Health: By tracking and supporting pollinators, IoT helps ensure better crop pollination, which is essential for food production and ecosystem stability.
- Early Detection of Biodiversity Threats: IoT offers real-time data on biodiversity changes, enabling farmers to take quick action against threats like invasive species or habitat loss.
- Habitat Preservation: Continuous monitoring of ecosystems helps preserve and enhance habitats for wildlife, fostering a more biodiverse and resilient farm ecosystem.

3.3.4.1.2 AI in Soil Health and Biodiversity Optimization

AI algorithms process an enormous amount of data from IoT sensors and other sources, offering predictive insights and optimization strategies for soil health management. Key strategies include the following:

- Predictive Soil Health Management: AI uses sensor data and weather forecasts to predict potential soil health issues such as malnutrition, compaction,

or erosion. Timely prediction allows farmers to take preventive actions like adjusting fertilization schedules or applying soil amendments before problems arise.
- Optimizing Fertilizer and Irrigation Schedules: AI systems analyze soil nutrient data alongside environmental conditions (e.g., rainfall forecasts, temperature) to create dynamic and adaptive fertilization and irrigation plans. These systems ensure that the right amount of inputs is applied at the optimal time, improving crop yields while minimizing input costs.
- Regenerative Farming Practices: AI promotes regenerative practices, including crop rotation and agroforestry. AI models simulate long-term outcomes of different management strategies, allowing farmers to adopt practices that regenerate soil health and enhance its resilience over time.
- Biodiversity Modeling and Simulation: AI can simulate the effects of various farming practices on biodiversity. For example, AI can model how crop rotations, the introduction of cover crops, or changes in pesticide use affect plant and animal species diversity.
- Predictive Biodiversity Management: AI models predict changes in biodiversity due to environmental factors such as weather patterns, pesticide use, or habitat destruction. By predicting these changes, AI allows farmers to proactively implement conservation strategies, such as adjusting planting patterns or creating wildlife corridors.
- Integrated Pest Management: AI can help optimize IPM strategies by analyzing pest populations and recommending the right balance of biological control agents, chemical treatments, and cultural practices to protect crops while preserving biodiversity.

Benefits

- Data-Driven Precision: AI provides actionable insights from large datasets, allowing for more informed decision-making regarding soil health management.
- Cost Reduction: By optimizing fertilizer, water, and input usage, AI reduces costs and enhances resource efficiency, which is especially valuable for small and large-scale farmers.
- Soil Regeneration: AI-supported strategies help implement practices that improve soil organic matter, increase water retention, and enhance biodiversity within the soil, supporting soil health and productivity.
- Better Ecosystem Modeling: AI helps farmers design farming systems that enhance biodiversity and ecosystem services by simulating the impact of different agricultural practices.
- Proactive Pest and Disease Management: AI supports biodiversity by reducing reliance on harsh chemicals and pest control measures, which deplete biodiversity.
- Optimized Habitat Conservation: By analyzing habitat data, AI can recommend farm design changes that enhance biodiversity, such as planting more diverse crops or creating buffer zones for wildlife.

3.3.4.1.3 ML in Soil Health and Biodiversity Optimization

ML enables systems to learn from historical and real-time data, improving predictions and recommendations over time.

- Soil Degradation Prediction: ML models analyze historical soil health data to identify patterns and predict areas at risk of degradation (e.g., erosion, compaction, or nutrient depletion). Farmers can then take early action to protect those areas, such as rotating crops, adding organic matter, or improving drainage.
- Soil Carbon Sequestration: ML algorithms can assess the potential for soil carbon sequestration under various farming practices, helping farmers implement methods that increase soil organic carbon levels.
- Automated Soil Quality Assessment: ML models can automate soil quality assessments by analyzing sensor data, providing real-time feedback on soil health, and helping farmers adjust practices to optimize conditions.
- Species Monitoring and Classification: ML algorithms can analyze images from cameras or drones to automatically identify and track species in the field. This helps monitor both beneficial species (e.g., pollinators, earthworms) and pest species, allowing for targeted interventions.
- Diversity Indices: ML models can calculate biodiversity indices, which measure the diversity of plant and animal species in a given area. By tracking these indices over time, farmers can assess the success of biodiversity-enhancing practices and make adjustments as needed [12].
- Adaptive Ecosystem Management: ML continuously analyzes data on biodiversity, pest populations, and environmental conditions to recommend adaptive strategies for improving farm ecosystems. This may involve adjusting land use, planting diverse crops, or altering pest management practices to enhance biodiversity.

Benefits

- Proactive Management: ML's predictive capabilities allow for proactive soil health management, preventing problems before they escalate.
- Improved Long-Term Planning: By analyzing long-term soil health trends, ML helps farmers make decisions that support the soil's long-term sustainability.
- Climate Resilience: ML can help adapt farming practices to changing climate conditions, ensuring that soil remains healthy and resilient to droughts, floods, and other environmental stresses.
- Continuous Improvement: ML's ability to learn from data over time ensures that biodiversity strategies evolve to become more effective, leading to a healthier, more resilient ecosystem.
- More Accurate Monitoring: By automating the identification and tracking of species, ML improves the accuracy of biodiversity assessments, enabling farmers to make informed decisions quickly.

3.4 ADVANTAGES OF RESOURCE OPTIMIZATION IN SUSTAINABLE AGRICULTURE

Optimizing resources in agriculture yields a wide range of benefits, both for farmers and the environment:

1. Environmental Sustainability
   - Reduced Pollution: By minimizing the use of synthetic chemicals (fertilizers, pesticides) and adopting practices such as precision application and organic farming, resource optimization reduces water and air pollution.
   - Carbon Sequestration: Policies such as agroforestry, conservation tillage, and organic farming can sequester carbon in the soil, contributing to climate change mitigation.
   - Biodiversity Conservation: Resource-efficient practices like crop diversification, agroecology, and organic farming support ecosystems and wildlife, helping preserve biodiversity in farming landscapes.
2. Economic Benefits
   - Cost Savings: Efficient use of water, fertilizers, and energy directly reduces input costs. Additionally, by improving yields and reducing waste, farmers can increase their profitability [16].
   - Risk Reduction: Resource optimization helps create more resilient farming systems that are better able to tolerate climate change, extreme weather events, and market fluctuations.
   - Market Access: Consumers increasingly demand sustainably produced food. Farmers who implement resource-efficient and environmentally friendly practices are better positioned to tap into these growing markets.
3. Social and Community Benefits
   - Improved Food Security: By optimizing the use of land, water, and nutrients, resource-efficient farming practices can increase agricultural productivity, supporting local food security and rural economies.
   - Healthier Communities: Sustainable practices, such as reduced pesticide use, healthier soils, and diversified farming systems, can improve food quality and public health, benefiting both producers and consumers.
   - Rural Development: Resource optimization creates opportunities for small-scale farmers and rural communities to innovate and improve livelihoods while protecting the environment.
4. Long-Term Sustainability
   - Soil Health and Fertility: Resource optimization helps to enhance soil health, which is critical for long-term agricultural sustainability. Crop rotation, reduced tillage, and organic matter addition can prevent soil degradation by maintaining soil fertility and increasing soil organic carbon content [20]. Healthy soils are more resilient to climate change, pests, and diseases, ensuring that agriculture can continue to produce food in the future.
   - Conservation of Natural Resources: Resource optimization reduces the pressure on natural resources, ensuring that they are available for future generations. Efficient irrigation systems and water management reduce

water consumption, which is vital as freshwater resources become scarcer in many parts of the world.
- Conserving biodiversity: Practices like agroforestry and reduced pesticide use support biodiversity by providing habitats for wildlife and beneficial organisms.
- Protecting air and water quality: By minimizing pesticide use and optimizing fertilizer application, resource optimization reduces pollution, helping to protect air and water quality.

5. Global Benefits and Alignment with Sustainable Development Goals
Resource optimization in agriculture aligns with several United Nations Sustainable Development Goals (SDGs), which aim to eradicate poverty, protect the planet, and ensure prosperity for all. Key SDGs supported by resource optimization include the following:
   - Zero Hunger – By optimizing resources, farmers can increase food production, improve food security, and ensure access to nutritious food for all.
   - Clean Water and Sanitation – Efficient water management and the reduction of water pollution contribute to clean water availability for communities.
   - Responsible Consumption and Production – Sustainable farming practices that optimize resource use help reduce food waste, lower environmental impact, and contribute to responsible production.
   - Climate Action – Resource optimization helps mitigate climate change by reducing greenhouse gas emissions, improving carbon sequestration, and enhancing agricultural resilience to climate impacts.

3.5 CHALLENGES TO RESOURCE OPTIMIZATION

While resource optimization offers clear benefits, there are several challenges to widespread adoption:

- Knowledge and Technical Barriers: Farmers may lack knowledge of sustainable practices or access to the tools and technologies needed for resource optimization. Education and extension services are crucial to overcoming this barrier.
- Upfront Investment: Transitioning to resource-efficient practices may require significant upfront investment in new technologies, infrastructure, or training. Financial support, subsidies, or low-interest loans can help offset these initial costs.
- Market Access and Policy Support: Government policies, such as subsidies for sustainable farming practices, research into new technologies, and improved market access for sustainably produced goods, can help encourage resource optimization.
- Climate Uncertainty: While resource optimization can help farmers adapt to changing weather patterns, unpredictable climate events and extreme weather continue to pose challenges (2). Flexible systems and long-term strategies are essential to address these risks.

3.6 CONCLUSION

The need for resource optimization in sustainable agricultural practices is urgent, as it addresses challenges such as resource depletion, climate change, and food insecurity. By optimizing land, water, nutrients, and energy use, farmers can increase productivity, reduce environmental harm, and ensure the long-term viability of their operations. This approach not only benefits individual farmers but also contributes to biodiversity conservation goals. Resource optimization in smart sustainable agriculture, powered by IoT, ML, and AI, holds immense potential to revolutionize farming practices. IoT sensors provide real-time data on soil conditions, weather patterns, and crop health, enabling precision irrigation, fertilization, and pest management. ML and AI algorithms analyze this data to predict optimal farming practices, adjust inputs dynamically, and forecast yield outcomes with remarkable accuracy. While challenges such as high initial costs, digital literacy, and technological infrastructure need to be addressed, the integration of these advanced technologies can pave the way for a more robust, conservation-focused, and sustainable agricultural system [21]. By embracing IoT, ML, and AI, farmers can not only optimize resource use but also contribute to global sustainability goals, ensuring food security for future generations in an environmentally responsible manner.

REFERENCES

1. Boix-Fayos C, De Vente J. Challenges and potential pathways towards sustainable agriculture within the European Green Deal. *Agric Syst.* [Internet]. 2023;207(December 2022):103634. Available from: https://doi.org/10.1016/j.agsy.2023.103634
2. Qazi S, Member S, Khawaja BA, Farooq QU. IoT-equipped and AI-enabled next generation smart agriculture: A critical review, current challenges and future trends. *IEEE Access*. 2022;10:21219–35.
3. Alwis SD, Hou Z, Zhang Y, Na MH, Ofoghi B, Sajjanhar A. A survey on smart farming data, applications and techniques. *Comput Ind.* [Internet]. 2022;138:103624. Available from: https://doi.org/10.1016/j.compind.2022.103624
4. Ur A, Alamoudi Y, Khalid HM, Morchid A, Muyeen M, Abdelaziz AY. Smart agriculture technology: An integrated framework of renewable energy resources, IoT-based energy management, and precision robotics. *Clean Energy Syst.* 2024;9(July):100132. Available from: https://doi.org/10.1016/j.cles.2024.100132
5. Maraveas C, Piromalis D, Arvanitis KG, Bartzanas T, Loukatos D. Applications of IoT for optimized greenhouse environment and resources management. *Comput Electron Agric.* 2022;198(April):106993. Available from: https://doi.org/10.1016/j.compag.2022.106993
6. Lee SW, Sarp S, Jeon DJ, Kim JH. Smart water grid: The future water management platform. *Desalin Water Treat.* 2015;55(2):341–8. Available from: http://doi.org/10.1080/19443994.2014.917887
7. Sharma V, Kumar A, Mittal H. Technological revolutions in smart farming: Current trends, challenges & future directions. *Comput Electron Agric.* 2022;201(June):107217. Available from: https://doi.org/10.1016/j.compag.2022.107217
8. Lytos A, Lagkas T, Sarigiannidis P, Zervakis M, Livanos G. Towards smart farming: Systems, frameworks and exploitation of multiple sources. *Comput Networks.* 2020;172(January):107147. Available from: https://doi.org/10.1016/j.comnet.2020.107147

9. Smaiot UI, Jani KA. A novel model for optimization of resource utilization in smart agriculture system. *IEEE Internet Things J.* 2022;9(13):11275–82.
10. Rai AK, Bana SR, Sachan DS, Singh B. Advancing sustainable agriculture: A comprehensive review for optimizing food production and environmental conservation. *Int J Plant Soil Sci.* 2023;35(16):417–25. Available from: https://doi.org/10.9734/IJPSS/2023/v35i163169
11. Chaudhary P, Singh S, Chaudhary A. Overview of biofertilizers in crop production and stress management for sustainable agriculture. *Front Plant Sci.*. 2022;13:930340. Available from: https://doi.org/10.3389/fpls.2022.930340
12. AlZubi, AA, Galyna K. Artificial intelligence and internet of things for sustainable farming and smart agriculture. *IEEE Access* 2023;11:78686–92. Available from: https://doi.org/10.1109/ACCESS.2023.3298215
13. Arya K. Machine learning for smart agriculture: Use of ML in precision farming, crop yield prediction, and resource optimization. 2024. Available from: http://dx.doi.org/10.2139/ssrn.5025948
14. Ramírez-Márquez C, Posadas-Paredes T, Raya-Tapia AY, Ponce-Ortega JM. Natural resource optimization and sustainability in Society 5.0: A comprehensive review. *Resources.* 2024;13(2):19. Available from: https://doi.org/10.3390/resources13020019
15. Khura TK, Lande SD, Parray R, Kushwaha H. Farm machinery use for energy use optimization under integrated crop management. *Integrated Crop Management Practices.* 2018:133. Available from: ISBN 978-93-83168-32-3
16. Tian Z, Wang JW, Li J, Han B. Designing future crops: Challenges and strategies for sustainable agriculture. *Plant J.* 2021;105(5):1165–78. Available from: https://doi.org/10.1111/tpj.15107
17. Adewusi AO, Asuzu OF, Olorunsogo T, Iwuanyanwu C, Adaga E, Daraojimba DO. AI in precision agriculture: A review of technologies for sustainable farming practices. *World J Adv Res Rev.* 2024;21(1):2276–85. Available from: https://doi.org/10.30574/wjarr.2024.21.1.0314
18. Powlson DS, Gregory PJ, Whalley WR, Quinton JN, Hopkins DW, Whitmore AP, et al. Soil management in relation to sustainable agriculture and ecosystem services q. *Food Policy.* 2011;36:S72–87. Available from: http://doi.org/10.1016/j.foodpol.2010.11.025
19. Ali A, Hussain T, Tantashutikun N, Hussain N, Cocetta G. Application of smart techniques, internet of things and data mining for resource use efficient and sustainable crop production. 2023;13(2):397. Available from: https://doi.org/10.3390/agriculture13020397
20. Said E, Belal AA, Abd-Elmabod SK, El-Shirbeny MA, Gad A. The egyptian journal of remote sensing and space sciences smart farming for improving agricultural management. *Egypt J Remote Sens Sp Sci.* 2021;24(3):971–81. Available from: https://doi.org/10.1016/j.ejrs.2021.08.007
21. Bahadur B, Dhanalakshmi R. Recent advancements and challenges of Internet of Things in smart agriculture: A survey. *Futur Gener Comput Syst.* 2022;126:169–84. Available from: https://doi.org/10.1016/j.future.2021.08.006

4 Revolutionizing Crop Protection
An Innovative Smart System to Deter Birds

Shivakumara T, Rajshekhar M Patil, Shantakumar B Patil, Muneshwara M S, Ananda R, and Rakshith M

4.1 INTRODUCTION

Agriculture is a cornerstone of the global economy, yet it faces constant threats, including bird predation. Birds damage crops extensively, leading to financial losses for farmers. Traditional methods like scarecrows and nets often fall short due to their limited effectiveness and scalability. With advancements in technology, smarter solutions are now within reach. This chapter presents the development and implementation of a smart crop protection system designed to deter birds dynamically. Built around a Raspberry Pi, the system combines hardware and software components to deliver real-time monitoring and deterrence. The system captures images using a camera module and processes them with machine learning algorithms to detect birds. Upon detection, it activates randomized alarm sounds to scare them away. Its flexibility, affordability, and ease of use make it suitable for diverse agricultural contexts.

The system's primary features include the following:

- **Dynamic Deterrence**: Randomized alarm sounds to prevent bird habituation.
- **Customizability**: Farmers can upload personalized alarm sounds in .wav format.
- **Cost-Effectiveness**: Leveraging the affordability of Raspberry Pi.
- **Sustainability**: Avoidance of harmful ultrasonic frequencies.

Recent advancements in agriculture highlight the increasing role of Internet of Things (IoT), Artificial Intelligence (AI), machine learning, and drone technologies in protecting crops from animal intrusions, bird damage, and environmental threats. These systems aim to enhance crop security, reduce financial losses, and improve agricultural productivity, particularly for rural and smallholder farmers. Several studies have proposed IoT-based solutions integrating sensors, microcontrollers, and communication modules to detect intrusions and automate responses such as activating

DOI: 10.1201/9781003594093-4

sprinklers, alarms, or deterrents. For example, Mohite et al. (2023) designed a system that uses smoke sensors and sprinklers to detect fire and animal intrusion, aiming for future integration with image processing and wireless networking [1]. Animal and bird encroachments are a major concern in agriculture. Raju et al. (2023) proposed a low-cost ultrasonic deterrent system to repel birds and wildlife using eco-friendly technology, with potential future integration of drone-based pest detection [2]. Similarly, Aranganathan et al. (2023) incorporated automated irrigation alongside intrusion alerts, demonstrating a cost-effective and non-invasive system suitable for rural India [3]. Patel et al. (2020) compared deep learning models like Single Shot Detector (SSD) and R-CNN for animal detection, finding SSD more effective, with future plans for app-based farmer alerts [4].

Meanwhile, Shaik et al. (2022) focused on intelligent systems that safeguard fields from wild animals using sensor networks [5, 6]. Tamilselvi, M., et al. (2023) aim to enhance agricultural productivity by integrating IoT, smart sensors, and drone technology for remote monitoring, disease detection, and crop growth prediction [7]. M Siva Sahithya (2022) proposes a system using IoT, machine learning, and AI to monitor weather, soil, and crop conditions, detecting diseases, pests, and animal attacks to improve crop production. The system provides alerts and insights to farmers, enabling better decision-making and increased crop yields [8]. Bird repellent technologies have also seen innovation. Wan Mohamed et al. (2020) and Wang ZiHao et al. (2019) explored drone-based bird deterrents using visual and auditory cues, demonstrating the effectiveness of combining airborne methods with psychological strategies [9, 10]. Condran et al. (2022) and Bertoglio et al. (2021) emphasize the transformative role of AIoT (AI + IoT) and digital agricultural technologies. These works identify challenges such as model interpretability, scalability, and cybersecurity, and advocate for sustainable, intelligent agricultural practices [11–13]. On the machine learning frontier, Mamdouh and Khattab (2021) utilized YOLO models for pest detection in olive crops, achieving high accuracy for real-time applications [14].

Emerging AI-based approaches like reinforcement learning (RL), convolutional neural networks (CNNs), and deep learning enhance system accuracy and responsiveness. Lee et al. (2021) introduced an RL-based repelling method that effectively delays bird invasion by using unfamiliar sound patterns [15]. A Hafeez et al. (2022) explores the role of drone technology in modernizing agriculture to meet the increasing food demands of a growing global population. It discusses advancements in drone applications, including crop monitoring, pesticide spraying, and AI-powered remote monitoring for precision agriculture [16]. Durgun (2023) developed an edge-computing acoustic system that uses machine learning to identify harmful bird species and play species-specific deterrent sounds [17]. Chaturvedi et al. (2024) proposed the Bird@Edge system using DenseNet-201 and multimodal edge AI for real-time bird identification [18]. In similar efforts, Simla et al. (2023) employed lightweight M2M protocols and deep learning to outperform traditional intrusion detection systems [19]. Naveenkumar et al. (2022) used DenseNet-201 to classify animals and activate sound-based deterrents, also sending location data to farm owners [20]. In terms of connectivity and protocols, Jerrin Simla et al. (2022) compared IoT topologies, concluding that mesh topologies using MQTT were optimal for intrusion data transmission [21].

4.2 SYSTEM ARCHITECTURE

The Adaptive Smart Crop Protection System (ASCPS) is designed to provide an automated and intelligent solution for bird deterrence in agricultural fields. The architecture of the system consists of four primary stages. First, a camera module captures real-time images or video of the field. This visual data is then passed to a CNN model that has been trained specifically to detect birds. Once a bird is detected, the system activates a deterrent mechanism by playing a randomly selected alarm sound from a preloaded library. This randomness helps prevent birds from becoming desensitized to specific noises. Finally, farmers can interact with the system through a user interface, which allows them to monitor field conditions, customize alarm settings, and adjust system parameters according to their preferences or field requirements.

The data flow within the ASCPS supports continuous and automated operation. The camera module is responsible for capturing uninterrupted video footage of the agricultural field. This video feed is processed in real time by the image recognition software running on a Raspberry Pi or similar embedded device. Each frame of the video is analyzed to detect the presence of birds. If a bird is identified, the system immediately selects a deterrent sound at random from a stored collection and plays it through a speaker to drive the bird away. If no birds are detected in the frame, the system continues to monitor the video stream without triggering the alarm, thereby conserving energy and minimizing unnecessary disturbances. The Smart Crop Protection System operates as a key component within the broader agricultural ecosystem. It is designed to work in harmony with environmental and human elements to effectively reduce crop losses caused by bird activity. The system's architecture ensures it can respond dynamically to field conditions and farmer inputs, making it adaptable to different agricultural contexts.

The scope of the ASCPS is primarily restricted to the field area monitored by the camera module. However, the system does interact with several external entities. Environmental factors such as weather conditions, bird population density, and crop type can significantly impact system performance. Additionally, farmers interact directly with the system to monitor activity and make configuration changes. Agricultural extension services may also utilize the data generated by the system for broader analyses or recommendations, enhancing the overall value of the system beyond immediate deterrence. The primary input to the system is the live video feed captured by the camera module. This video serves as the raw data for the detection algorithms. The outputs generated by the system include the audible deterrent sounds that are emitted when birds are detected, as well as visual data or status displays provided to users via the interface. These outputs ensure that both automated and manual components of the system are working in sync.

Despite its usefulness, the ASCPS faces several operational constraints. Hardware limitations, particularly those related to the Raspberry Pi, such as limited processing power, memory capacity, and power consumption, can affect performance. Environmental conditions like rain, fog, or poor lighting can also hinder the system's ability to capture clear images and make accurate detections. Furthermore, the effectiveness of the deterrent sounds may vary depending on bird species and behavioral patterns. Some birds may become accustomed to the sounds over time, reducing the long-term efficacy of the deterrence mechanism. To function effectively, the system is based on several assumptions. It is assumed that the device is deployed in an outdoor setting with sufficient natural lighting for proper image capture. The system must have access to a stable power source, which may include renewable options like solar panels for continuous operation in remote areas. Additionally, it is presumed that the farmers or designated operators have a basic level of technical knowledge required to install, configure, and maintain the system. These assumptions are critical to ensuring reliable and uninterrupted system performance in real-world agricultural environments.

In Figure 4.1, the context diagram illustrates the interactions and relationships between the Smart Crop Protection System and its associated entities. The components and their interactions are detailed as follows.

The Adaptive Smart Crop Protection System (ASCPS) is designed to continuously monitor the field environment, specifically focusing on detecting the presence of birds, which pose a significant threat to crops. Birds are treated as environmental inputs that trigger the system's core functions. The system identifies bird activity through image recognition and analysis and responds accordingly. This creates a feedback loop where the presence of birds informs the system to initiate protective measures, such as activating deterrent sounds. This real-time responsiveness ensures proactive and timely protection of crops. A stable and reliable power supply is a critical input for the effective operation of the Adaptive Smart Crop Protection System. It ensures uninterrupted functionality of the system's components, including the camera module, processing unit, and output devices such as speakers. Any disruption in the power source can impact the system's ability to detect threats and execute protective actions. Consequently, the efficiency and overall reliability of the system are heavily dependent on a consistent power supply. In many cases, the system may rely on traditional grid power, backup batteries, or renewable sources like solar panels to maintain operational continuity.

At the core of the setup is the Smart Crop Protection System itself, which serves as the central processing unit that coordinates all operations. It assimilates inputs from both the environment (bird activity) and the power supply to make intelligent decisions regarding crop protection. The system processes visual data to detect

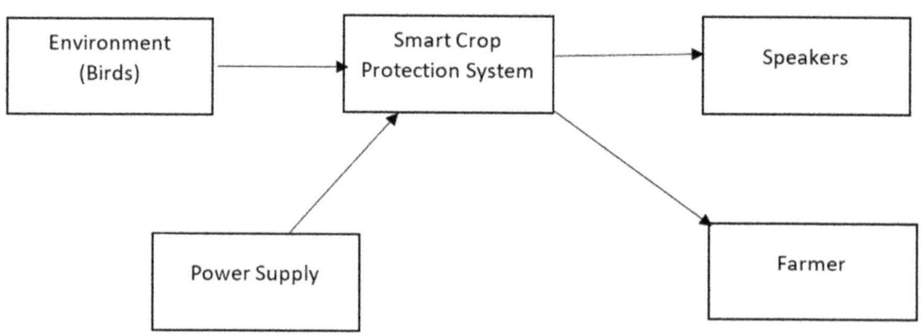

FIGURE 4.1 Context diagram.

threats and then triggers appropriate responses. It also delivers real-time data and actionable insights to farmers through an integrated user interface. These outputs help farmers monitor crop safety and adjust system settings as needed. Additionally, the system sends output commands to external devices such as speakers to deter birds effectively.

Speakers are integral output devices used by the ASCPS to carry out bird deterrence. When the system detects a bird, it activates the speakers to emit preloaded deterrent sounds designed to scare birds away from the field. This automatic response minimizes the need for manual intervention by the farmer. While the primary function of the speakers is to deter birds, their activation indirectly supports the farmer's objective of crop protection by ensuring fewer birds come into contact with the crops, thereby reducing the risk of damage. Farmers are the end-users and key beneficiaries of the Smart Crop Protection System. The system provides them with vital information about bird activity and system performance through the user interface. This allows farmers to stay informed and react promptly to emerging threats. Furthermore, the insights and notifications provided by the system support data-driven decision-making, empowering farmers to implement timely and effective crop protection strategies. By automating bird deterrence and offering constant monitoring, the system significantly reduces manual labor and enhances crop security for the farmer.

4.3 TOOLS AND TECHNOLOGIES USED

4.3.1 Hardware Requirements

The development and deployment of the Smart Crop Protection System rely on a blend of essential hardware components that ensure robust and efficient functioning. At the heart of the system is the Raspberry Pi, which acts as the central processing unit responsible for image analysis, sound playback, and overall system control. To support efficient multitasking and smooth operation, a RAM of 8 GB or above is recommended, along with a storage capacity of at least 16 GB, typically through a memory card. A camera module is used to capture real-time video footage of the field, which is essential for bird detection. Upon detecting a bird, a connected speaker is triggered to emit deterrent sounds, thus preventing birds from damaging the crops. A consistent and stable power supply ensures the system remains operational without interruptions, while internet connectivity facilitates real-time interaction, remote monitoring, and system updates if needed.

4.3.2 Software Requirements

The software infrastructure of the Smart Crop Protection System is built primarily on Python, selected for its readability, versatility, and the broad availability of scientific and machine learning libraries. Among the key libraries used are OpenCV, which facilitates image processing and computer vision tasks; NumPy, for handling numerical operations and data structures; and PyAudio, which manages audio input and output functionalities.

The system operates on Raspberry Pi OS (formerly known as Raspbian), a lightweight, Linux-based operating system optimized for the Raspberry Pi hardware, ensuring stability and compatibility with the required software stack.

4.4 SCOPE AND OBJECTIVE

The Smart Crop Protection System is designed with adaptability in mind, allowing it to serve a variety of users involved in the agricultural ecosystem. Its functionality is tailored to reduce bird-related crop losses and provide meaningful data for both operational and analytical purposes. The primary users of the system are farmers, who aim to protect their crops from bird damage and minimize losses. Additionally, agricultural extension workers can use the system to monitor field activity and provide data-driven recommendations to farmers. Another important user group includes researchers, who benefit from analyzing the system-generated data to study bird behavior, deterrent effectiveness, and system performance trends across different environmental conditions. For farmers, the main goal is to protect crops in an efficient, automated manner while optimizing the use of available resources. Extension workers, on the other hand, utilize the data and monitoring capabilities of the system to provide timely and accurate advice to farmers, thus enhancing the overall agricultural decision-making process.

4.5 DATA FLOW DIAGRAM

The data flow diagram (DFD) for the Smart Crop Protection System illustrates the sequential processes involved in bird detection and deterrence. The system begins with the camera module, which constantly captures real-time video of the field area to monitor bird activity. Images are captured at defined intervals and sent to the Raspberry Pi, which serves as the central data processor. Here, the image data undergoes analysis using a CNN and OpenCV library.

The image analysis process involves two major steps: pre-processing, where images are resized and filtered to optimize them for analysis, and bird detection, where the CNN model is applied to identify the presence of birds. Upon detecting a bird, the system activates its response mechanism by randomly selecting an alarm sound from a preloaded audio library and playing it through the speaker to scare away the birds. Simultaneously, the system sends real-time alerts to the user interface, allowing the farmer to monitor system activity and modify settings if needed. If no bird is detected, the system continues monitoring by processing the next set of captured images without triggering an alarm.

The user interaction component of the system allows farmers to upload custom .wav files through the user interface, tailoring alarm sounds to local bird species for greater deterrence effectiveness. The interface also provides visibility into the system's operational status and logs of bird detection events, enabling the farmer to assess and adjust strategies. This entire workflow operates in a continuous feedback loop, with real-time image input and user customizations guiding the system's ongoing responses to bird activity (Figure 4.2).

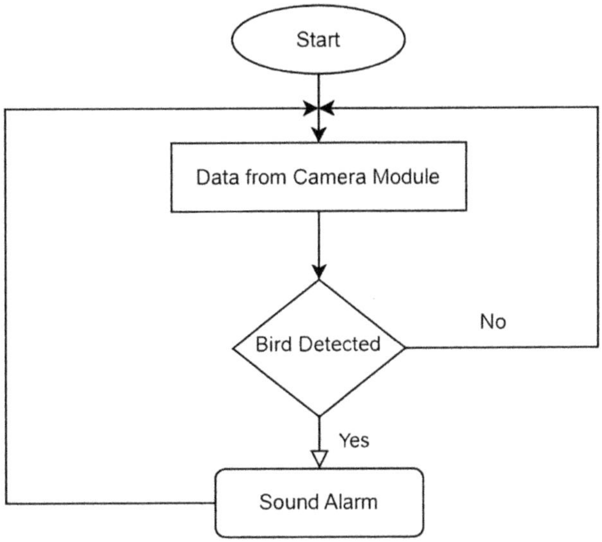

FIGURE 4.2 Data flow diagram.

4.6 ACTIVITY DIAGRAM

The activity diagram encapsulates the core operational workflow of the Smart Crop Protection System, highlighting its repetitive and adaptive nature. The process begins with the initialization of the system, which includes loading necessary software components and establishing connections with hardware elements like the camera and speaker. Once initialized, the system enters a monitoring mode where it detects birds using image data captured by the camera module.

If a bird is detected in the image, the system proceeds to play a randomly selected alarm sound through the speaker to deter the bird from damaging the crops. This response is both immediate and automated. The system then loops back to continue monitoring for further activity, ensuring continuous and real-time protection of the crop area. This looped workflow forms the basis for an effective, autonomous bird deterrent mechanism designed to operate with minimal human intervention while offering adaptability through user inputs and environmental feedback (Figure 4.3).

The Bird Detection Decision stage plays a critical role in the system's workflow. During this phase, the bird detection algorithm performs a detailed analysis of the visual data captured by the camera module to determine whether birds are present. This initiates a conditional flow: if no bird is detected, the system returns to a monitoring state, continuously gathering visual data and reprocessing it in an ongoing cycle. However, if a bird is detected, the system transitions to the next phase, initiating a deterrent action.

In the Play Sound phase, the system triggers its alarm mechanism. A random preloaded alarm sound is selected and played through connected speakers. This sound generation process is designed to produce noises specifically aimed at deterring birds. The auditory disruption startles the birds, discouraging them from entering or staying in the crop area, thereby protecting the field from damage.

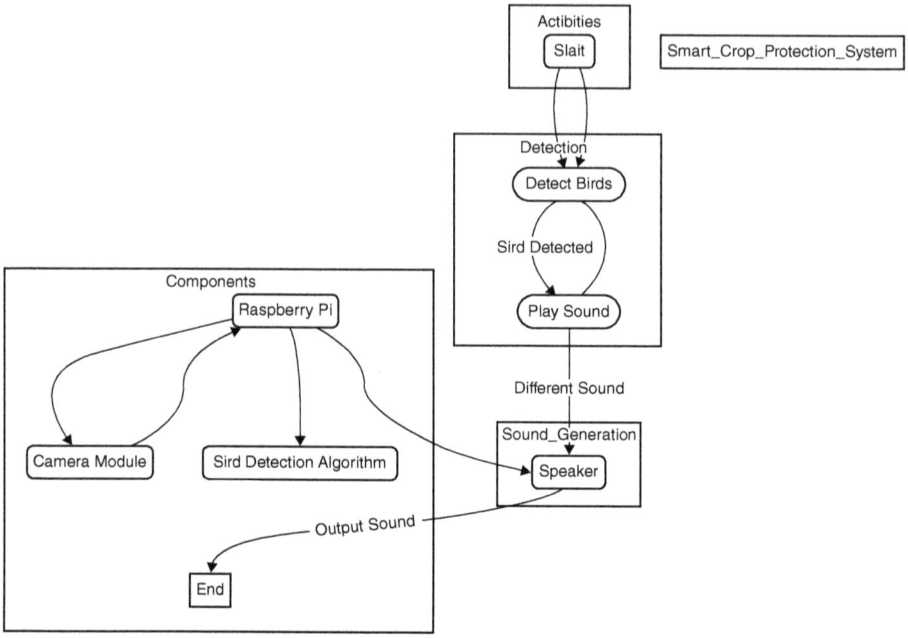

FIGURE 4.3 Activity diagram.

The final step is marked as the End of the activity workflow, where the selected sound has been played, completing the process. The cycle then restarts, ensuring continued vigilance and real-time responsiveness.

System Components supporting this workflow include the following:

- **Camera Module**: Captures real-time images of the environment.
- **Raspberry Pi**: Acts as the processing unit running the detection algorithm.
- **Bird Detection Algorithm**: Analyzes images to detect bird presence.
- **Speaker**: Emits deterrent sounds upon detection.

The process benefits are significant. The system offers real-time monitoring, ensuring immediate detection of any bird activity through continuous video input. The automated response mechanism greatly reduces the need for manual observation, offering convenience to the farmer. Additionally, effective bird deterrence is achieved through prompt and randomized alarm sounds, enhancing the protection of crops against bird-related damage.

4.7 USE CASE DIAGRAM

The Use Case Diagram offers a comprehensive and high-level overview of the interactions between the various users and system components within the Smart Crop Protection System. It serves to clearly visualize how different actors engage with the system, highlighting the functional relationships and delineating the scope of its behavior. This diagram is particularly useful for understanding the roles, inputs, and responses that define system operations. In this system, there are two primary actors. The first is the Farmer, who is the main end-user responsible for interacting with the system. The farmer's role includes system customization, such as uploading personalized alarm sounds – these could be recordings of local predator calls or loud, disruptive noises intended to effectively scare away birds. The second actor is the Bird, which, although not a user in the conventional sense, is treated as an external entity that interacts indirectly with the system. The presence of a bird in the monitored field triggers the detection and response workflow, initiating the core functions of the system.

The use cases describe the specific actions and interactions that occur between these actors and the system. One such use case is "Detect Bird," where the system leverages image processing through a trained CNN to identify birds appearing in the camera feed. Another is "Capture Image," which refers to the continuous operation of the camera module, capturing frames of the crop field at regular intervals. These images serve as input data for further analysis. Next, the "Process Data" use case involves the Raspberry Pi, which functions as the central processing unit of the system. It runs the detection algorithm that processes the captured image data to evaluate whether birds are present. If bird activity is confirmed, the system proceeds to the "Sound Alarm" use case, wherein it activates a speaker system to play a deterrent sound and scare the birds away from the crops. This structured use case framework provides a clear understanding of how each system component and user action contributes to the overall goal of crop protection. By defining these interactions, it becomes easier for stakeholders – including farmers, developers, and researchers – to assess the system's functionality and potential areas for enhancement (Figure 4.4).

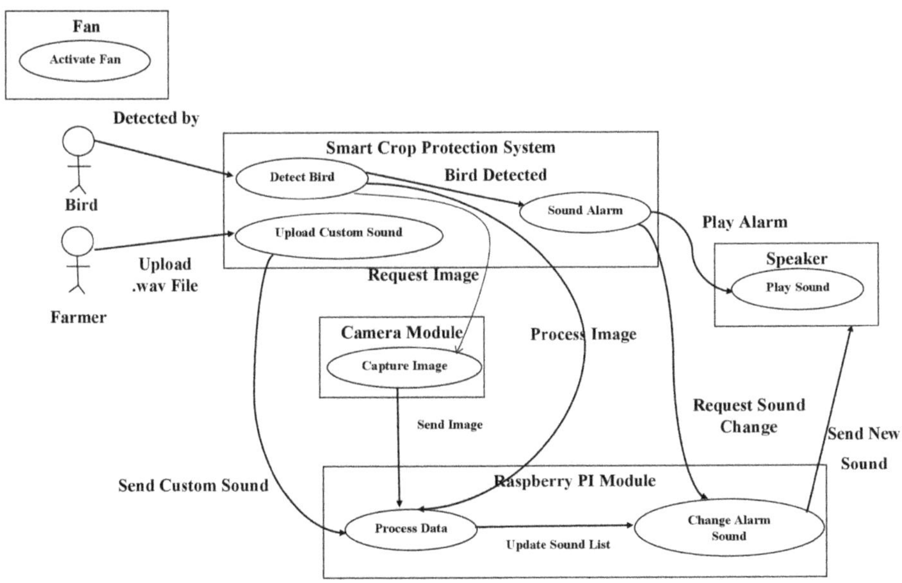

FIGURE 4.4 Use case diagram.

The system is designed with several features aimed at deterring birds effectively. One of the key functionalities is the ability to change the alarm sound. This ensures that birds do not become accustomed to a single sound, maintaining its effectiveness in repelling them. The system also includes an option to play a sound through a speaker. When triggered, the alarm sound plays to deter birds from the area. This can be activated automatically based on the system's settings or manually by the user. Farmers have the flexibility to upload custom alarm sounds to the system. They can upload .wav sound files, which allows for personalization and the use of different tones or sounds that may be more effective in their specific environment. Additionally, if the system includes a fan, it can be activated as part of the bird deterrence mechanism. The fan works in conjunction with the alarm sound, helping to create an environment that discourages birds from settling in the area.

4.8 SEQUENCE DIAGRAM

In Figure 4.5, the sequence diagram illustrates the interactions within the Smart Crop Protection System when a bird is detected and the process for uploading custom sounds. The Bird Detection Sequence begins when a bird enters the vicinity, which triggers the Camera Module. The Camera Module captures an image of the area and sends this data to the Smart Crop Protection System for further analysis. Upon receiving the image data, the system forwards it to the Raspberry Pi Module, where the data is processed to determine if a bird is detected in the frame. If the Raspberry Pi Module identifies a bird, the system then requests it to determine the appropriate alarm sound. The Raspberry Pi Module selects the alarm sound based on pre-configured settings or specific parameters. Once the sound is selected, the

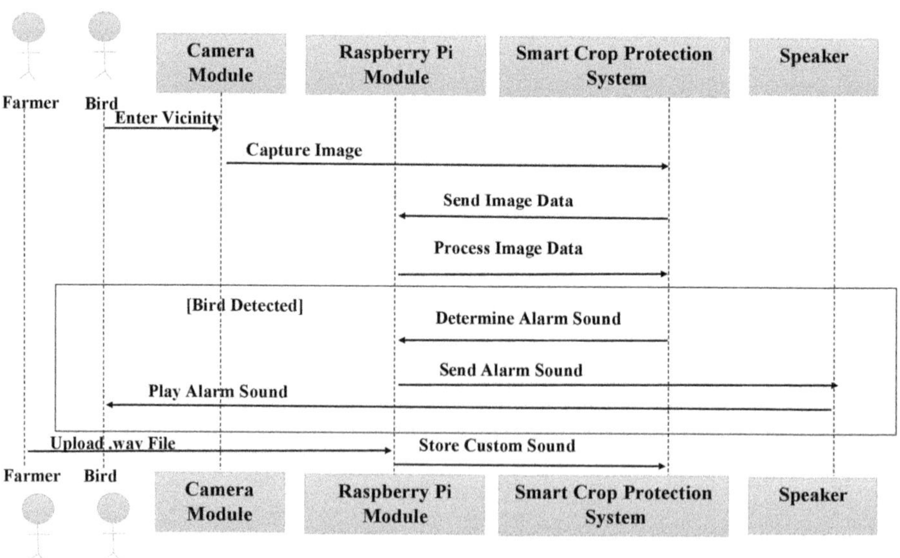

FIGURE 4.5 Sequence diagram.

module sends the sound data to the Speaker, which then plays the alarm to scare away the bird and prevent it from causing damage. In addition to the automated alarm sounds, there is a Custom Sound Upload Sequence. In this sequence, the farmer has the ability to upload a custom wav file to the Raspberry Pi Module. Once uploaded, the Raspberry Pi Module stores the custom sound in the Smart Crop Protection System, making it available for future use in bird deterrence. This feature allows the farmer to use personalized sounds that may be more effective in protecting their crops.

4.9 EXPERIMENTAL SETUP AND RESULTS

Figure 4.6 shows the Raspberry Pi interphase. Here we can see the output screen displaying the birds being detected through the camera module. In Figure 4.7, the system setup for the Smart Crop Protection System is illustrated. The circuit diagram showcases the hardware components and their connections within the Smart Crop Protection system, which is designed to detect birds and deter them using sound.

The bird deterrence system is comprised of several key components that work in tandem to protect crops from bird damage. At its core, the system is powered by a portable power supply, allowing it to operate independently in the field. The Raspberry Pi module serves as the system's processing unit, running a sophisticated bird detection algorithm and controlling the various components. A camera module, connected to the Raspberry Pi, captures real-time images of the crop area, providing the visual data necessary for detecting bird presence. Once a bird is detected, the system triggers speakers to play deterrent sounds, effectively scaring the birds away. Additionally, a fan is integrated into the system to cool the Raspberry Pi, preventing overheating during prolonged operation. The system's operational workflow is designed for seamless functionality. The camera module continuously

FIGURE 4.6 Successful detection of birds.

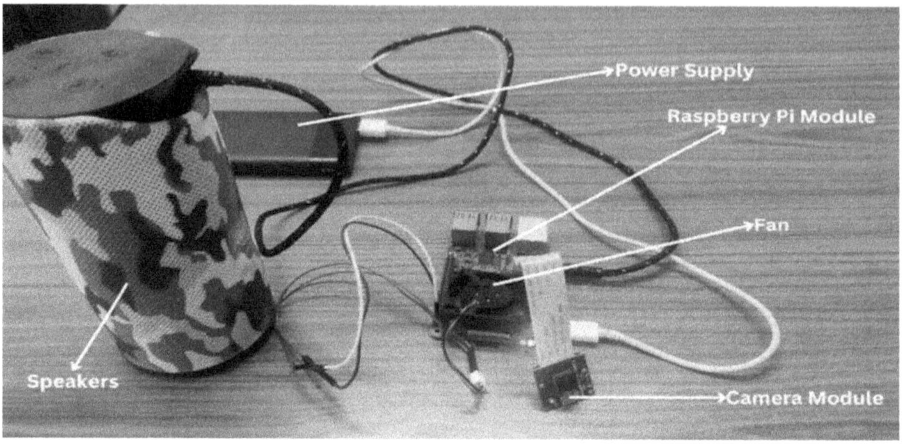

FIGURE 4.7 System setup.

captures images of the crop area, sending the data to the Raspberry Pi for analysis. When the bird detection algorithm identifies a bird, the Raspberry Pi promptly triggers the speakers to play the deterrent sounds. This autonomous operation not only reduces the need for constant human supervision but also provides an effective solution for protecting crops from bird damage. By leveraging advanced technology and real-time data analysis, the system offers a reliable and efficient means of safeguarding agricultural yields.

4.10 CONCLUSION

The smart crop protection system offers a practical solution for mitigating bird-related crop losses. By integrating image recognition, IoT, and machine learning, the system represents a significant advancement in agricultural technology. Its ability to dynamically detect and deter birds through randomized alarm sounds makes it a highly effective and adaptable tool for farmers. Additionally, the system's cost-effectiveness, simplicity, and customizable features ensure its usability across various farming contexts. Initial testing has showcased promising results, highlighting the system's potential to improve crop yields and reduce economic losses. However, further refinement is needed, particularly in expanding its scalability, integrating solar power for sustainability, and optimizing detection algorithms for broader environmental conditions. Field trials across diverse agricultural environments will provide valuable insights into the system's long-term performance. Future enhancements may include incorporating multi-sensor setups, advanced data analytics for bird behavior prediction, and remote monitoring capabilities. Ultimately, the smart crop protection system aligns with global efforts toward sustainable agriculture, addressing the persistent challenge of crop losses due to bird predation while promoting eco-friendly practices. This innovation exemplifies how technology can empower farmers, ensuring food security and improving livelihoods in a rapidly changing world.

REFERENCES

[1] Mohite, R. B., Thorat, Z. V., Chinchawade, A. J., Mane, V. B., Waghmode, U. P., and Pawar, V. E., "Smart Farm Protection System Using IoT," (2023, March 23). *IEEE Conference Publication* | IEEE Xplore, 2023. https://ieeexplore.ieee.org/document/10125709

[2] Raju, R., Sreedharan, N. K., Krishnan, A., Kumar, N., and Sreejith, R., "Innovative Low-Cost Field Protection (LCFP) Technique: Safeguarding Agricultural Land and Crops through Advanced Technology Integration," (2023, May 9). *IEEE Conference Publication*. IEEE Xplore, 2023. https://ieeexplore.ieee.org/document/10425242

[3] Aranganathan, A., et al., "Smart Agricultural Security System: Integrating Arduino Technology for Intruder Detection and Crop Protection," *2023 International Conference on Innovative Computing, Intelligent Communication and Smart Electrical Systems (ICSES)*, Chennai, India. IEEE, 2023.

[4] Patel, R. S., Amin, D. D., Patel, S., Rahevar, M., Patel, C., and Nayak, A., "Smart crop protection against animal encroachment using deep learning," *International Journal of Innovative Technology and Exploring Engineering (IJITEE)*, vol. 9, no. 10, August 2020. https://ieeexplore.ieee.org/document/10167118

[5] Shaik, M. F., Mounika, R., Prasad, A. D., Raja, I. R., Sekhar, B. P., and Sampath, D., "Intelligent Secure Smart Crop Protection from Wild Animals," *2022 8th International Conference on Advanced Computing and Communication Systems (ICACCS)*, IEEE, 2022. https://doi.org/10.1109/icaccs54159.2022.9785265

[6] Shaik, M. F., "Intelligent Secure Smart Crop protection From Wild Animals," *2022 8th International Conference on Advanced Computing and Communication Systems (ICACCS)*, Coimbatore, India, IEEE, 25 March 2022. https://doi.org/10.1109/ICACCS54159.2022.9785265

[7] Tamilselvi, M., et al., "Internet of Things Enabled Energy-Efficient Flying Robots for Agricultural Field Monitoring Using Smart Sensors," *Intelligent Technologies for Sensors*. Conference Proceedings (ICACCS-22). New York: Apple Academic Press, 2023. 59–73.

[8] Siva Sahitya, M., "An Investigation on Smart Crop Protection System Utilizing Internet of Things," *2022 8th International Conference on Signal Processing and Communication (ICSC)*, Noida, India, IEEE, 1 December 2022. https://doi.org/10.1109/ICSC56524.2022.10009442

[9] Mohamed, W. M. W., Naim, M. N. M., and Abdullah, A., "The Efficacy of Visual and Auditory Bird Scaring Techniques Using Drone at Paddy Fields," *IOP Conference Series: Materials Science and Engineering*. Vol. 834. No. 1. IOP Publishing, 2020. https://doi.org/10.1088/1757-899X/834/1/012072

[10] Wang, Z., Griffin, A. S., Lucas, A., and Wong, K. C., "Psychological warfare in vineyard: Using drones and bird psychology to control bird damage to wine grapes," *Crop Protection*, vol. 120, pp. 163–170, 2019. https://doi.org/10.1016/j.cropro.2019.02.025

[11] Condran, S., Bewong, M., Islam, M. Z., Maphosa, L., and Zheng, L., "Machine learning in precision agriculture: A survey on trends, applications and evaluations over two decades," *IEEE Access*, vol. 10, pp. 73786–73803, 2022. https://doi.org/10.1109/ACCESS.2022.3188649

[12] Bertoglio, R., Corbo, C., Renga, F. M., and Matteucci, M., "The digital agricultural revolution: A bibliometric analysis literature review," *IEEE Access*, vol. 9, pp. 134762–134782, 2021. https://doi.org/10.1109/ACCESS.2021.3115258

[13] Mukhopadhyay, S. C., Tyagi, S. K. S., Suryadevara, N. K., Piuri, V., Scotti, F., and Zeadally, S., "Artificial intelligence-based sensors for next generation IoT applications: A review," *IEEE Sensors Journal*, vol. 21, no. 22, pp. 24920–24932, 2021. https://doi.org/10.1109/JSEN.2021.3055618

[14] Mamdouh, N. and Khattab, A., "Yolo-based deep learning framework for olive fruit fly detection and counting," *IEEE Access*, vol. 9, pp. 84252–84262, 2021. https://doi.org/10.1109/ACCESS.2021.3088075
[15] Lee, C. W., Muminov, A., Ko, M.-C., Oh, H.-J., Lee, J. D., Kwon, Y.-A., Na, D., Heo, S.-P., and Jeon, H. S., "Anti-adaptive harmful birds repelling method based on reinforcement learning approach," *IEEE Access*, vol. 9, pp. 60553–60563, 2021. https://doi.org/10.1109/ACCESS.2021.3073205
[16] Hafeez, A., Husain, M. A., Singh, S., Chauhan, A., Khan, M. T., Kumar, N., Chauhan, A., and Soni, S., "Implementation of drone technology for farm monitoring & pesticide spraying: A review," *Information Processing in Agriculture*, 2022. https://doi.org/10.1016/j.inpa.2022.02.002
[17] Durgun, M., "An Acoustic Bird Repellent System Leveraging Edge Computing and machine Learning Technologies," *2023 Innovations in Intelligent Systems and Applications Conference (ASYU)*. IEEE, 2023. https://doi.org/10.1109/ASYU58738.2023.10296574
[18] Chaturvedi, K., et al., "Deep learning-based identification of birds in agricultural fields," *2024 International Conference on Computing and Intelligent Reality Technologies (ICCIRT)*. IEEE, 2024. https://doi.org/10.1109/ICCIRT59484.2024.10921862
[19] Simla, A. J., Chakravarthi, R., and Megalan Leo, L., "Agricultural intrusion detection (AID) based on the internet of things and deep learning with the enhanced lightweight M2M protocol," *Soft Computing*, vol. 27, Issue: 03, pp. 1–12, 2023. https://doi.org/10.1007/s00500-023-07935-1
[20] Naveenkumar, M., Manoj, R., Nandhakumar, B, and Rahul, R., "DenseNet201 for animal detection and repellent system", *International Conference on Electronic Systems and Intelligent Computing (ICESIC)*, pp. 213–218, 2022. https://doi.org/10.1109/ICESIC53714.2022.9783582
[21] Jerrin Simla, A., Chakravarthy, R., and Megalan Leo, L., "An experimental study of IoT-based topologies on MQTT protocol for agriculture intrusion detection, measurement: sensors," *Science Direct*, vol. 24, p. 100470, 2022. ISSN 2665–9174.

5 Climate Change and Agriculture

Emerging Challenges and Far-Reaching Impacts

Poonam Chaudhary, Sneha Kandacharam, Swastika, and B. Santhana Krishnan

5.1 INTRODUCTION

There is an urgent and escalating problem that interests everyone because it influences ecosystems and weather patterns. The United Nations defines climate change as gradual changes in the environment. This is a long process that is associated with shifts in temperature and precipitation levels. They are largely caused by human actions, for instance, the emission of greenhouse gases (GHGs), including carbon dioxide and methane, which lead to observable changes in the Earth's climate. Through natural variation over millions of years, the Earth's climate has reacted. But many of the more significant recent growths in global warming for the past hundred years were produced anthropogenically – mainly from fossil fuel combustion to meet energy consumption and requirements for clearing forests. The burning resulting from industries, agricultural practice, among others, hence raise carbon dioxide concentrations in the air. It creates the greenhouse effect blanket trapping air in an attempt by sunlight to break through toward space. This increases the air's internal temperature globally.

Rising climate change is a common problem the entire human race is facing, and it is only possible to overcome it through coordinated efforts all over the world. It urges the substitution of non-renewable and clean energy sources, the use of sustainable land budget planning, the promotion of energy efficiency, and the making of policies that reduce GHG emissions. Additionally, it is essential that mitigation strategies be designed to accommodate already-occurring consequences. Logical solutions to climate change challenges can be developed only when we have in-depth knowledge of the complexities of climate change and its consequences. In the process of this research, we should also explore different aspects related to climate change in order to understand the whole problem. This will also act as the foundation for any other actions that can be performed to ensure the Earth's better and healthier future.

DOI: 10.1201/9781003594093-5

5.1.1 Impacts of Climate Change

Mastering the implications of climate change is an essential step in the process of revealing the spectrum of the consequences that this phenomenon involves for our planet and our lives. Climate change can be attributed to human activities to a large extent, and the impact can be felt profoundly across ecosystems, economies, and people. One of the most observable impacts of climate change is the alteration in weather patterns and the increased frequency of extreme weather events. The increased severity and frequency of heat waves, hurricanes, floods, droughts, and wildfires pose a danger to the community, infrastructure, agriculture, and economy. These events often result in loss of lives, displacement of populations, damage to property, and disruptions in food and water supplies.

In addition to the rising temperatures, sea level rise is another major effect of climate change. It mostly occurs as a result of melting ice masses and glaciers at the poles, which threatens the marine territories and coastal lowlands, so it is the source of the river deposits, the coasts, and the flooding of inhabited areas. Islands and highly inhabited marine regions will be the ones to suffer the most, as these places will be at the forefront of the threats; hence, the authorities of these places will be forced to devise strategies to keep both their communities and infrastructure safe. In addition, due to shifts in climate patterns, agriculture and food security are heavily influenced. The effect of temperature and rainfall on the crops causes a variation in the yield of the crops and their quality, which ultimately leads to the availability and quality of the food. It is thus a big problem because it can worsen the conditions in the area of food production worldwide and could contribute to the increasing hunger and malnutrition in the vulnerable population. Ecosystems are also under immense stress due to climate change. Species are not able to adapt quickly to the sudden changes in their habitats, which results in the movement of those habitats, disruption of migration routes, and threats to biodiversity. On the other hand, climate change also affects human health, bringing an increased threat of heat-related illness, the spread of vector-borne disease, and respiratory issues that result from degraded air quality. Understanding these impacts emphasizes the urgency of addressing climate change through collective global action. With all-inclusive research on climate change, we get a better picture of the interworking process. We could be formulating ways for combating a disaster or educating communities that are at risk concerning surviving the new circumstances. As the scope of this subject narrows, we realize that collaborative actions and creative answers play a critical role in life support of our planet and the creation of a sustainable environment for the upcoming generations.

5.1.2 Role of Technology in Addressing Climate Change

Computers and technology are the most powerful tools to address climate change because of their impacts and contributions across various domains. The complex computing models and simulations driven by high-performance computing help scientists in simulating the complicated global warming scenarios and also enable them to forecast the more precise future climate patterns and their possible impacts. These models give us the ability to observe every complex aspect of how sea level rise, temperature fluctuations, and GHG emissions contribute to climate change, which, as a result, helps the government to make informed decisions and formulate mitigation and adaptation strategies.

Along with those, machine learning models, big data analytics, enable researchers to conduct trials using a huge amount of environmental data. They shed light on what patterns are the most prominent in air temperature changes, modifications in biodiversity, and possible influence of human activity on our planet. It facilitates the understanding of the consequences, associations, and solutions that mitigate climate risks and enhance sustainability. Advanced technologies support the development and implementation of renewable energy sources by enhancing efficiency and reducing costs through innovations in solar, wind, and hydroelectric power generation. They play a key role in facilitating the transition to a low-carbon economy by providing energy storage solutions like smart grids and energy-efficient devices.

The use of remote sensing technologies such as satellites and sensors provides real-time monitoring and data collection on multiple environmental parameters. These tools monitor changes in land use, calculate carbon emissions, track deforestation, and record variations in weather. This variety of information aims to help with the early detection of environmental changes and prompt the implementation of conservation, crisis management, and land use planning, among other initiatives. Additionally, global scientific, policy, and community collaboration is encouraged by digital platforms, devices, and technology, which boosts the development of innovative green solutions to combat global warming on a global scale.

5.2 AI-ENABLED CLIMATE CHANGE MITIGATION

5.2.1 AI-Driven Smart Grid and Energy Management Solutions

Nowadays, smart grids have been widespread. The smart grid is an electrical network that applies digital innovations to enhance the efficiency, reliability, and sustainability of the power grid. Contrary to one-way communication, they enable bi-directional contact between producers and consumers. They facilitate the immediate data exchange across the power grid parts for better monitoring and supervision. Through smart metering, there is minute-by-minute energy tracking, which gives consumers enough knowledge to reflect on their usage patterns and potentially change their habits, if needed. Using smart grids, one can integrate renewable energy at peak times while at the same time optimizing energy consumption based on real-time needs and grid conditions. Here, it implies promoting consumption away from peak periods or regulating demand to spare the peak periods.

5.2.2 AI-Based Carbon Capture, Utilization, and Sequestration Technologies

The root cause of climate change and global warming is the emission of GHG, particularly carbon dioxide, to the atmosphere as it depletes the ozone blanket surrounding the atmosphere, which protects it from the sun's heat. Carbon capture, utilization, and sequestration (CCUS) technologies aim to stop the carbon dioxide generated from power plants, industries, and fossil fuel burning from reaching the atmosphere. CCUS plays an important role in decarbonization, although it's not a direct mitigation strategy, but combined with other efforts, it can significantly contribute to mitigating climate change [1]. There's a new jargon nowadays, called Direct Air Capture (DAC), which aims to capture carbon dioxide from the air. The computation abilities of artificial intelligence (AI) and machine learning can be leveraged to find the most suitable chemical sorbent from a variety of chemicals for carbon capture.

The Paris Climate Agreement aims to limit global warming to 2°C, with a recommendation to reduce it to 1.5°C in COP-26. CCUS is being increasingly recognized as a measure to mitigate climate change. Initiatives like the Global CCS Institute, Decarbonization Technology, and Flexible Carbon Capture and Storage (FLECCS) are promoting the adoption of CCUS technologies. Efficient implementation of cost-effective CCUS systems requires robust evaluation of available technological alternatives. The CO_2 capture is still the most expensive component of the CCUS network. Such federal and provincial initiatives like the Energy Innovation Program play a great role. But the efficiency in costs could be improved with more comprehensive modeling, simulation, and optimization techniques for techno-economic and environmental considerations related to CCUS systems. Technologies such as artificial intelligence and big data analytics seem to have promising possibilities in developing such kinds of tools. An enhancement to current CCUS systems or new designs must be adopted in order to decrease design, construction, and operational costs. For the medium and long-term time horizon, deep reinforcement learning agents could be implemented and be either open-source simulation platforms or commercial simulators for addressing economic and technological uncertainties in optimizing the CCUS system design and operation. Besides, in the long term, AI can help design sophisticated capture materials and catalysts to convert CO_2 into high-value products [2].

5.2.3 AI-Powered Carbon Footprint Calculation and Personalized Recommendations

Personalized information is very effective in promoting behavior change. The first is the "norm activation theory," which states that people need to be aware of the consequences of their actions, such as rewards or sanctions, in order to act on personal norms. The second reason is that personalized strategies will provide information and behavior change guidance relevant to the circumstances of the individual and, thus, more relevant and feasible for them to adopt. The third reason presumes that personalized feedback might motivate people to maintain or extend positive behavior by showing how much progress they have made toward their goals. Participants can also compare their carbon impact across different behavioral domains against average values by calculating emissions based on various behavioral questions and providing individualized recommendations to reduce carbon footprints. This approach may inspire actions in areas where visible results are attainable, helping to counter feelings of helplessness and activate self-transcendent values [3].

The cutting-edge use of AI is indeed transforming the handling of carbon emissions quantification through to carbon footprint reduction. AI can precisely measure carbon footprints in real time by using machine learning algorithms, Internet of

Things (IoT) devices, and blockchain technology. Moreover, AI will enable optimization, effective supervision of the supply chain, and proactive emission reduction. This approach benefits businesses and the environment through case studies and examples, while addressing ethical considerations and social equity concerns. With the growing power of AI, its integration for carbon footprint measuring and mitigating procedures will lead to a more sustainable world where environmental duties and technology-based solutions are parallel concepts. Through the assessment and handling of the carbon metrics, stakeholders will be able to make informed decisions based on appraisals and design strategies to mitigate the impacts of climate change. As a result, AI becomes of paramount importance for environmental protection of the planet and for creating a more sustainable future [4].

5.3 AI-DRIVEN DECARBONIZATION OF TRANSPORTATION

The transport sector is one of the major contributors to the emission of GHGs, responsible for 14% of all direct and indirect emissions, which in turn contribute to climate change [5]. AI presents a powerful opportunity to optimize transportation systems and accelerate decarbonization. AI can revolutionize electric vehicle charging, public transportation, and autonomous vehicles. AI models are able to predict electricity demand for EVs based on historical data, climatic conditions, and the amount of traffic. This factor aids placement of charging stations at strategic places where demand for an electric car is high, hence, making sure drivers with EVs have easy access to charge their vehicles. AI algorithms can determine charging schedules for individual owners of an electric car, which includes two factors: grid stability and reduced car occupant's waiting time. This involves billing during off-peak hours, integrating renewables like solar energy, and giving priority to such sources while supplying electricity. AI models can leverage this data along with real-world comparisons to spot trends. The turns of passenger demand can also be predicted by the analysis of historical ridership data, special events, and weather. Short-term passenger demand forecasting is crucial for on-demand ride service platforms, as it helps incentivize vacant cars to move from oversupply to over-demand regions [6]. Also, public vehicles (buses, trains, subways) between them can be integrated, and journeys with the most efficient combination can be recommended. Therefore, AI can observe the transportation management and consequently help shorten road travel time, minimize fuel usage, and enhance the experience for pedestrians and electro mobile vehicle (EMV) drivers – more comfortable, reliable, and predictable one with appropriate resource allocation. Some examples related to this implementation include Google's DeepMind project, which has reduced fuel consumption by 15% and commuting times by analyzing traffic data from sensors, and Tesla's Autopilot system, which uses AI to optimize routes, speed, and battery usage.

5.4 AI FOR INDUSTRIAL PROCESS OPTIMIZATION

AI can also be leveraged to optimize industrial processes and manufacturing activities, minimizing energy consumption and reducing GHG emissions per unit of production, as well as cutting down our costs. AI can be used to amalgamate sustainable practices with economic objectives, with an emphasis on maximizing the financial and environmental impact of corporate operations. Streamlining of procedures, energy usage, and practices – all of which can help enterprises achieve net-zero emissions by tracking, tracing, and optimizing resources including emissions, energy, water, trash, materials, and safety. AI-driven real-time sustainability metrics can automatically detect unsafe behavior among large operations and help to implement environmentally viable and deployable sustainable practices, the developed framework makes the sustainable initiatives cut the costs, and the generated real-time sustainability metrics have a great value compared to the regular information when used in decision-making processes within a fraction of time [7].

Some of the past implementations include a regression model developed to optimize waste product utilization in the metal processing industry, use of software in designing metallurgical recycling process plans, regression analysis to come up with the best energy-efficient light sources for a manufacturing plant facility, and significant energy savings without having to replace the entire facility. In addition, procedures have been developed in order to optimize maintenance by instant monitoring of images of potential manufacturing defects and thereby easy corrections, reduction of scrap, and improvement in the efficiency of production. Flick et al. presented an examination of energy efficiency in car manufacturing systems using regression analysis and principal component analysis (PCA) in order to identify the determining parameters of energy consumption. The logistic regression method has been used in various application areas, such as real-time health monitoring of industrial components, the analysis of sound patterns from rotating cutting tools, resource efficiency in industrial printing, and predictive strategies in maintenance. Additionally, Doreswamy used Naive Bayes clustering and decision trees to classify datasets related to different materials used in manufacturing in order to make decisions concerning the choice of material to be used in performing specific tasks or on specific manufacturing components, among other applications [8].

5.5 AI FOR RENEWABLE ENERGY DEVELOPMENT (RED) AND OPTIMIZATION

The basis of clean energy lies in using renewable energy, which is the key to diminishing the dire outcome of global warming. By 2030, it is intended that half of the electricity generated must be from renewable sources, comprising wind, solar, hydrogen, or the like, to be achieved faster by smartly employing the tools: AI and Industry 4.0. The main goal will be to develop industrial robots and other communication technologies that can cut down on waste, reduce production cycles, and increase energy efficiency, using byproduct-less production methods. They decrease inequalities, improve the quality of products for trading, promote eco-friendly advancements, and decrease carbon emissions. Currently, climate finance (CF) plays a vital role in helping developing nations combat climate change and lower GHG emissions. It helps balance economic growth with reducing GHG emissions by offering rewards for using low-carbon paths. It enhances the ability of certain regions to withstand climate change and accelerates the transition to an eco-friendlier economy [9]. Intelligent energy management automation also heavily depends on AI. Consisting of multiple advanced algorithms, it enables demand response, energy trading, energy forecasting, and scheduling. Support vector machines and neural networks are two popular methods for forecasting [10].

5.6 AI-DRIVEN CLIMATE CHANGE ADAPTATION

5.6.1 AI-Driven Adaptive Agriculture and Land-Use Planning

AI helps in encouraging climate-smart farming practices. A greater degree of sustainable farming that can meet the challenges from climate change is made possible by technologies like precision agriculture, remote sensing, IoT, AI, and biotechnology. In precision agriculture, farmers can select very carefully the usage of resources like water, fertilizer, and pesticides. They can thus reduce the negative outcomes that come from the environment and increase yield. Genetically modified crops that are resistant to pests or drought are examples of biotechnological advancements that show promise for increasing crop yields in unfavorable climates. These technologies can enable climate-smart agriculture through their adoption, reduction of resource waste, and eco-friendliness by the implementation of the agricultural functions [11].

AI can revolutionize agriculture when it comes to figuring out how to utilize renewable energy and the challenges that underdeveloped countries face. To renew aging and already developed agricultural sites, one needs to install cutting-edge devices and equipment or implement them into the current landscape. AI-based conservation prioritization has been made possible through the creation of spatial conservation planning, wherein specific areas are determined as priority areas for conservation through the use of focus protection, restoration planning, impact avoidance, as well as loss reduction. AI plays a dominant role in solving issues like climate change, water scarcity, and environmental pollution, so AI is being used as a prominent tool in the development of sustainable agriculture. AI algorithms may automate most tasks that are labor intensive, such as crop harvest, pruning, and plowing, and also may enable better irrigation and use less water. These cutting-edge agricultural tools, which are made up of the GPS, sensors, and analytics, help to make the right decision at the right time, reduce the unnecessary use of resources, and achieve higher yields. The energy scheme employs renewable energy sources to produce electricity. Thus, this assists in reducing the energy cost and carbon emission [12].

IoT-based solutions will create a path to data-driven decision-making, which will take place based on the integrated system of devices and sensors. The employment of AI in the agricultural production cycle makes it more efficient in several places, such as crop irrigation, soil content detection, crop monitoring, and weeding. The execution of irrigation, providing the exact allocation of the resources, and measurement of soil temperature and moisture levels, is the set of sensing and remote machine-to-machine technologies. The computer systems that are AI-based can detect illnesses, supervise cattle's health on the go, and know how to treat existing diseases. AI can apply machine learning algorithms to improve animal health and use electricity more wisely in dairy farms. Measurable data that is used to guide the decision-making of precision farming increases crop quality and agricultural productivity, leading to higher profitability for farmers. The integration of high-tech sensors and devices to aid data collection and enable automated farming, use of renewable energy sources, and vertical farming all constitute the core of AI-based robots' innovation in farm mechanization. AI and nanotechnology may enhance smart plant sensors, plant nutrition, and precision agriculture through their technologies [12].

5.6.2 AI-Supported Water Resource Management and Conservation

AI is fast becoming an integral part of applications in urban water resource management toward sustainability. Adaptive Intelligent Dynamic Water Resource Planning (AIDWRP) is an applied AI that improves efficiency in water use, rationalizing data with numeric AI technology integrated with the expertise of the human mind. Applying the Markov Decision Process (MDP), AIDWRP addresses hard constraints, especially yearly use and discharge restrictions – location specific. This sensitivity-driven method optimizes efficient environmental planning and management policies, resulting in substantial improvements in local economic efficiency. AIDWRP aims to develop water infrastructure to increase the overall hydraulic return, enhancing the quality and cost of services provided. By incorporating future climate scenarios in water resource planning and development, AIDWRP can mitigate adverse effects and ensure sustainable water resource development [13].

The 21st century brings along with it a great task for handling the scarce natural resources to cater to human needs without causing damage. The struggle for resources, especially those relating to water, grows more heightened with increasing population, increasing economic activities, and changing patterns of living. Agriculture still is the largest consumer of water in water-stressed countries; however, the other sectors are also using it in increasing amounts. Water scarcity and deterioration of water quality are mainly due to this reason in the developing world. The International Water Management (IWRM) was developed in the 1980s with the goal of improving water management, including balancing water use in different sectors such as agriculture, industry, and conservancies. On the other hand, climate change that can alter water balance and cycle may have an impact on water supply. The management strategies for climate change demand the acknowledgment of both short- and long-term needs. These strategies should integrate adaptation to climate change into existing water management institutions and theories. An overview of smart water management is described in Figure 5.1. IWRM is a strategic and decision-making tool that tends to minimize uncertainties and adjust to emerging challenges [14].

5.6.3 AI Models to Predict Tipping Points and Their Downstream Consequences

Tipping points are critical convergence points in Earth's system where small changes can trigger large, potentially irreversible shifts. Predicting these tipping points and their cascading effects is crucial for proactive mitigation strategies. AI can play a vital role in this endeavor. AI models can be created to illustrate the intricate physics of the climate system of Earth, integrating the relationships between

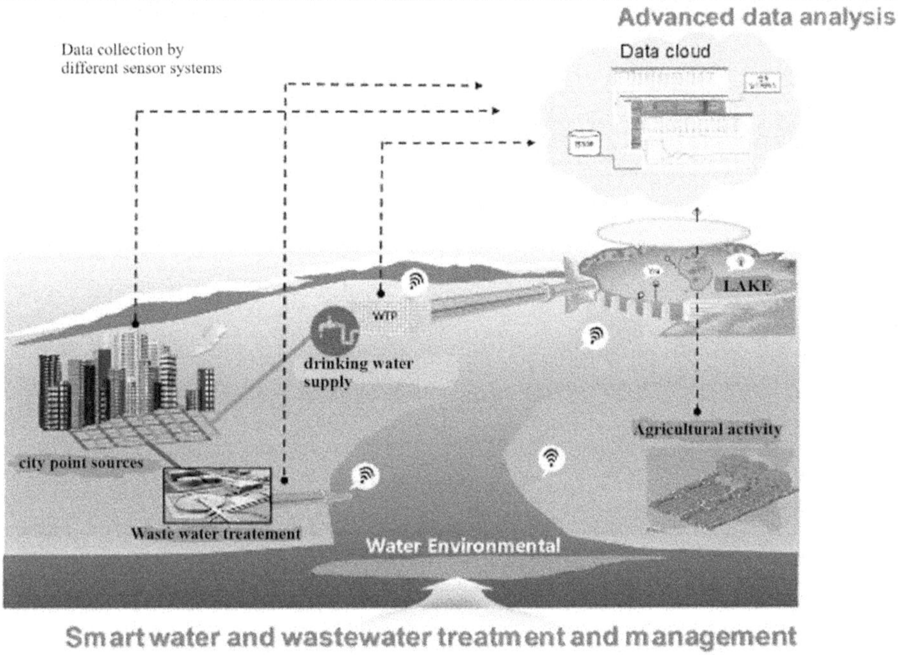

FIGURE 5.1 An overview of smart water management.

atmosphere, oceans, ice sheets, and the living world. These models can be developed and trained on extensive datasets that include climate data, historical records, and paleoclimate simulations, as well as techniques like deep learning and neural networks can be utilized to identify the interrelationships and connections that might be missed by conventional methods [15]. AI models are not limited to only one factor like ocean temperature but analyze the whole climate along with various other significant climate indicators, like ice sheet melt rate, and atmospheric CO_2 concentration. AI engines can subtly reflect these markers, showing that the likelihood of a tipping point is very high. This way of border information and strength is a salient one to enable government and relevant decision-makers to begin the curative measures in a timely fashion. The AI models could be utilized for the possible effects on the natural world through the simulation of a tipping point passing. This process requires modeling of the natural environment transformations, including weather variations, sea level rising, ecosystem disruptions, and the socio-economic impacts. Therefore, by studying the impact in such parts, allocation of the efforts of mitigation and formulating working strategies of adaptation to the vulnerable communities are possible. Examples of AI applications that are now in place include the early warning system for coral reefs' distressed events, computer-based estimation of the coming collapse of the Atlantic Meridional Overturning Circulation (AMOC), and the modeling of the trigger-effect cascades that will occur due to the permafrost thaw on global methane emissions. Prediction of the tripping point is shown in Figure 5.2.

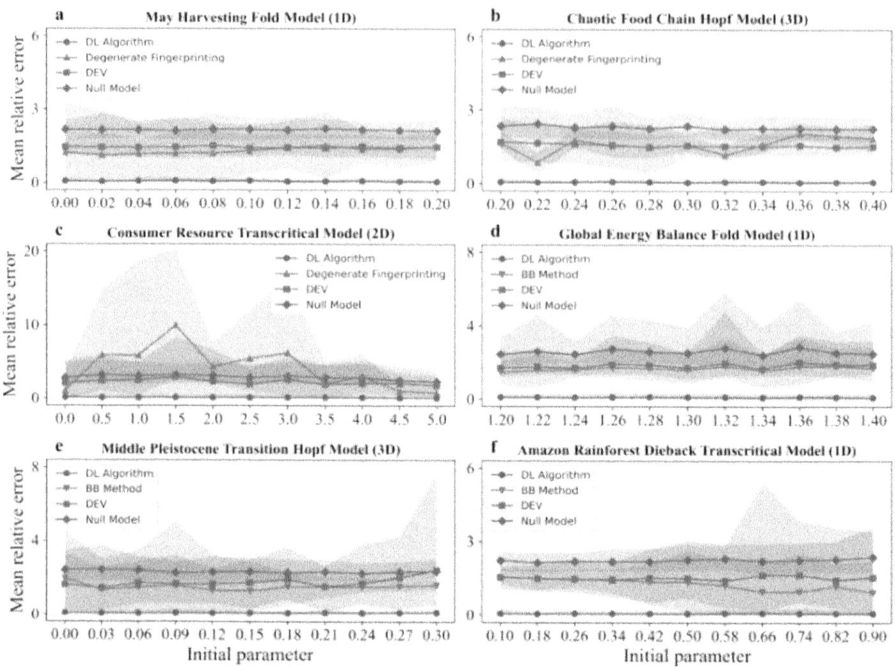

FIGURE 5.2 Predicting the tripping point using several models.

5.7 AI FOR CLIMATE-RESILIENT INFRASTRUCTURE DESIGN

AI can be used to improve the existing infrastructure design and construction plans that are capable of withstanding extreme events such as floods, storms, and high sea levels as a result of climate change. AI and machine learning algorithms can detect risks associated with the state of earthquake fault lines, weather modeling, and environmental chemical exposure, capturing a huge pool of data regarding disaster risks. This information can be used to enhance not only delicate and wise but also forecasting analysis, which might consequently differentiate disaster-resilient infrastructure constructions. The AI development in the buildings helps reduce the climate adaptive effort of the building, including its resilience and its capacity for load-bearing. This is achieved through monitoring the buildings against climate's impact through weather phenomena. Enhancement of computational power is accompanied by the use of improved sensors in satellites or drones for images with higher throughput and real-time analysis. For example, drones can be utilized to capture various areas for imaging, and the gathered data through these observations can then be evaluated for efficiently mapping risk-vulnerable parts and deciding on appropriate mitigation methods that could be adapted. The investigation and development of the algorithms that could make such data evaluation possible on larger scales is being conducted by different research groups as well as industrial world institutions across the globe [16].

Extreme weather conditions such as earthquakes, floods, pandemics, and droughts may be happening rapidly due to climate change, resulting in emergencies. Such events result in devastation to the infrastructure, agriculture, and possibly financial systems. Cloud platforms allow multiple tenants (organizations such as agencies and NGOs, as well as public and private entities) to be able to pool the storage capacity, while AI enables them to commonly analyze the data that existed before, during, and after the disaster. Public systems fail to get through vital data, such as understanding of different emergency shelter options, fatality, government policies, medicines, and diagnostic and treatment options. It is indispensable that the community that is at risk can have uninterrupted access to the precise and credible data that is made public by agencies and governments in a transparent way. The stakeholders of emergency, extreme weather, and disaster relief non-state acts are government institutions, non-governmental organizations, R&D centers, and companies. Streamlining information sharing is very important for appropriate coordination, and the system is widely usable and demonstrated in real time. Leveraging AI and cloud-based solutions, the systems that integrate and analyze the data provided by every participant can ultimately rescue more lives through quicker realization and recovery. AI and cloud storage make it possible to store massive amounts of data that citizens receive through different informative channels. Instead, the AI can assist in the unhindered coordination of people involved in disasters that occur from the pre-disaster to post-disaster phases. These technologies help in disaster management by the quick and accurate dissemination of information among the stakeholders toward the smooth coordination of inter-organizational activities, time use for damage assessments, and quicker deliveries of aid and supplies in extremely inhospitable conditions. AI needs to build a strong database that can assist in better predictions and recommendations; this can be done with AI and crowd-sourced tools, which can convert speech into text and vice versa. Cloud systems act as an application space for emergency or severe weather attention, whereas mobile applications allow end users to access services, enable more usage of applications, and receive upgrades from the source of information. Data coming in different forms (text, video, audio, count, and images) in emergency and extreme weather scenarios can be analyzed using various tools (e.g., AI), which require continuous synchronization of data and, subsequently, easy sharing through stakeholders' networks [17].

5.8 SYNERGISTIC APPROACHES: AI INTEGRATION WITH OTHER TECHNOLOGIES

5.8.1 AI and Citizen Science

A wide range of definitions and conceptions exist for the concept of citizen science, including generally the involvement of the public, voluntary contribution to the knowledge-generating process, and the creation of new knowledge. Citizen science resonates with the concept of Public Participation in Scientific Research (PPSR), defined as the intentional cooperation between the public and scientists for the generation of novel, science-based knowledge. There are five types of PPSR projects, including contractual, contributory, collaborative, co-created, and collegial projects.

Citizen science includes any initiative that produces scientific knowledge through volunteer participation, such as community-based monitoring, participatory research, participatory action research, citizen-generated data, crowdsourcing, volunteered geographic information, and participatory sensing [18].

It is over the past two decades that citizen science has rapidly gained ground in almost all domains, though its application to climate change research and studying its impacts is a more recent development. Most of the work currently underway aims at bettering scientific knowledge about climate change, like changes in phenology within ecosystems, atmospheric measurements collection, weather pattern classification, storms, and relevant data. In fact, despite all these factors, there has been relatively little emphasis on how citizen science might aid in climate mitigation or adaptation at the community level, or on its role in supporting governance and decision-making for climate action [19]. Citizen science is a growing domain and aims to achieve scientific advancement, improve scientific knowledge, increase public participation in decision-making, and facilitate personal discussion on complex, modern topics. Citizen observatories represent effective methods for monitoring the progress toward Sustainable Development Goals (SDGs) by involving the public in developing and evaluating trusted, accurate, complete, and standardized data. These observatories engage citizen scientists in data collection and analysis across disciplines, including environmental monitoring, biodiversity studies, and astronomy. While some concerns exist regarding the quality of data generated by citizen-based efforts and the lack of strict quality control measures that are necessary for effective decision-making and management of the environment, citizen-focused observatories foster longer-term participation, which often results in better quality data generation [20].

5.9 AI, BLOCKCHAIN, AND LLM FOR CLIMATE CHANGE

The use of blockchain and Large Language Models (LLM) technologies could really increase as key contributors to curbing climate change. Fostering public trust and proper climate action requires them to emphasize transparency, reliability, and openness of climate science. Blockchain as a technology guarantees transparent and manipulation-free record of transactions and impacts on the environment, which enables trust among stakeholders and a highly participatory environment in the fight against climate change. LLMs function in a synergistic way with blockchain in bringing together deep insight and actionable intelligence, and are therefore a desirable evidence-based policymaking instrument. The blend of the two technologies guarantees better handling of data, more realistic climate models, and the enabling of efficient climate action initiatives. On the one hand, though, the main issues revolving around scalability and energy consumption of blockchain technologies and a lack of high-quality data for LLMs are considered. The alliance of blockchain and LLMs can imply an unusual way to fight climate change by making climate data more precise, transparent, and more protected. This process is Mercury Fusion, which is the desired long-term solution for data management and analysis, and would solve one of the highest-ranked challenges. Blockchain is providing a new way of dealing with climate change through enhancing transparency, traceability, and efficiency in key fields, namely carbon credit trading, renewable energy certificates, and sustainable supply chain management. It is incapable of altering transactions, commands trust, and has a better verification process, hence making the carbon credit trading system reliable and efficient. The blockchain technology also offers efficiency through traceability in renewable energy certificates, where the certificates are narrowed directly to their source of renewable energy, which is hardly capable of fraud and links directly to the growth of the renewable energy market. Furthermore, it enhances the process of supply chain management for sustainable practices, which fosters fair sourcing and production methods. The symbiotic manifestation of blockchain technology and LLMs is robust for data integrity enhancement, decentralized data sharing that is secure and transparent, and smart contract development among environmental governance figures. This integration not only increases efficiency but also enforces the environmental standards and becomes a tool for a new era of environmental stewardship. Through relating these aspects to specific mitigation measures and research lines, a road more reassuringly feasible can be built for the synergy between blockchain and LLMs to make a positive impact on climate mitigation [21].

5.10 AI IN COMBINATION WITH IOT FOR CLIMATE MONITORING AND DATA COLLECTION

It would also encompass smart solutions within an IoT: devices furnished with sensors, software, and computing to enable network- and internet-enabled communication for data connection and sharing. For the time being, environmental challenges and problems – such as provision of pure clean water, landfill avoidance through waste, redeforestation, and curbing or minimizing emission-related air pollutants – can indeed benefit from new opportunities that technological advancements can bring about within this IoT scope. As the world market is projected to increase, large companies can already foresee the possible uses of IoT for both human lives and health in the environment.

IoT devices collect robust data from a number of various sources, including metrics that might include temperature, humidity, and precipitation, to offer unparalleled precision in monitoring ocean temperatures and sea levels. Now, most projects incorporate IoT in reducing emissions, pollution, and illegal logging, which reduces environmental damage. The way we relate to the environment has changed; these changes in the sense are made quite potent when used in the urban parts, where approximately 70% of the global population is expected to be concentrated by 2050. Since climate change remains a globally sensitive issue, it should be solved with IoT. It can help by organizing cities smartly to organize the navigation for easy transport, reducing congestion, and bringing down the emissions level. For instance, IoT-based transportation systems reduce carbon emissions by directing drivers to available parking spaces while avoiding congested roads. IoT devices can further monitor and alert on deforestation and illegal logging activities, thus reducing labor, time, and cost.

For example, in Los Angeles, IoT technology synchronizes traffic lights, which reduces GHGs and saves more than 35 million gallons of gasoline annually. This is the same as how utilities are supported by IoT in creating responsive energy networks to use predictive analytics in making sure that the energy supply matches the demand, waste is minimized, and load balancing is done. Similarly, smart meters track the consumption of building energy. It helps in load balancing, waste reduction, and reduction of reliance on fossil fuels, with carbon footprints decreased thereby [22].

5.11 AI AND BIG DATA ANALYTICS FOR CLIMATE PATTERN ANALYSIS AND TREND IDENTIFICATION

Climate data is the key to unraveling the climate system interactions and reveals the changes in the climate that may occur due to global warming. Thus, the Earth observation technology has been successfully integrated with the multi-dimensional big data system, which is still developing, and bringing the capability of observing and monitoring a wide variety of Earth and climate parameters all over the world. The climate data sources comprise on-the-spot reading, telemetry, model outputs, as well as paleoclimate data. The 5Vs of data science – volume, variety, velocity, veracity, and value – are the main reasons that make interpreting, predicting, and optimizing huge data difficult. Big data researchers are armed with many resources at hand, for example, the Global Climate Observing System (GCOS), the Earth System Grid Federation, the National Center for Atmospheric Research, the United Nations Global Pulse, the Climate Data Guide, NASA Global Climate Change, and the NASA Center for Climate Simulation. These resources have enabled us to gain knowledge of what big climate data are, how they are collected, and what the data can be used for [23].

These are some of the very efficient frameworks, including HDFS, cross-MapReduce, YARN, Google BigTable, Spark, Mahout, and Flume. HDFS is an analytical storage framework for enormous amounts of data that have been generated from Earth observation and climate simulation. It ensures simplicity by allowing data to be read into distributed arrays without creating a single point for data conversion. Cross-MapReduce is a technological framework designed to improve document search and processing of geo-distributed data. It has played a very important role in climate change detection, with large datasets of weather information and machine learning used for decision-making. YARN serves as a batch-processing framework orchestrating the components of the Hadoop ecosystem that enables parallel execution of multiple distributed climate big data applications. This framework has improved multi-tenancy, cluster utilization, scalability, and compatibility of the climate big data for management and processing.

High storage solutions for climate big data play a significant role in managing unstructured data. Geospatial-temporal processing and storing data is especially important as Google BigTable and Cassandra allow dealing with data of weather, climate, and air quality through time and space. Spark is a high-speed, large-scale framework that processes climate data for graphX, MLlib for machine learning, and the stream Application Programming Interfaces (APIs). This processes large climate datasets in-memory and enables fast response, along with a platform for streaming data analysis. Finally, Mahout offers a scalable, user-friendly, and extensible library application for machine learning using big data analytics. These have largely been adopted for the effective processing and streaming of climate change-related data [24].

5.12 CHALLENGES AND ETHICAL CONSIDERATIONS

5.12.1 AI Footprint

Resources required by AI system operations, from AI research to AI equipment manufacturing, negatively affect climate change. AI research has a carbon footprint, which means it is important to be systematic about the GHG emissions produced in AI research, as it consumes energy and resources. AI has a plethora of skills, for instance, the power to process a lot of unstructured data in determining high-dimensional climate datasets and forecasting prospective trends. Apart from that, AI brings up ethical issues that are quite important to our environmental area. Also, implementing AI systems comes with computational cost, which is directly linked to climate change. Refinement of deep learning has boosted the amount of energy consumed, and the associated issue of environmental impact has raised concerns about the emergence of negative effects. Large cloud providers, including Microsoft Azure and Google Cloud, are doing their best to lessen their carbon footprint by implementing renewable energy and investing in carbon offsetting policies, as well as other schemes that help make their cooling systems more efficient and recycle waste heat. Several approaches to monitoring and estimating GHG emissions of AI research activities have been offered, including Henderson et al.'s "experiment-impact-tracker" and Lacoste et al.'s Machine Learning Emissions Calculator.

Energy consumption issues, which may not allow SOTA hardware to be accessed for training large-scale machine learning models, or leave hardware performance behind the computing requirements, will also hamper AI progress research. The algorithmic development are shown to be effective in the time and effort of operations; on the other hand, studies centering on new hardware-software-neural network combinations and algorithms have been put on hold. The challenge of limiting both computational load and energy use in research is being overcome as researchers continue to build ever more efficient models using various techniques such as random search over hyperparameters, pruning, transfer learning, and early stopping for underperforming models [25]. To maintain the pace while reducing the carbon footprint, AI research needs to shift away from intense, compute-heavy studies and review performance metrics that emphasize accuracy. Another significant barrier for entry is that the focus on precision over speed currently enables only well-funded research teams to afford the rising costs of computationally intensive studies [26]. A balance has to be struck between energy usage and efficiencies that are both achievable and acceptable for researchers and regulatory agencies alike. Guidelines for carbon assessment and transparent disclosure can incentivize AI providers to use green metrics and report their emissions footprints more understandably than ever before [27].

5.12.2 Data Quality and Availability Challenges for AI Implementation in Climate Science

Climate change is a pressing issue that can be resolved only when we have a robust data infrastructure, including well-populated datasets, servers, and data centers. However, the availability and proper maintenance of climate data for AI models can be hindered by the presence of geographical diversity. Developed countries and urban regions generally have their own systems for how data is collected, but third-world countries may have a higher chance of being victims of climate change. It turns out that such infamous data collection in developing areas makes AI less needed and not appropriate for global climate investigations. The solution to this is thus to provide data accessibility through inter-ministerial and inter-departmental data sharing platforms while also taking care of privacy concerns. This will help mitigate the risk of harm from AI-driven climate change research and innovation [28].

AI algorithm-driven systems are based on the availability and quality of ample amounts of data for making predictions or suggesting recommendations. While this certainly streamlines the process, if the data from which the AI algorithms are trained is biased, the results may also be biased. To illustrate this, imagine, for instance, that the data considered in training an AI system is only historical for a region or a population group, and may not be accurate enough to present the experiences of people from other origins. Such biases might lead to wrong forecasts and insights that could discriminatingly affect the weak groups of society. Sensor-based datasets, such as radar-derived hail estimates, could be used to address data issues. However, these datasets must be interpreted to create a full dataset, as they may contain overprediction biases and incomplete data due to sensor errors or missing sensors. Sensor placement can also be influenced by geological features and economic factors, such as crowd-sourced sensors. For example, the PurpleAir low-cost air quality sensors, deployed in whiter, higher-income areas, need access to less-privileged communities to democratize air pollution data [29].

The Indian government has put in place numerous reforms that seek to enhance collaboration and climate action. The National Data Governance Framework Policy focuses on the creation of efficient methods of data sharing and the liberation of data for both the private sector and startups. The Digital India Act will be about regulating non-personal anonymous data; consequently, the influence it can have within AI applications will be regulated. Government-backed programs for data collection can promote private sector participation in climate-relevant datasets. Data task forces in crucial sectors, such as energy, transportation, and manufacturing, can be established to set standards for data and promote data collection. An exchange of experience in different sectors is needed too, and the Paris Agreement and Indo-US Science and Technology Forum are examples of such collaboration on innovative technologies for sustainable development. It is very useful because the technology transfer is supported by the funding in research institutions, joint workshops, and centralized coordination centers [28].

5.12.3 Addressing Bias and Fairness in AI-Driven Climate Solutions

AI systems sometimes are complicated and hard to understand, which can cause problems when it comes to spotting a bias or error. Lack of transparency in algorithm choosing and teaching models, as well as in the possibility (source) of bias formation, could become difficult obstacles to overcome. AI systems have the potential of not displaying a reflective movement on the many kinds of perspectives that are necessary for the building of effective climate change adaptation approaches. To illustrate, if the development team remains exclusive in diversity, it may overlook communal or group-specific needs. AI systems need huge amounts of data that include personal information, which can be different in nature. To benefit from the ethos of data, privacy, and individual details should be well preserved. Ethical factors are among the significant issues that a researcher has to deal with in the design and development of AI technology to fight climate problems. There are several of them, such as getting varied and represented data, making the AI algorithms assisted by humans transparent, gathering input from stakeholders of diverse backgrounds by involving them in the work, and regularly doing ethical audits and complying with personal and sensitive data. AI may deepen the issue of bias with the occurrence of ethnic discrimination or discrimination against other outcast groups, creating situations where the AI system neglects the infrastructure. Data selection and preprocessing should be given careful attention to avoid this. Furthermore, AI-aided climate change adaptation policies must take into account the likelihood of harming poor, vulnerable communities. Examples include gentrification (redeveloping neighborhoods) or displacement. With this in mind, AI can be applied to use it wisely, all-inclusive, transparent, and usable nature for the benefit of all [30].

5.12.4 Ensuring Transparency and Explainability in AI Models – XAI

Explainable AI (XAI) is a technology that allows people to decode complex AI algorithms and models in a comprehensible language. It involves the use of interpretability and explainability, which are often interchangeable terms. XAI helps in understanding causation through giving a depiction of the inner-working mechanisms of complicated AI models. For the hydroclimatic natural hazard, interpretability or explainability is about understanding what compels different weather variables to occur. Google's whitepaper argues for the explainability of noticeable transparency, educated human decisions, debugging, fair auditing, and generalization capabilities. XAI is critical in the development of ML-based methods of studying natural hazards because machine Learning (ML) models may give false conclusions [31].

Artificial neural networks are widely used in Earth sciences, including marine science, solid Earth science, climate science, and meteorology. Prediction of climate is an important task in the field of climate science, with significant implications for the economy, water security, and ecosystem management. It usually uses sea surface temperature (SST) values from the surface of the sea to generate either deterministic or statistical models. Although neural networks (NNs) are useful at predicting climate at different scales primarily through XAI, it is not the only way. XAI helps to uncover the traces of human activities in the climatic changes and observations/simulations. Detecting climate change signals is a signal-to-noise issue, where the warming signal that arises from the slow human-caused changes of GHG concentrations contradicts the background noise of natural climate variability. However, these models have limitations due to incomplete knowledge of initial conditions, uncertainties, and computational resources. To address this, data-driven machine learning methods have been suggested, leveraging information from the entire globe. The state-of-the-art approach to quantify or account for natural variability within the climate community is the utilization of large ensembles of climate model simulations [32].

The ML models, including the XAI ones, are getting more important in assessing and understanding hydrometeorological events as well as in the study of natural disasters. Recent technological breakthroughs in machine learning have led to a better understanding of natural hazards, but trust and fairness are critical for proper real-world decision-making. XAI can ease future problems and develop plans for effective hazardous event management, response, recovery, and resilience. Nevertheless, conventional ML model generation may use XAI in areas like data availability and dimensionality reduction techniques in search of the most important variables, irrespective of bias, and data explanation. XAI can help researchers and ML developers not only to create more profound models but also to further AI system development. XAI is a great platform in computer science that can be applied to recognize the significant features inside the datasets, delete the redundant variables, and enhance the decision-making process. Ethical aspects, exemplified by transparency and accountability, are indispensable, because XAI models should be constructed while the proper data collection and management policies are considered to avoid the representation and perpetuation of biases. XAI can likewise tackle model transferability, revealing how a model reaches its conclusions, but utmost care must be taken to ensure the features of an explainable model are retained [31].

5.12.5 Potential Risks and Unintended Consequences of AI Applications in Climate Change Effort

AI-driven climate change adaptation strategies could bring about indirect consequences that might end up being negative for communities, or for ecosystems in general. An instance of this would be if the system suggested a seawall compound to protect against the rising sea level, which in turn would have a negative impact on the marine ecosystem [30]. With the emergence of the XAI domain, new AI approaches that are transparent and verifiable according to human reasoning have been developed, but the field is still in the early stages. A concern of over-reliance on AI algorithms whose level of performance is variable is that they increase bias risks as well as the wider performance risks. Apart from hacking, malicious repurposing of algorithms created with good intentions could also be another security issue, raising new threats to global safety. Good governance is required to build explainability, transparency, and validity into algorithms, and to prevent misuse. Adversarial attacks can also be used to deceive machine learning models, posing new risks. Autonomy risks are inherent in AI systems taking decisions on their own and coexisting with each other, which develop complex loops that may result in unintended effects. These issues could be proactively controlled by the creation of controls, monitoring, and safeguards to make sure they don't turn into bigger problems. Along with tax-base erosion, the digitized globalized economy causes difficulties for the current system to be competitive. Social challenges include the displacement of workers from the transportation, manufacturing, agriculture, and service sectors; joblessness; and a wider income gap. Autonomous weapons carry a profound risk of damaging the environment in the long run. Ethical risks include outsourcing big data, increased reliance on algorithms, and a slow reversal of human involvement in processes. These generate questions of fairness, responsibility, equality, and human rights. Privacy concerns also have an impact, as the effect of AI on individuals and communities goes beyond individual levels [33].

Maladaptation, which raises the vulnerability to climate change, may lead to limitations on affected stakeholders' freedoms. Integrated assessment models (IAM) can be used to explore how different sectors and decision-makers are affected by climate change. Adaptation may result in positive outcomes but at the same time can cause negative impacts due to, for instance, data constraints, management- and information-sharing arrangements, and stakeholders' awareness. In a number of cases, the provision of adaptation concerns within national policy is often difficult because of low capacity of staff, poor data on adaptation options, lack of information sharing and management across sectors, and limited awareness among stakeholders and the general public [34].

5.12.6 Ethical and Environmental Implications of Geoengineering Solutions and Their Impacts on Global Climate Patterns and Regional Weather Systems

Carbon removal, a popular geoengineering solution for climate change mitigation, has been linked to negative co-impacts, such as interference with water security and aggravating water scarcity. If carbon removal is not carefully applied, then its side effects could be negative, such as consuming scarce water, which is further needed for farmers. Bioenergy with Carbon Capture and Storage (BECCS) and Direct Air Capture (DAQ) technologies also employ large-scale water consumption, leading to security issues related to water scarcity globally. Deterring mitigation often puzzles NGOs and civil society to undertake mitigation, as they may feel that carbon removal compensates emissions burden without reduction. BECCS and afforestation at a large scale take up large land areas, which somewhat affects land use and the food system. This can result in land appropriations and land enclosures that are the foremost cause of anthropogenically driven extinction. Besides, the bottom stream of carbon dioxide removal (CDR) is polluted, which in the end affects land availability and contributes to the food insecurity issue. Carbon offsetting is another dangerous outcome in the category of negative co-effects. The entities may assert they have a zero net carbon footprint and are carbon neutral, but they in fact are still using carbon and emitting GHGs. In addition to this, ethical questions like the nature of our planet and climate, which cause problems, result in such negative responses to solar geoengineering. Solar geoengineering techniques are scrutinized for their possible use to serve as weather modifiers, but the subject has to comply with international law and declining social approval [35–37].

Climate engineering has three dimensions: direct environmental impacts, indirect environmental impacts, and indirect impacts on climate mitigation politics and policies. Direct environmental impacts include sulfate release from Solar Radiation Management (SRM) projects, volcanic eruptions, and ocean fertilization projects. Indirect environmental impacts involve climate engineering, lowering global mean temperature but potentially changing local temperatures and precipitation levels, harming agriculture and ecosystems. Regional disparities in benefits and risks complicate the situation. Fast-food climate engineering projects only focus on lowering global mean temperature, leaving environmental problems from conventional fossil energy usage unfixed. Indirect impacts on politics and policies involve governments changing their future development strategies, such as investing in renewable energy instead of green transformation [38].

Latest environmental hazards emerge by way of either a direct or an indirect impact brought about by scientific controversy, regional specificity, and the termination problem. Achieving a pre-industrial global surface temperature equilibrium may happen within just a couple of years after the deployment, but if the method fails or is cancelled, a rapid transition to a warmer climate could occur. The geographic specificity of some SRM methods can also create disruptions in the uniform temperature distribution in different regions. On the contrary, the global temperatures could fall and might decrease the respiration rates of plants and consequently increase the net CO_2 uptake of the land biosphere, thus leading to the marine ecosystems' acidification. Human intrusion in the natural biological and chemical processes may, in turn, result in a diverse range of issues in biodiversity and ecosystems. The ocean iron fertilization and other ocean-based CDR methods harm the fact that they increase carbon uptake within the ocean and can subsequently lead to dissolved CO_2 in the water and the acidification of the oceans [39].

5.13 HUMAN EVOLUTION IN RESPONSE TO CLIMATE CHANGE

Every living organism influences the entirety of the Earth system, which comprises land, air, and water. The shifting of the Earth system is the reason why species scatter, modify, or lose their entire existence. For example, the polar bear range may have shrunk, and the animal population decreased because of the Arctic's warming temperature, which makes the process of hunting almost impossible. The Earth system is also changed due to the evolutionary variation of organisms, migration, and extinction. Humankind has, throughout time, adapted to ecological problems and advantages in an effort to survive, accumulate food, build shelters, and stay away from carnivores, as well as learn how to overcome social difficulties [40].

5.14 FUTURE DIRECTIONS AND RECOMMENDATIONS

On the way to specifying opportunities for the application of AI in climate change mitigation and adaptation, there are certain aspects that need to be addressed. The first one is to keep doing research and development of AI as it applies to climate challenges. Here, the existing AI systems and models need to be optimized to capture more complicated dynamics to lead to more precise predictions. Individual approaches need to be made, including hyper-personalized suggestions for carbon footprint reduction based on the lifestyle and consumption patterns of every individual and boost Carbon Capture and Storage (CCS) technologies and develop technologies for removing GHGs from the atmosphere. After this, AI tools and datasets must be developed and released open to everybody for the climate research community to enable innovation and collaboration. The collaboration between the scientists of AI, climatologists, policymakers, and industry actors is a propitious way to create solutions that are effective, practical, with an ethical background. In addition, designing

AI systems that will automatically collect and process the data from citizen science initiatives, providing the public with the ability to contribute to valuable data for use in climate observation and adaptation planning. Efforts jointly taken by educational institutions, people, industrial bodies, and government in the sphere of AI can effectively speed up the uptake and transition to AI-powered climate solutions in implementation on a global level.

Accurate ethical consideration and accountability presence are a must in the artifact of AI systems in climate action. Such measures entail building the regulatory context for safe, objective, and responsible algorithms to detect the possibilities of bias, discrimination, and unintended consequences linked with the usage of AI applications and provide fair and even outcomes for the inhabitants of different communities in the world. Attention must be on building XAI solutions for climate solutions, such that stakeholders will comprehend the intelligence's decisions, along with being able to correct them. Besides this, sensible and proper planning of research work plays a great role in this regard to develop the framework of carbon footprint reduction by considering the consumption of energy and its influence on the environment when constructing and running AI systems. Developing educational and capacity-enhancing programs for stakeholders at all levels (scientists, policymakers, industry professionals, and citizens) through which all can be taught the right skills and knowledge to leverage the use of AI for climate-related issues is critical. Through the adoption of these principles and co-operatively advancing common objectives, the AI transformational power can be mobilized for fighting climate change with regard to meaningful and lasting results.

5.15 CONCLUSION

AI has in hand a very promising future in relation to climate change, as it brings up better and great ideas to mitigate the impacts of climate change and adapt to the challenges. Different AI-dependent strategies must be explored that get the work done during the process in sectors like energy and agriculture but also infrastructure and water resources. These solutions bring in a new opportunity, through their approaches, to confront rising levels of carbon emissions efficiently, bringing in innovations to cope with the needs of this fast-changing world. However, it seems that while looking into the future, in order to realize the full potential of AI with respect to climate change, all stakeholders have a role to play. They are required to stay on the right track, keep inputting efforts in research and development, which involves engulfing different disciplines and creating innovative technologies in order to unveil entirely new ways of using AI in the context of climate change. Also, ethical issues must always be considered, and intelligent machine-driven solutions must obey principles of justice, openness, and responsibility. Ensuring that AI is unbiased, data is correct, and the unintended consequences are addressed will help to create a belief that the application of AI against the climate crisis is well grounded and can be used to best effect without negative consequences. Apart from this, the synergistic interplay of AI with other emerging technologies gives exceptional chances to improve our capacity for climate monitoring, prediction, and decision-making. Having AI, blockchain, IoT, big data processing, and digital twins work together, complete and innovative solutions that cope with the multi-level and interconnected issues in the climate area evolve. In the end, the capability of the AI to change climate flow through mitigation and adaptation depends on the general motivation and commitment of all of us. Through promoting an environment of teamwork, innovation, creativity, and responsible custodianship, the amazing power of AI can be used to empower a better and more adaptable society in the future. Together, let us embrace the challenge and opportunity of leveraging AI in the fight against climate change, forging a path toward a brighter and more sustainable tomorrow.

ACKNOWLEDGMENTS

We express our appreciation to the academic and scientific community for their pioneering research and innovative insights that have laid the foundation for the advancements in climate change and AI discussed in this chapter.

REFERENCES

1. Munawar, H. S., Mojtahedi, M., Hammad, A. W. A., Kouzani, A. Z., & Mahmud, M. A. P. (2022, February 1). Disruptive technologies as a solution for disaster risk management: A review. *Science of the Total Environment.* https://doi.org/10.1016/j.scitotenv.2021.151351
2. Al-Sakkari, E. G., Ragab, A., Dagdougui, H., Boffito, D. C., & Amazouz, M. (2024, March 1). Carbon capture, utilization and sequestration systems design and operation optimization: Assessment and perspectives of artificial intelligence opportunities. *Science of the Total Environment.* https://doi.org/10.1016/j.scitotenv.2024.170085
3. Gupta, S., Modgil, S., Kumar, A., Sivarajah, U., & Irani, Z. (2022, December 1). Artificial intelligence and cloud-based collaborative platforms for managing disaster, extreme weather and emergency operations. *International Journal of Production Economics.* https://doi.org/10.1016/j.ijpe.2022.108642
4. Bhuiyan, M. A. H. (2024, January 1). Carbon footprint measurement and mitigation using AI. *Social Science Research Network.* https://doi.org/10.2139/ssrn.4746446
5. Perez-Dávila, O., & Fernández, R. L. (2023, March 1). Optimization algorithm applied to extended range fuel cell hybrid vehicles. Contribution to road transport decarbonization. *Energy (Oxford).* https://doi.org/10.1016/j.energy.2022.126519
6. Ke, J., Zheng, H., Yang, H., & Chen, X. M. (2017, June 20). Short-term forecasting of passenger demand under on-demand ride services: A spatio-temporal deep learning approach. Retrieved April 12, 2024, from https://arxiv.org/pdf/1706.06279.pdf
7. Suboyin, A., Eldred, M., Thatcher, J., Rehman, A., Gee, I., & Anjum, H. (2023, January 19). Environomics framework for sustainable business practices: Industrial case studies on true impact reduction and process optimization through AI. https://doi.org/10.2118/214459-ms
8. Waltersmann, L., Kiemel, S., Stuhlsatz, J., Sauer, A., & Miehe, R. (2021, June 12). Artificial intelligence applications for increasing resource efficiency in manufacturing companies—A comprehensive review. *Sustainability.* https://doi.org/10.3390/su13126689
9. Zhao, C., Dong, K., Wang, K., & Nepal, R. (2024, May 1). How does artificial intelligence promote renewable energy development? The role of climate finance. *Energy Economics.* https://doi.org/10.1016/j.eneco.2024.107493
10. Li, J. H., Herdem, M. S., Nathwani, J., & Wen, J. Z. (2023, January 1). Methods and applications for artificial intelligence, big data, internet of things, and blockchain in smart energy management. *Energy and AI.* https://doi.org/10.1016/j.egyai.2022.100208

11. Singh, A., & Kumar, P. (2024, January 1). *Tech-Driven Agriculture: Adapting to Climate Challenges*. Springer eBooks. https://doi.org/10.1007/978-3-031-52708-1_10
12. Mana, A., Allouhi, A., Hamrani, A., Rahman, S. M., Jamaoui, I. E., & Jayachandran, K. (2024, February 1). Sustainable AI-based production agriculture: Exploring AI applications and implications in agricultural practices. *Smart Agricultural Technology*. https://doi.org/10.1016/j.atech.2024.100416
13. Xiang, X., Li, Q., Khan, S., & Khalaf, O. I. (2021, January 1). Urban water resource management for sustainable environment planning using artificial intelligence techniques. *Environmental Impact Assessment Review*. https://doi.org/10.1016/j.eiar.2020.106515
14. Ludwig, F., Van Slobbe, E., & Cofino, W. (2014, October 1). Climate change adaptation and integrated water resource management in the water sector. *Journal of Hydrology*. https://doi.org/10.1016/j.jhydrol.2013.08.010
15. Gnanadesikan, A., Brett, J., Sleeman, J., & Chung, D. (2023, August 1). Using AI to detect climate tipping points- or why it's hard to understand rapid changes in the earth system. *OpenReview*. https://openreview.net/forum?id=JCyVm71Wyv
16. Munawar, H. S., Mojtahedi, M., Hammad, A. W. A., Kouzani, A. Z., & Mahmud, M. A. P. (2022, February 1). Disruptive technologies as a solution for disaster risk management: A review. *Science of the Total Environment*. https://doi.org/10.1016/j.scitotenv.2021.151351
17. Gupta, S., Modgil, S., Kumar, A., Sivarajah, U., & Irani, Z. (2022, December 1). Artificial intelligence and cloud-based collaborative platforms for managing disaster, extreme weather and emergency operations. *International Journal of Production Economics*. https://doi.org/10.1016/j.ijpe.2022.108642
18. Fraisl, D., Campbell, J., See, L., Wehn, U., Wardlaw, J., Gold, M. M., Moorthy, I., Arias, R., Piera, J., Oliver, J. L., Masó, J., & Penker, M. (2020, July 2). Mapping citizen science contributions to the UN sustainable development goals. *Sustainability Science*. https://doi.org/10.1007/s11625-020-00833-7
19. Bremer, S., Haque, M. M., Aziz, S. B., & Kvamme, S. (2019, April 1). 'My new routine': Assessing the impact of citizen science on climate adaptation in Bangladesh. *Environmental Science & Policy*. https://doi.org/10.1016/j.envsci.2018.12.029
20. Woods, S. M., Daskolia, M., Joly, A., Bonnet, P., Piera, J., Liñán, S., Woods, T., Piera, J., & Ceccaroni, L. (2022, March 30). How networks of citizen observatories can increase the quality and quantity of citizen-science-generated data used to monitor SDG indicators. *Sustainability*. https://doi.org/10.3390/su14074078
21. Heston, T. F. (2024, January 1). A blockchain AI solution to climate change. *Social Science Research Network*. https://doi.org/10.2139/ssrn.4736849
22. El-Mawla, N. A., Badawy, M., & Arafat, H. (2019, April 8). IoT for the failure of climate-change mitigation and adaptation and IIoT as a future solution. *World Journal of Environmental Engineering*. https://doi.org/10.12691/wjee-6-1-2
23. Hassani, H., Huang, X., & Silva, E. S. (2019, February 2). Big data and climate change. *Big Data and Cognitive Computing*. https://doi.org/10.3390/bdcc3010012
24. Ikegwu, A. C., Nweke, H. F., Mkpojiogu, E. O., Anikwe, C. V., Igwe, S., & Alo, U. R. (2024, February 7). Recently emerging trends in big data analytic methods for modeling and combating climate change effects. *Energy Informatics*. https://doi.org/10.1186/s42162-024-00307-5
25. Riaz, K., McAfee, M., & Gharbia, S. (2023, February 28). Management of climate resilience: Exploring the potential of digital twin technology, 3D city modelling, and early warning systems. *Sensors*. https://doi.org/10.3390/s23052659
26. Ye, X., Du, J., Han, Y., Newman, G., Retchless, D., Zou, L., Ham, Y., & Cai, Z. (2022, November 29). Developing human-centered urban digital twins for community infrastructure resilience: A research agenda. *Journal of Planning Literature*, 38(2), 187–199. https://doi.org/10.1177/08854122221137861

27. Cowls, J., Tsamados, A., Taddeo, M., & Floridi, L. (2021, October 18). The AI gambit: Leveraging artificial intelligence to combat climate change—opportunities, challenges, and recommendations. *AI & Society*. https://doi.org/10.1007/s00146-021-01294-x
28. Landscaping the use of AI for Climate Change: Addressing Challenges and Risks. (2023, June 14). https://www.ikigailaw.com/article/17/landscaping-the-use-of-ai-for-climate-change-addressing-challenges-and-risks
29. McGovern, A., Ebert-Uphoff, I., Gagne, D. J., & Bostrom, A. (2022, January 1). Why we need to focus on developing ethical, responsible, and trustworthy artificial intelligence approaches for environmental science. *Environmental Data Science*. https://doi.org/10.1017/eds.2022.5
30. Jain, H., Dhupper, R., Shrivastava, A., Kumar, D., & Kumari, M. (2023, July 17). AI-enabled strategies for climate change adaptation: protecting communities, infrastructure, and businesses from the impacts of climate change. *Computational Urban Science*. https://doi.org/10.1007/s43762-023-00100-2
31. Dikshit, A., Pradhan, B., Matin, S., Beydoun, G., Santosh, M., Park, H., & Maulud, K. N. A. (2024, July 1). Artificial Intelligence: A new era for spatial modelling and interpreting climate-induced hazard assessment. *Geoscience Frontiers*. https://doi.org/10.1016/j.gsf.2024.101815
32. Mamalakis, A., Ebert-Uphoff, I., & Barnes, E. A. (2022, January 1). Explainable artificial intelligence in meteorology and climate science: Model fine-tuning, calibrating trust and learning new science. *Lecture Notes in Computer Science*. https://doi.org/10.1007/978-3-031-04083-2_16
33. Pwc. (n.d.). *Fourth Industrial Revolution for the Earth Harnessing Artificial Intelligence for the Earth*. Pwc.
34. van Zyl-Bulitta, V. H., Patt, A., Mudombi, S., & Fabricius, C. (2023, September 20). Unintended consequences of climate change adaptation: African case studies and typologies on pitfalls and windfalls. *Development Southern Africa*, 41(1), 130–163. https://doi.org/10.1080/0376835x.2023.2254801
35. Sovacool, B. K., Baum, C. M., & Low, S. (2023, February 1). Beyond climate stabilization: Exploring the perceived sociotechnical co-impacts of carbon removal and solar geoengineering. *Ecological Economics*. https://doi.org/10.1016/j.ecolecon.2022.107648
36. Taebi, B., Lenzi, D., Buhr, L., Claassen, K., Gerola, A., Hofbauer, B., Paiusco, E., & Rijssenbeek, J. (2023, September 5). 4. Climate engineering and the future of justice. *Ethics of Socially Disruptive Technologies*, 83–112. https://doi.org/10.11647/obp.0366.04
37. Resnik, D. B., & Vallero, D. A. (2011). Geoengineering: An idea whose time has come?. *Journal of Earth Science & Climatic Change*, S1, 001.
38. Liu, Z., & Chen, Y. (2015, September 1). Impacts, risks, and governance of climate engineering. *Advances in Climate Change Research*. https://doi.org/10.1016/j.accre.2015.10.004
39. Proelß, A., & Steenkamp, R. C. (2022, November 15). *Geoengineering: Methods, Associated Risks and International Liability*. Springer eBooks. https://doi.org/10.1007/978-3-031-13264-3_9
40. National Research Council (US) (2010). *Committee on the Earth system Context for Hominin Evolution. Understanding Climate's Influence on Human Evolution*. National Academies Press (US); 1, Introduction. Available from: https://www.ncbi.nlm.nih.gov/books/NBK208097/

6 Leveraging Advanced Technologies for Postharvest Management in Horticulture

The Role of AI, ML, and IoT

Shuprajhaa T, Suresh Kumar P, Dhayalini K, and C. Sivananth

6.1 INTRODUCTION

In the modern era, swift technological progress has profoundly impacted horticulture, which includes both perennial and seasonal crops, spanning a diverse array of fruits, vegetables, flowers, spices, plantation crops, nuts, and exotic high-value crops. Although horticulture occupies a smaller land area compared to other agricultural sectors, it stands out for its high economic and nutritional value. Over the past decade, the area dedicated to horticulture in India has increased by 2.6% annually, while production has risen by 4.8%. Horticulture occupies 10% of the land in India, contributes over 34.4% of the agricultural GDP, and supports nearly 20% of the agricultural labor force [1]. The importance of postharvest management in horticultural crops cannot be overstated. Unlike food grains, horticultural crops such as fruits, vegetables, and ornamental plants are highly perishable and susceptible to rapid quality deterioration after harvest. Proper postharvest management is critical for minimizing losses, maintaining quality, ensuring food safety, and extending the shelf life of these products [2]. Globally, postharvest losses are a significant issue, often accounting for 30–40% of production.

Conventional postharvest management systems face several challenges that limit their effectiveness. Many traditional methods rely heavily on manual handling, subjective quality assessment, and rudimentary storage practices, which can lead to inconsistent quality control and significant postharvest losses. Additionally, traditional methods are labor-intensive and costly, limiting scalability, especially in regions with limited access to modern technology. Inadequate infrastructure, such as cold chain systems and packaging technologies, further exacerbates these losses, where access to advanced storage and transportation facilities may be limited. Other issues include the lack of real-time monitoring, inadequate pest and disease control,

DOI: 10.1201/9781003594093-6

and the difficulty in predicting optimal harvesting times, which can lead to over-ripening or under-ripening of crops. With increasing global demand for food and rising pressures on natural resources, it is critical to adopt more efficient, sustainable, and scalable solutions [3].

6.2 DIGITAL TECHNOLOGIES IN HORTICULTURE

Digitalization involves the integration of digital technologies into horticultural practices, allowing stakeholders to collect and analyze data in real time [4]. The emergence of digital technologies has revolutionized postharvest management, introducing innovative tools to overcome challenges associated with traditional systems. Technologies such as the Internet of Things (IoT), artificial intelligence (AI), blockchain, and big data analytics facilitate precise, real-time monitoring of crop conditions during storage and transportation, enabling timely interventions to preserve quality and reduce losses. The IoT connects various devices and sensors, enabling continuous monitoring of crops and environmental conditions. This steady stream of data empowers farmers to make informed decisions on sowing, planting, and intercultural operations like irrigation, fertilization, and pest control, ultimately enhancing quality and minimizing losses. AI and machine learning (ML) process vast datasets from multiple sources, identifying patterns and making predictions about crop yields, pest infestations, and disease outbreaks. For instance, ML algorithms can analyze historical weather data and crop performance to determine optimal harvest times, ensuring crops are harvested at peak quality to improve postharvest management. Additionally, cloud computing offers a centralized platform for storing and analyzing large datasets related to postharvest processes. Meanwhile, big data analytics enables stakeholders to detect trends, monitor performance, and optimize operations, further enhancing efficiency and sustainability in postharvest management. Blockchain enhances transparency and traceability in horticultural supply chains. By creating a secure, decentralized ledger of transactions, blockchain allows all stakeholders – producers, distributors, retailers, and consumers – to track the journey of crops from farm to table. This is particularly beneficial in postharvest management, as it ensures the integrity of product quality and safety, enabling stakeholders to verify the origin and handling of horticultural crops throughout the supply chain. Automation plays a significant role in the postharvest phase, especially in packing and sorting. Robotic systems can handle delicate fruits and vegetables with precision, ensuring minimal damage during the packing process. Automated sorting systems equipped with AI can analyze color, size, and quality, categorizing products for optimal market distribution. This reduces human error and enhances the efficiency of postharvest operations. Drones are increasingly utilized in postharvest management for tasks such as monitoring storage facilities and assessing the health of crops before they are harvested. They can provide high-resolution imagery and data on crop conditions, allowing for better planning of postharvest processes. Additionally, drones can be used for targeted applications of protective treatments during the postharvest phase, ensuring that crops remain disease-free during storage and transport. The integration of digital technologies in horticulture forms the backbone of modern postharvest management. Among these, IoT-enabled smart sensors, AI-driven decision-making, blockchain for supply chain transparency, and advanced manufacturing methods like 3D food printing work together to enhance efficiency and reduce losses. The following sections explore these technologies and their applications in detail.

6.3 SMART SENSORS-BASED SOLUTIONS FOR POSTHARVEST SUPPLY CHAIN MANAGEMENT

6.3.1 Quality and Grading of Fruits and Vegetables

Traditional methods of quality assessment often rely on human inspectors, which can be time-consuming, inconsistent, and subjective. While manual inspections have been the cornerstone of produce grading for years, they are inherently limited by human error, fatigue, and variations in judgment, especially when large volumes of fruits and vegetables must be assessed within short timeframes. This inefficiency often results in inaccurate grading, poor-quality produce reaching markets, and financial losses for farmers and distributors. To address these limitations, AI and ML techniques are increasingly being integrated into postharvest management systems, offering solutions that enhance the accuracy, efficiency, and consistency of quality assessment processes [5, 6].

AI, particularly convolutional neural networks (CNNs), has proven to be highly effective for image recognition tasks and automated grading of fruits and vegetables [7, 8]. By analyzing visual attributes such as size, color, surface texture, and shape, CNN models can detect defects and anomalies with remarkable precision. For instance, any ML model can be trained on large datasets of labeled images showing both high-quality and defective produce. Once trained, these models can automatically assess incoming images of fruits and vegetables, categorizing them based on pre-defined quality parameters [9–13]. In the case of mangoes, ML algorithms can detect ripeness levels by analyzing the peel's color while simultaneously identifying defects such as bruises, black spots, or irregular shapes. Similarly, in bananas, CNNs can distinguish between perfectly ripened fruits and those affected by mechanical damage, disease, or over-ripening [14]. This level of automation ensures faster processing and minimizes human involvement, which is particularly useful for large-scale sorting facilities that need consistent grading standards to meet export requirements.

Beyond CNNs, ML algorithms such as support vector machines, random forests, and decision trees are widely employed to detect physical defects. These algorithms analyze image data to identify imperfections like bruises, blemishes, discoloration, and shape irregularities. ML techniques are capable of processing pixel-level data, allowing them to identify defects that may not be visible to the naked eye, such as internal bruising or the onset of early spoilage. By combining image processing with ML, it is also possible to quantify the severity of defects, further enhancing the accuracy and reliability of the grading process. In addition to supervised learning methods, unsupervised learning techniques, such as clustering algorithms, are gaining prominence in quality assessment systems. Unlike supervised methods that require labeled datasets, unsupervised techniques analyze data patterns and detect outliers or anomalies that deviate from expected norms. For instance, clustering algorithms can be applied to identify unusually sized fruits, irregular textures, or early spoilage symptoms without extensive pre-labeled data. This makes them particularly effective for handling diverse horticultural produce where variations and anomalies are frequent.

The integration of imaging technologies such as hyperspectral imaging with ML has further expanded the scope of automated quality detection. Hyperspectral imaging captures spectral data across a range of wavelengths, allowing for the assessment of both external and internal quality attributes [15–17]. This technology can detect internal defects such as browning, measure sugar content, or assess ripeness levels in fruits like apples, tomatoes, and bananas. By processing hyperspectral data, ML algorithms provide a comprehensive evaluation of produce quality, ensuring that only products meeting the highest standards are distributed to markets. Researchers deployed IoT-enabled sensors to continuously monitor environmental parameters such as temperature and humidity, crucial for maintaining fruit quality for real-time monitoring and regulation during off-season grafting and postharvest storage of citrus fruits. The system provided timely alerts when conditions deviated from optimal ranges, enabling immediate corrective actions. This approach not only reduced postharvest losses but also improved the overall quality of the citrus produce, demonstrating the effectiveness of IoT in enhancing postharvest management practices [18].

The adoption of AI and ML in quality assessment brings significant advantages to postharvest management. These technologies reduce the dependency on human labor, minimize errors, and enable faster and more accurate grading. By ensuring uniformity in the assessment process, AI-driven systems can help exporters meet stringent quality standards demanded by international markets, thereby enhancing the value of produce. Moreover, the ability to detect defects and spoilage at an early stage reduces postharvest losses and improves profitability for farmers and distributors. By addressing inconsistencies in grading, these technologies ensure that high-quality fruits and vegetables consistently reach consumers, meeting both domestic and international market demands.

6.3.2 Smart Sensors for Real-Time Monitoring and Cold Chain Logistics and Processing

Monitoring cold chain logistics is a critical component in the postharvest management of horticultural crops, ensuring that products maintain their quality and safety from the point of harvest to the final consumer. Smart sensors are integral to modern postharvest management, providing accurate and timely data that allow for informed decision-making. A smart sensor is a device that captures input from the physical environment, utilizes built-in computing capabilities to execute pre-defined functions upon detecting specific stimuli, and processes the data before transmitting it further. In smart storage facilities, a network of IoT devices is employed to continuously monitor critical environmental parameters such as temperature, humidity, and gas composition. Sensors are strategically placed throughout storage areas to gather real-time data, allowing for precise control over the storage environment [19, 20].

Temperature and humidity sensors are deployed within storage facilities and transport containers to continuously monitor environmental conditions. Typically, these sensors use thermocouples or thermistors to measure temperature and hygrometers to gauge humidity levels. These sensors transmit data via wireless protocols (e.g., Zigbee, Wi-Fi) to a centralized monitoring system. The system can analyze the data in real time and trigger alerts if conditions deviate from pre-defined thresholds [21]. For instance, maintaining a temperature of 0–4°C for fruits like strawberries can significantly prolong shelf life. If the temperature rises above 4°C, automated notifications can prompt corrective actions, such as adjusting cooling systems or redistributing products to maintain optimal conditions. Additionally, the integration of Global Positioning System (GPS) tracking with cold chain monitoring systems enhances visibility and control throughout the transportation process. By combining location data with temperature and humidity readings, logistics managers can track not only where the shipment is but also the conditions it is experiencing at any given moment.

Gas composition sensors utilize electrochemical or infrared detection technologies to monitor gases such as oxygen (O_2), carbon dioxide (CO_2), and nitrogen (N_2) in controlled atmosphere storage. These sensors are installed in storage chambers where the atmospheric composition is adjusted to slow down respiration and ripening processes. Real-time adjustments to gas mixtures can be made based on sensor feedback, optimizing conditions for different types of produce. Ethylene sensors, often based on metal oxide semiconductor (MOS) technology, detect concentrations of ethylene gas emitted by ripening fruits. By integrating these sensors into storage and packing areas, stakeholders can manage ripening schedules more effectively. Furthermore, analyzing historical data collected from these shipments can provide insights into patterns and trends that inform future logistics strategies. Predictive maintenance alerts can be generated if temperature fluctuations occur frequently, indicating potential equipment malfunctions before they lead to costly failures (Figure 6.1).

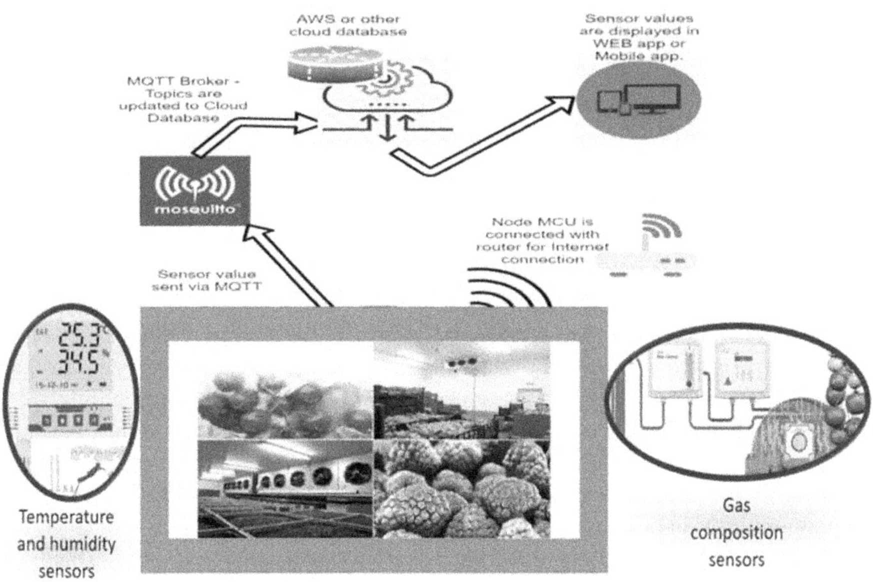

FIGURE 6.1 Sensor- and IoT-based monitoring of postharvest quality of fresh produces.

6.3.3 Intelligent Packaging Solutions

Intelligent packaging solutions are revolutionizing postharvest management by incorporating active and smart features that interact dynamically with their environment, ensuring that fruits and vegetables remain fresh, safe, and of high quality throughout the supply chain. Unlike traditional packaging, which serves merely as a protective barrier, intelligent packaging systems actively monitor and respond to changes in environmental conditions, providing a proactive approach to minimizing postharvest losses. These solutions integrate advanced technologies such as sensors, indicators, and communication systems to offer real-time feedback on the condition of the produce, allowing stakeholders to make timely decisions and ensure product quality [22]. Time-temperature indicators (TTIs) are a prominent feature in smart packaging systems. They are designed to monitor temperature exposure and signal when products have been subjected to conditions outside safe limits. TTIs typically function by changing color when specific temperature thresholds are exceeded, providing a clear visual indication of potential spoilage risks. This is particularly valuable for perishable produce like mangoes, tomatoes, and bananas, where temperature fluctuations during transit or storage can accelerate ripening and decay. By alerting farmers, distributors, and retailers to breaches in temperature control, TTIs enable corrective action, reducing the likelihood of spoilage and food waste [23].

Freshness indicators, another critical component of intelligent packaging, monitor the chemical changes associated with produce spoilage. Technologies such as electronic noses (E-nose) detect spoilage-related volatile organic compounds (VOCs) emitted during the breakdown of fruits and vegetables. These indicators are often coupled with visual cues, such as color changes or display panels, to signal the decline in product quality. For example, in tomatoes, VOC emissions increase as the produce begins to spoil, and the freshness indicators can provide an early warning, allowing the produce to be redirected to alternative markets or consumption channels before complete deterioration occurs. In addition to tracking freshness and environmental conditions, intelligent packaging plays a crucial role in enhancing traceability and transparency across the supply chain. Incorporating advanced technologies such as QR codes and Near Field Communication (NFC) allows consumers to access comprehensive information about a product's journey with a simple scan. This includes details about its origin, processing and handling conditions, storage history, transportation records, quality certifications, and freshness status. Such transparency not only builds consumer trust but also helps ensure food safety, reduce counterfeiting, and support more informed purchasing decisions. Moreover, intelligent packaging enables real-time data sharing among stakeholders, improving supply chain efficiency and minimizing losses due to spoilage or mishandling. By simply scanning a QR code or tapping an NFC-enabled package with a smartphone, consumers gain insights into the produce's journey from farm to market [24, 25]. This level of transparency not only builds trust among consumers but also encourages accountability among retailers and suppliers, ensuring adherence to quality and safety standards. For exporters, this traceability is particularly critical to meet stringent international requirements and maintain credibility in global markets [26].

IoT-enabled sensors are another game-changing advancement integrated into intelligent packaging systems. These sensors continuously monitor critical parameters such as temperature, humidity, and gas concentrations, providing real-time updates on environmental conditions during storage and transit. The collected data is transmitted to cloud-based platforms, where it is analyzed and visualized for actionable insights [27]. For example, if a sensor detects a sudden rise in temperature inside a cold storage container, an immediate alert can be sent to stakeholders, prompting intervention to restore optimal conditions. Similarly, sensors monitoring oxygen and carbon dioxide concentrations can detect abnormal gas levels indicative of spoilage or improper ventilation, allowing corrective measures to be taken before significant losses occur.

Beyond real-time monitoring, the data collected by IoT-enabled packaging solutions contributes to supply chain optimization. Advanced analytics and predictive models leverage this data to anticipate potential quality issues and adjust logistics strategies proactively. For instance, by analyzing trends in temperature fluctuations or humidity levels, predictive algorithms can recommend adjustments to transportation routes, storage durations, or packaging materials to better preserve produce quality. This data-driven approach not only minimizes spoilage but also reduces operational costs, making the supply chain more efficient and sustainable.

By integrating TTIs, freshness indicators, IoT sensors, and traceability technologies, stakeholders can ensure that horticultural produce such as bananas, mangoes, onions, and tomatoes maintain their quality across extended supply chains. These innovations are particularly valuable for smallholder farmers, as they provide access to affordable tools for improving postharvest management and increasing profitability. Additionally, intelligent packaging can enhance India's competitiveness in global markets by meeting international standards for food safety, quality, and traceability. Smart sensors play a pivotal role in real-time monitoring of postharvest conditions, ensuring optimal storage and transport of fresh produce. However, data alone is not enough – ML algorithms process this information to generate predictive insights that can drive better decision-making. The next section delves into how ML enhances postharvest operations through predictive analytics.

6.4 MACHINE LEARNING FOR POSTHARVEST DECISION-MAKING

ML has emerged as a powerful tool in postharvest decision-making, significantly enhancing the management of horticultural crops. By leveraging predictive models, quality assessment techniques, and defect detection algorithms, ML can optimize storage practices, improve product quality, and reduce waste, which are particularly relevant in the context of rapidly evolving agricultural markets and increasing consumer demand for high-quality produce [28].

6.4.1 Predictive Models for Shelf-Life Estimation

Shelf-life prediction models are gaining significant importance in postharvest management, as they utilize both historical and real-time data to estimate the remaining freshness and usability of fruits and vegetables. By analyzing patterns and trends, these models provide valuable insights that help optimize storage conditions and reduce postharvest losses. Advanced ML techniques, including regression analysis, support vector machines (SVM), and neural networks, are extensively used to assess and forecast shelf life. These algorithms process key influencing factors such as temperature, humidity levels, ethylene concentration, and respiration rate to generate accurate predictions. By integrating such predictive models, stakeholders can make informed decisions regarding storage, transportation, and distribution, ultimately enhancing product quality and reducing wastage in the supply chain. These algorithms analyze vast datasets to identify trends and relationships between environmental conditions and produce quality, allowing for accurate predictions. For instance, linear regression models can provide insights into how changes in temperature or humidity accelerate spoilage, while SVM models offer robust classification capabilities to categorize produce based on their remaining shelf life. More advanced models, particularly artificial neural networks (ANNs), are gaining prominence due to their ability to handle complex, non-linear relationships between multiple variables. ANNs are particularly suited for tasks involving large volumes of multi-dimensional data, where traditional models may struggle to deliver accurate predictions. By training these models on extensive datasets that capture the effects of varying environmental parameters across different types of produce, stakeholders can accurately forecast how long a product will remain marketable. This predictive capability allows farmers, distributors, and retailers to optimize inventory management, reduce spoilage, and make well-informed decisions regarding storage durations and transportation schedules. For example, in temperature-sensitive crops such as bananas, tomatoes, or grapes, ANN-based models can predict the onset of spoilage under different transit conditions, enabling timely interventions to maintain freshness and extend shelf life.

The CNN-XGBoost approach stands out for its ability to provide an efficient and reliable method for non-destructive classification of fruit ripening stages, particularly in bananas. CNNs are highly proficient in image recognition tasks, making them ideal for analyzing key visual indicators of ripening, such as changes in color, texture, and surface patterns. When integrated with eXtreme Gradient Boosting (XGBoost), a powerful ensemble learning technique, this hybrid model significantly enhances classification accuracy, even when working with limited datasets. This is a notable advancement since deep learning models typically require large amounts of training data to achieve high precision. The CNN-XGBoost model effectively addresses this challenge, making it particularly suitable for real-world applications where data availability is restricted [29]. To validate its performance, the model is often compared with other ML algorithms, including k-nearest neighbor (KNN), multilayer perceptron (MLP), logistic regression (LR), random forest (RF), decision tree (DT), and AdaBoost, ensuring comprehensive evaluation and benchmarking. Each of these models has its strengths and limitations [30]. KNN is simple and effective for smaller datasets but may struggle with computational efficiency as data volume grows. MLP, a type of feedforward neural network, is capable of learning non-linear relationships but often requires fine-tuning of hyperparameters to perform optimally. Logistic regression, while widely used for binary classification tasks, may not capture the complex non-linear patterns required for shelf-life prediction. Random forests and decision trees offer interpretability and robustness but may suffer from overfitting if not properly managed. AdaBoost, an ensemble technique, improves prediction accuracy by combining multiple weak learners but may require significant computational resources for larger datasets.

The integration of CNN and XGBoost, therefore, offers a robust and scalable solution for ripening classification and shelf-life estimation, combining the feature extraction capabilities of CNNs with the gradient boosting efficiency of XGBoost. This hybrid approach ensures precise predictions with minimal error margins, making it a valuable tool for stakeholders across the supply chain. By enabling non-destructive assessments, the CNN-XGBoost method ensures that fruits and vegetables can be analyzed without compromising their integrity, a critical requirement for commercial operations where product quality must be maintained [14]. By accurately forecasting shelf life and ripening stages, these technologies empower farmers and distributors to optimize the timing of harvest, storage, and transportation. This reduces spoilage, improves resource utilization, and enhances profitability for smallholder farmers. Moreover, predictive models integrated with IoT-enabled sensors provide real-time data on environmental conditions, further improving the accuracy of shelf-life predictions. As a result, the deployment of ML techniques, including CNN-XGBoost and other comparative models, represents a significant advancement in postharvest management, paving the way for smarter, data-driven decision-making and enhanced supply chain efficiency (Figures 6.2 and 6.3).

In Tamil Nadu, India, researchers developed an IoT-based smart technology system aimed at optimizing postharvest handling in the agro-industry, particularly focusing on tomatoes. The system utilized IoT sensors to collect data on environmental conditions such as temperature, humidity, and precipitation. By applying spatial

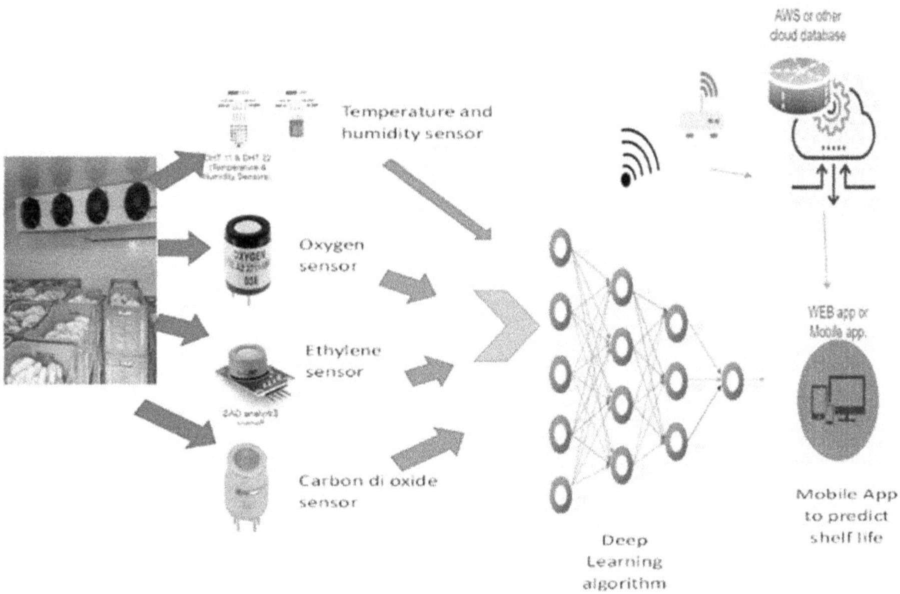

FIGURE 6.2 Predictive models for shelf-life estimation in horticultural produces.

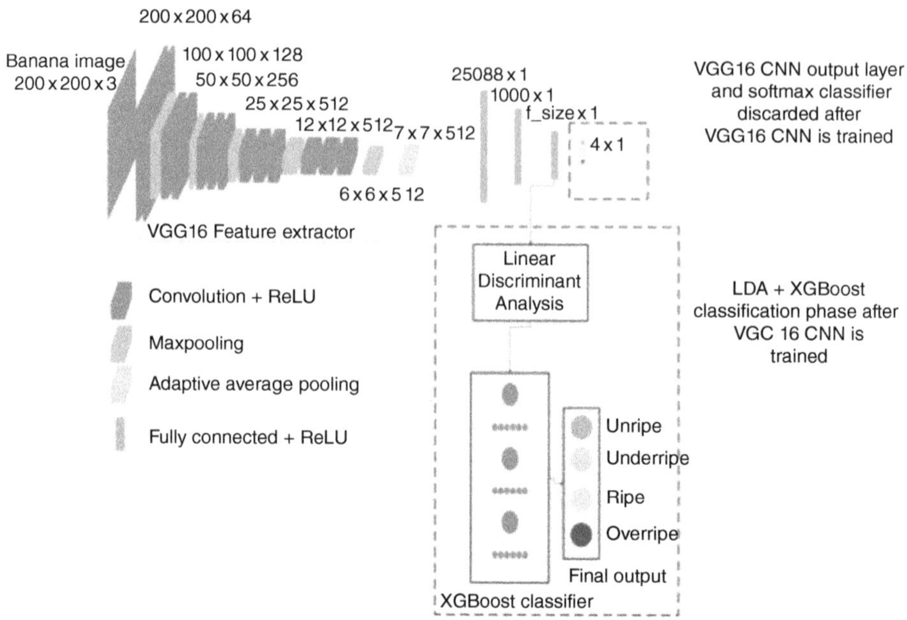

FIGURE 6.3 CNN-Xgboost-based ripening identification in banana.

analysis techniques, the system could predict harvest quantities and production capacity with an accuracy rate of 89.35%. This predictive capability enables farmers to make informed decisions regarding postharvest handling, storage, and marketing strategies, thereby reducing losses and improving profitability [31]. While sensors provide continuous data, ML models transform this raw information into actionable insights, such as predicting shelf life and detecting spoilage patterns. However, transparency and traceability in these AI-driven predictions are crucial, particularly in global supply chains. This is where blockchain technology becomes essential, ensuring that recorded data remains tamper-proof and verifiable.

6.5 BLOCKCHAIN FOR TRACEABILITY AND SUPPLY CHAIN TRANSPARENCY IN FRESH AND PROCESSED PRODUCTS

Blockchain technology has emerged as a powerful solution for improving traceability and transparency in the supply chain, addressing longstanding issues in postharvest management for both fresh and processed horticultural products. Traditional traceability systems are often fragmented, with data scattered across various databases and recorded in inconsistent formats. This lack of integration makes it difficult to effectively track the journey of produce and identify areas where losses occur. Blockchain technology overcomes these limitations by creating a decentralized, immutable, and verifiable digital ledger that records every event and transaction in the supply chain. This not only ensures greater accountability but also establishes a transparent record of the produce's journey from farm to fork [32].

For example, as produce is harvested, information such as the date, time, location, method of harvest, and even farmer details can be recorded on the blockchain. As the produce moves through the supply chain – being sorted, graded, packed, transported, and stored – additional information, including temperature, humidity, and handling conditions, can be continuously updated and verified on the digital ledger [33]. Since blockchain is decentralized, all stakeholders in the supply chain – including farmers, processors, distributors, retailers, and consumers – have access to the same verified data, eliminating inconsistencies, fostering collaboration, and reducing disputes regarding product quality or handling. This level of transparency is particularly important in ensuring that fresh produce meets safety standards and quality expectations throughout its lifecycle. By providing tamper-proof records, blockchain enables stakeholders to identify and trace the source of any quality issues quickly. This is particularly valuable for addressing contamination events, identifying sources of spoilage, or verifying compliance with food safety standards [34, 35]. If a batch of produce is found to be substandard or contaminated, blockchain can pinpoint exactly where in the supply chain the issue occurred, enabling swift corrective action and reducing wastage.

The integration of blockchain technology with Internet of Things (IoT) devices further enhances its utility in postharvest management. IoT sensors and Radio Frequency Identification (RFID) tags are capable of tracking and transmitting real-time data on environmental conditions such as temperature, humidity, and gas concentrations during transportation and storage [36]. For instance, perishable produce like bananas, tomatoes, or mangoes requires specific temperature and humidity conditions to maintain freshness. IoT sensors monitor these parameters continuously, and any deviations from the optimal range are recorded on the blockchain. This allows for timely alerts to be sent to logistics managers, enabling immediate interventions to prevent spoilage and reduce losses. Additionally, IoT-enabled RFID tags can provide real-time location tracking, offering granular visibility into the movement of produce across the supply chain. This ensures that stakeholders can monitor not only the condition of the produce but also its precise location, helping optimize logistics and minimize delays [37, 38]. Such visibility is critical for time-sensitive products, where even minor delays or mishandling can result in significant losses. Furthermore, these advanced systems can identify and segregate spoiled or low-quality produce during transportation, ensuring that only the best-quality products reach the market. Technologies such as electronic noses (E-nose) integrated with IoT sensors can detect spoilage-related VOCs, further strengthening quality control mechanisms.

The enhanced traceability provided by blockchain, combined with IoT technologies, also plays a key role in building consumer trust and meeting regulatory requirements, especially for export markets. For example, international buyers often require detailed traceability to ensure compliance with stringent food safety standards. Blockchain provides an end-to-end, verifiable record of a product's journey, from the point of harvest to the retail shelf, including data on growing conditions, storage, handling, and transportation. This level of transparency assures consumers of the product's quality, safety, and authenticity while also adding value to the produce in competitive markets. For consumers, blockchain-enabled systems offer greater confidence in the quality and origin of the produce they purchase. Features such as QR codes or NFC tags on product packaging allow customers to access detailed traceability data by scanning the code with a smartphone. This data can include the product's origin, harvest date, farmer details, storage history, and transportation conditions. Such transparency fosters greater trust in food systems, encourages informed purchasing decisions, and promotes brand loyalty. Blockchain technology secures and streamlines the movement of fresh and processed products across the supply chain, ensuring authenticity and quality assurance. As digital tracking becomes more prevalent, advancements in food processing, such as 3D printing, are also gaining momentum, offering innovative ways to add value to perishable produce. The next section explores how AI and deep learning facilitate customization in 3D food printing.

6.6 3D FOOD PRINTING FOR VALUE-ADDED FRUIT PRODUCTS

3D food printing technology uses layer-by-layer deposition techniques to create edible structures from fruit-based materials such as purees, pastes, gels, or powders. The technology is often supported by extrusion-based systems, where fruit-based inks are shaped into intricate designs with high precision [39]. This approach has revolutionized the creation of visually appealing and functional foods while reducing waste by utilizing byproducts like fruit peels and pulps as raw materials. In fruit processing, 3D printing finds applications in designing products such as fruit-based snacks, decorative dessert elements, and functional foods fortified with vitamins or fibers. For instance, mango or banana purees are often used as base materials for creating complex geometries in snacks or dessert toppings. Furthermore, freeze-dried fruit powders combined with binding agents can produce nutritious and lightweight snacks ideal for space missions or long-term storage. The ability to blend multiple fruit ingredients during printing allows for creating multi-flavored, multi-colored, or multi-layered products that cater to evolving consumer preferences. This has led to the development of hybrid fruit products, such as strawberry-banana snacks or layered apple-kiwi desserts, enhancing their sensory and nutritional appeal.

6.6.1 DL and AI for Customization in 3D Food Printing: Texture, Flavor, and Nutritional Profiles

Deep learning (DL) and AI are key enablers of customization in 3D food printing. These technologies allow for precise adjustments to parameters such as texture, flavor, and nutritional content based on individual preferences or dietary requirements. DL models analyze user data, such as health metrics or taste preferences, to create customized recipes that meet specific needs. For example, a consumer requiring low-sugar options might receive a 3D-printed fruit snack with reduced natural sugar content while retaining optimal sweetness through the addition of alternative fruit-based sweeteners [40]. AI algorithms enable real-time control of printing parameters like nozzle speed, layer thickness, and material composition to achieve the desired texture. For example, softer textures suitable for elderly consumers or children can be achieved by adjusting the viscosity of the fruit ink. AI-powered sensors also ensure consistency in texture across batches by analyzing the rheological properties of the material in real time. Flavor customization is achieved by blending fruit components in varying ratios. AI models use flavor profiling techniques to predict and optimize flavor combinations during the printing process. For instance, AI systems have been used to create customized fruit jelly shapes with unique flavor blends like orange-berry or mango-lime, tailored to individual taste profiles. Nutritional profiles can also be fine-tuned with the help of AI. For example, fortifying 3D-printed fruit snacks with calcium, iron, or fiber is possible by adding microencapsulated nutrients to the fruit ink. AI-driven nutritional modeling ensures that the final product meets the consumer's dietary goals, such as high-protein snacks for athletes or low-calorie treats for weight management [40].

6.6.2 Innovations in Creating Fruit-Based Products through 3D Printing

3D food printing has enabled the creation of entirely new categories of fruit-based products that were previously difficult or impossible to manufacture using conventional methods. One such innovation is the development of complex edible structures, such as hollow or lattice-like designs, which enhance the visual and sensory appeal of the product. For example, apple or pear purees have been used to create intricately shaped fruit sculptures for gourmet desserts, offering a premium consumer experience. Similarly, 3D printing has enabled the creation of multi-layered fruit bars with alternating layers of different flavors, textures, or colors, such as a strawberry-kiwi-mango bar with distinct flavor zones. Another innovative application is the incorporation of probiotics or prebiotics into fruit-based products through 3D printing. For instance, a fruit-based yogurt snack can be fortified with live probiotic cultures during the printing process, enhancing its functional benefits. Edible packaging for fruit products is another area of innovation. Using biodegradable fruit-based inks, 3D printers can produce edible wraps or containers that reduce plastic waste while adding a functional layer to the food. For example, an edible fruit cup made from banana puree and gelatin can serve as a container for fruit salads, eliminating the need for disposable plastic packaging.

6.6.3 Role of 3D Printing in Personalization and Nutritional Optimization of Processed Fruit Foods

One of the most transformative aspects of 3D food printing is its ability to personalize processed fruit products according to individual dietary needs. This personalization is particularly valuable in healthcare, where patients with specific dietary restrictions can receive tailored food solutions. For example, 3D-printed fruit gels can be designed for patients with dysphagia (difficulty swallowing). By modifying the texture and consistency of the gels, the risk of choking can be minimized while still delivering essential nutrients. Nutritional optimization is another significant advantage of 3D printing. For instance, fruit-based energy bars can be enriched with additional protein, vitamins, or electrolytes for athletes. AI algorithms calculate the precise nutritional requirements and adjust the composition of the printed product accordingly. Personalized fruit-based meals for children are also gaining popularity. Using 3D printing, playful and appealing shapes like animals or cartoon characters can be created to encourage healthy eating habits. A 3D-printed "fruit dinosaur" made from mango puree can make nutritious snacks more engaging for children. In addition, 3D printing enables real-time customization based on consumer feedback. AI systems analyze data from previous purchases or taste preferences and adjust recipes accordingly. This flexibility allows manufacturers to offer a wide range of products tailored to diverse consumer needs, from high-fiber snacks for digestive health to low-sugar desserts for diabetic consumers. 3D food printing enables the creation of customized fruit-based products by leveraging AI and deep learning to optimize texture, flavor, and nutritional content.

6.7 AI FOR PREDICTIVE LOGISTICS: MANAGING FRESH AND PROCESSED PRODUCT FLOW

AI-driven predictive logistics is essential for streamlining the movement of fresh and processed fruit products, ensuring efficiency across the supply chain. By leveraging historical data and real-time inputs, AI systems can anticipate demand variations, detect potential bottlenecks, and optimize inventory management. For instance, an AI-driven system can analyze seasonal patterns, weather conditions, and consumer purchasing behavior to accurately forecast demand for fresh strawberries. This enables suppliers and retailers to make data-driven decisions, reducing wastage, preventing stock shortages or surpluses, and ultimately improving supply chain resilience and profitability. This enables suppliers to adjust harvesting schedules, minimize overproduction, and prevent wastage. Predictive logistics also optimizes transport routes to ensure timely delivery while reducing costs and environmental impact. For instance, AI can recommend the fastest and most fuel-efficient delivery routes for refrigerated trucks carrying frozen fruit purees. By incorporating live traffic data and road conditions, these systems help reduce delays and maintain product freshness. Additionally, AI can manage multi-modal logistics for processed products, such as dried fruit snacks or fruit syrups, ensuring that inventory reaches distribution centers efficiently. This is particularly important for export operations, where delays can lead to spoilage or loss of market value [41]. AI-driven predictive logistics ensures seamless management of fresh and processed produce by optimizing storage and transport routes based on real-time data. As technology evolves, future innovations such as autonomous robots, nano-sensors, and quantum computing will further revolutionize postharvest management, making supply chains even more efficient and sustainable.

6.7.1 Real-Time Decision-Making for Reducing Losses and Optimizing Storage and Transport

Real-time decision-making powered by IoT, AI, and blockchain is critical for minimizing losses and ensuring the efficiency of storage and transportation systems. IoT-enabled storage units equipped with temperature, humidity, and gas sensors monitor conditions in real time. For example, smart cold storage facilities can detect deviations in temperature or humidity that might accelerate spoilage and automatically adjust settings to restore optimal conditions. AI algorithms analyze data from these sensors to identify potential risks and recommend corrective actions. For instance, if a refrigerated truck's cooling system malfunctions, an AI system can immediately alert the driver and suggest the nearest repair facility or alternative storage location. Blockchain further supports real-time decision-making by providing end-to-end visibility of the supply chain. For instance, if a shipment of processed fruit juice is delayed, blockchain can trace the delay's origin and allow stakeholders to make informed decisions about rerouting or prioritizing the shipment to prevent spoilage. Dynamic pricing is another application of real-time analytics. AI systems can analyze the quality and shelf life of products and adjust prices accordingly. For example, fruit nearing the end of its shelf life can be sold at a discounted price to prevent waste while still generating revenue.

6.8 FUTURE TRENDS AND EMERGING TECHNOLOGIES

6.8.1 Autonomous Robots for Harvesting and Postharvest Processing

A significant breakthrough in agricultural technology is the development of autonomous robots designed for harvesting and postharvest processing. These state-of-the-art robots are integrated with advanced sensors and AI, allowing them to execute intricate tasks with remarkable accuracy and efficiency. By leveraging machine vision and deep learning algorithms, these robots can identify optimal harvesting times, selectively pick ripe produce, and handle postharvest sorting and packaging with minimal human intervention. This innovation not only enhances productivity but also reduces labor dependency and postharvest losses, making it a game-changer for the horticulture industry. For example, robotic systems can be programmed to identify ripe fruits based on color, size, and other visual indicators, significantly reducing labor costs and the risk of damage associated with manual harvesting [42]. Furthermore, autonomous robots can operate continuously, allowing for timely harvesting that minimizes losses due to overripeness or spoilage. In postharvest processing, robots can automate sorting, grading, and packing, which enhances efficiency and consistency while freeing human workers for higher-value tasks [43].

6.8.2 Use of Nano-Sensors and Quantum Computing in Postharvest Monitoring

Nano-sensors, which can detect minute changes in environmental conditions, are being developed to monitor factors such as gas composition, humidity, and temperature at unprecedented levels of sensitivity. These sensors can be embedded in packaging materials or integrated into storage facilities, providing real-time data that can be analyzed for effective decision-making [44]. Meanwhile, quantum computing holds the potential to process complex datasets rapidly, enabling advanced predictive modeling and optimization of postharvest logistics. By harnessing the power of quantum algorithms, stakeholders can simulate and analyze multiple scenarios simultaneously, leading to more effective strategies for minimizing waste and enhancing product quality [45].

6.8.3 Integration of 5G and Edge Computing for Real-Time Decision-Making

The integration of 5G and edge computing technologies is set to revolutionize real-time decision-making in postharvest management. With the advent of 5G networks, the speed and capacity of data transmission will increase dramatically, allowing for seamless communication between IoT devices deployed throughout the supply chain. Edge computing, which involves processing data closer to the source rather than relying solely on centralized data centers, will facilitate real-time analytics and responsiveness. This real-time capability is crucial for maintaining the integrity of perishable products and ensuring that stakeholders can respond proactively to any challenges that arise during transportation and storage.

6.8.4 Digital Twin Technology

Digital twin technology is revolutionizing postharvest management in horticultural crops by offering a dynamic, data-driven approach to optimizing storage, transportation, and overall supply chain efficiency. At its core, a digital twin is a virtual replica of a physical system or object, continuously updated with real-time data to simulate, analyze, and enhance performance [46]. In postharvest operations, digital twins play a pivotal role in improving efficiency, minimizing waste, and maintaining produce quality throughout critical handling stages. Beyond real-time monitoring, they leverage predictive analytics to optimize workflows, ensuring that best practices are identified and implemented to reduce postharvest losses. By serving as intelligent decision-support systems, digital twins provide stakeholders with actionable insights, enabling them to make informed choices regarding storage conditions, transportation logistics, and inventory management. For exporters, retailers, and consumers, digital twins offer critical data such as the estimated shelf life of shipments, allowing for strategic logistics planning and targeted marketing approaches. Additionally, they help diagnose and predict potential disruptions in the supply chain that could compromise food quality and lead to losses. More importantly, digital twins can recommend proactive, shipment-specific interventions to mitigate waste at both retail and household levels, ensuring a more sustainable and efficient food supply system. The future of postharvest management is shaped by cutting-edge advancements, including autonomous robots for processing, nano-sensors for real-time quality assessment, and quantum computing for enhanced decision-making. Integrating these technologies with 5G and edge computing will further enable instantaneous monitoring and control.

6.9 CHALLENGES IN THE ADOPTION OF DIGITAL TECHNOLOGIES

The adoption of digital technologies for postharvest management of fruits and vegetables presents a myriad of benefits; however, it has several challenges, including technological barriers, data privacy concerns, cybersecurity, high initial costs, skills gaps, and scalability issues. Technological barriers and infrastructure limitations are among the most significant obstacles to the adoption of digital technologies in postharvest management.

Limited internet connectivity, inadequate power supply, and lack of access to modern storage facilities can impede the implementation of digital solutions. Stakeholders may also be reluctant to adopt technologies that require sharing data with third parties due to fears of data breaches or misuse. Additionally, as more devices become interconnected, the risk of cyberattacks increases. Cybersecurity vulnerabilities can lead to unauthorized access to sensitive data or disruption of operations, undermining trust among stakeholders. Besides, high initial costs and concerns regarding return on investment (ROI) represent another barrier to widespread adoption. The skills and knowledge gap among stakeholders poses a significant challenge in the successful implementation of digital technologies in postharvest management. Scalability and interoperability issues also complicate the adoption of digital technologies in the postharvest sector. Many digital solutions are designed as stand-alone systems, making it challenging to integrate them with existing technologies and processes. Furthermore, as businesses grow and operations expand, scaling up digital solutions can be complex and costly.

6.10 CONCLUSION

The integration of digital technologies in the postharvest management of fruits and vegetables has ushered in a new era of efficiency, quality, and sustainability within the horticultural supply chain. By leveraging advanced tools such as IoT devices, ML algorithms, and blockchain technology, stakeholders can significantly enhance various aspects of postharvest operations. One of the most significant benefits of digital technologies is the enhancement of efficiency and the reduction of losses throughout the postharvest process. With the deployment of IoT devices, real-time monitoring of environmental conditions, such as temperature and humidity, becomes feasible. This level of automation reduces the need for manual inspections and minimizes the likelihood of human error, leading to a more streamlined process and lower operational costs. Advanced quality assessment tools, such as ML algorithms and computer vision systems, enable precise grading of fruits and vegetables based on visual attributes like color, size, and surface defects. Moreover, blockchain technology enhances food safety by providing an immutable record of the product's journey, including details about handling practices and any treatments applied. Advanced analytics platforms allow stakeholders to model and simulate different storage conditions and transportation routes, identifying the optimal parameters for specific types of produce. By utilizing predictive analytics, stakeholders can forecast demand more accurately, allowing for better inventory management. ML models can analyze historical sales data, weather patterns, and market trends to predict the shelf life of various produce items. By aligning production with actual demand, farmers and distributors can minimize excess supply that leads to waste. Additionally, real-time monitoring systems can detect spoilage at an early stage, enabling quick responses to salvage affected batches or redirect them to processing facilities rather than allowing them to go to waste. The extensive data collected through various digital tools provide valuable insights into consumer preferences, market trends, and product performance. This ability to leverage data for strategic planning enhances market competitiveness and can significantly boost sales and profitability. By integrating sustainability into the core of digital strategies, the postharvest sector can contribute to broader climate goals while also meeting consumer demands for responsible sourcing.

REFERENCES

1. National Horticultural Board. (2022). Ministry of agriculture and farmers welfare, government of India.https://nhb.gov.in/StatisticsViewer.aspx?enc=MWoUJibk35dW2g36TUJWAoZqESmAYFi7h2irlsmjlINT
2. Kader, A. A. (2013). Postharvest technology of horticultural crops-An overview from farm to fork. *Ethiopian Journal of Applied Science and Technology*, *1*, 1–8.
3. Singh, R., Singh, R., Gehlot, A., Akram, S. V., Priyadarshi, N., & Twala, B. (2022). Horticulture 4.0: Adoption of industry 4.0 technologies in horticulture for meeting sustainable farming. *Applied Sciences*, *12*(24), 12557.
4. Addanki, M., Patra, P., & Kandra, P. (2022). Recent advances and applications of artificial intelligence and related technologies in the food industry. *Applied Food Research*, *2*(2), 100126.
5. Bora, G. C., Lin, D., Bhattacharya, P., Bali, S. K., & Pathak, R. (2015). Application of bio-image analysis for classification of different ripening stages of banana. *Journal of Agricultural Science*, *7*(2), 152.
6. El-Bendary, N., El Hariri, E., Hassanien, A. E., & Badr, A. (2015). Using machine learning techniques for evaluating tomato ripeness. *Expert Systems with Applications*, *42*(4), 1892–1905.
7. Jiao, W., Hao, X., & Qin, C. (2021). The image classification method with CNN-XGBoost model based on adaptive particle swarm optimization. *Information*, *12*(4), 156.
8. Khaing, Z. M., Naung, Y., & Htut, P. H. (2018) Development of control system for fruit classification based on convolutional neural network, *2018 IEEE Conference of Russian Young Researchers in Electrical and Electronic Engineering (EIConRus)*, Moscow and St. Petersburg, Russia, pp. 1805–1807.
9. Dimililer, K., & Olaniyi, E. O. (2015). Intelligent sorting system based on computer vision for banana industry. *International Journal of Scientific Research in Science, Engineering and Technology*, *6*(9), 332–337.
10. Mohapatra, A., Shanmugasundaram, S., & Malmathanraj, R. (2017). Grading of ripening stages of red banana using dielectric properties changes and image processing approach. *Computers and Electronics in Agriculture*, *143*, 100–110.
11. Mustafa, N. B. A., *et al.*, Image processing of an agriculture produce: Determination of size and ripeness of a banana, *2008 International Symposium on Information Technology*, Kuala Lumpur, Malaysia, 2008, pp. 1–7.
12. Surya Prabha, D., & Satheesh Kumar, J. (2015). Assessment of banana fruit maturity by image processing technique. *Journal of food science and technology*, 52, 1316–1327.
13. Zhang, L., Jia, J., Gui, G., Hao, X., Gao, W., & Wang, M. (2018). Deep learning based improved classification system for designing tomato harvesting robot. *IEEE Access*, *6*, 67940–67950.
14. Shuprajhaa, T., Raj, J. M., Paramasivam, S. K., Sheeba, K. N., & Uma, S. (2023). Deep learning based intelligent identification system for ripening stages of banana. *Postharvest Biology and Technology*, *203*, 112410.
15. Lorente, D., Aleixos, N., Gómez-Sanchis, J. U. A. N., Cubero, S., García-Navarrete, O. L., & Blasco, J. (2012). Recent advances and applications of hyperspectral imaging for fruit and vegetable quality assessment. *Food and Bioprocess Technology*, *5*, 1121–1142.
16. Lu, Y., Huang, Y., & Lu, R. (2017). Innovative hyperspectral imaging-based techniques for quality evaluation of fruits and vegetables: A review. *Applied Sciences*, *7*(2), 189.
17. Nanyam, Y., Choudhary, R., Gupta, L., & Paliwal, J. (2012). A decision-fusion strategy for fruit quality inspection using hyperspectral imaging. *Biosystems Engineering*, *111*(1), 118–125.

18. Lamsal, R. R., Acharya, U. K., Karthikeyan, P., Otero, P., & Ariza, A. (2024). Implementing internet of things for real-time monitoring and regulation of off-season grafting and post-harvest storage in citrus cultivation: A case study from the hilly regions of Nepal. *AgriEngineering*, *6*(3), 2082–2100. https://doi.org/10.3390/agriengineering6030122
19. Cil, A. Y., Abdurahman, D., & Cil, I. (2022). Internet of Things enabled real time cold chain monitoring in a container port. *Journal of Shipping and Trade*, 7(1), 9.
20. Wanganoo, L. & Shukla, V. K. Real-Time Data Monitoring in Cold Supply Chain Through NB- IoT, *2020 11th International Conference on Computing, Communication and Networking Technologies (ICCCNT)*, Kharagpur, India, 2020, pp. 1–6.
21. Campos, Y. and Villa, J. L. Technologies applied in the monitoring and control of the temperature in the Cold Chain, *2018 IEEE 2nd Colombian Conference on Robotics and Automation (CCRA)*, Barranquilla, Colombia, 2018, pp. 1–6.
22. Kuswandi, B., Wicaksono, Y., Jayus, Abdullah A., Heng, L. Y., & Ahmad, M. (2011). Smart packaging: Sensors for monitoring of food quality and safety. *Sensing and Instrumentation for Food Quality and Safety*, *5*, 137–146.
23. Liu, D., Yang, L., Shang, M., & Zhong, Y. (2019). Research progress of packaging indicating materials for real-time monitoring of food quality. *Materials Express*, *9*(5), 377–396.
24. Bhatia, S., & Albarrak, A. S. (2023). A blockchain-driven food supply chain management using QR code and XAI-faster RCNN architecture. *Sustainability*, *15*(3), 2579.
25. Yang, S. Y., Jan, H. C., Chen, C. Y., & Wang, M. S. (2023). CNN-based QR code reading of package for unmanned aerial vehicle. *Sensors*, *23*(10), 4707.
26. Abekoon, T., Buthpitiya, B. L. S. K., Sajindra, H., Samarakoon, E. R. J., Jayakody, J. A. D. C. A., Kantamaneni, K., & Rathnayake, U. (2024). A comprehensive review to evaluate the synergy of intelligent food packaging with modern food technology and artificial intelligence field. *Discover Sustainability*, *5*(1), 160.
27. Torres-Sánchez, R., Martínez-Zafra, M. T., Castillejo, N., Guillamón-Frutos, A., & Artés-Hernández, F. (2020). Real-time monitoring system for shelf life estimation of fruit and vegetables. *Sensors*, *20*(7), 1860.
28. Singh, A., Vaidya, G., Jagota, V., Darko, D. A., Agarwal, R. K., Debnath, S., & Potrich, E. (2022). Recent advancement in postharvest loss mitigation and quality management of fruits and vegetables using machine learning frameworks. *Journal of Food Quality*, *2022*(1), 6447282.
29. Shuprajhaa, T., Raj, M., Sheeba, K. N., & Dhayalini, K. Deep learning based mobile application for varietal identification and ripeness grading of traditional Indian banana varieties, *2023 Second International Conference on Electrical, Electronics, Information and Communication Technologies (ICEEICT)*, Trichirappalli, India, 2023, pp. 1–6. IEEE.
30. Shuprajhaa, T., Sekar, C., Suresh Kumar, P., Sheeba, K. N., & Dhayalini, K. (2024, May). Development of predictive model based mobile application for maturity stage identification of Indian traditional red bananas, *2024 International Conference on Advances in Modern Age Technologies for Health and Engineering Science (AMATHE)*, Shivamogga, India, pp. 1–9.
31. Prithviraj, S. R., Kumar, N., Niranjan, C., Praveen, R., & Ahmed, M. R. (2020). Technological advances in agriculture from pre-processing of land management to post-harvest management: A critical review. *International Journal of Advance Science and Technology*, *29*, 3055–3067.
32. Madumidha, S., Ranjani, P. S., Varsinee, S. S., & Sundari, P. S., Transparency and Traceability: In Food Supply Chain System using Blockchain Technology with Internet of Things, *2019 3rd International Conference on Trends in Electronics and Informatics (ICOEI)*, Tirunelveli, India, 2019, pp. 983–987.

33. Astill, J., Dara, R. A., Campbell, M., Farber, J. M., Fraser, E. D., Sharif, S., & Yada, R. Y. (2019). Transparency in food supply chains: A review of enabling technology solutions. *Trends in Food Science & Technology*, *91*, 240–247.
34. Majdalawieh, M., Nizamuddin, N., Alaraj, M., Khan, S., & Bani-Hani, A. (2021). Blockchain-based solution for secure and transparent food supply chain network. *Peer-to-Peer Networking and Applications*, *14*, 3831–3850.
35. Sunny, J., Undralla, N., & Pillai, V. M. (2020). Supply chain transparency through blockchain-based traceability: An overview with demonstration. *Computers & Industrial Engineering*, *150*, 106895.
36. Garrido-López, J., Jiménez-Buendía, M., Toledo-Moreo, A., Giménez-Gallego, J., & Torres-Sánchez, R. (2024). Monitoring perishable commodities using cellular IoT: An intelligent real-time conditions tracker design. *Applied Sciences*, *14*(23), 11050.
37. Azizi, Aydin. *Applications of artificial intelligence techniques in industry 4.0*. Berlin: Springer, 2019.
38. Zuo, J., Feng, J., Gameiro, M. G., Tian, Y., Liang, J., Wang, Y., et al. (2022). RFID-based sensing in smart packaging for food applications: A review. *Future Foods*, *6*, 100198.
39. Mantihal, S., Kobun, R., & Lee, B. B. (2020). 3D food printing of as the new way of preparing food: A review. *International Journal of Gastronomy and Food Science*, *22*, 100260.
40. Tomašević, I., Putnik, P., Valjak, F., Pavlić, B., Šojić, B., Markovinović, A. B., & Kovačević, D. B. (2021). 3D printing as novel tool for fruit-based functional food production. *Current opinion in food science*, *41*, 138–145.
41. Rane, N., Choudhary, S., & Rane, J. (2024). Artificial intelligence and machine learning for resilient and sustainable logistics and supply chain management. *Available at SSRN 4847087*.
42. Rajendran, V., Debnath, B., Mghames, S., Mandil, W., Parsa, S., Parsons, S., & Ghalamzan-E, A. (2024). Towards autonomous selective harvesting: A review of robot perception, robot design, motion planning and control. *Journal of Field Robotics*, *41*(7), 2247–2279.
43. Chauhan, A., Brouwer, B., & Westra, E. (2022). Robotics for a quality-driven post-harvest supply chain. *Current Robotics Reports*, *3*(2), 39–48
44. Zhong, X., Zhang, M., Tang, T., Adhikari, B., & Ma, Y. (2023). Advances in intelligent detection, monitoring, and control for preserving the quality of fresh fruits and vegetables in the supply chain. *Food Bioscience*, *56*, 103350.
45. Bharti, A., Jain, U., & Chauhan, N. (2024). From lab to field: Nano-biosensors for real-time plant nutrient tracking. *Plant Nano Biology*, 9, 100079.
46. Defraeye, T., Shrivastava, C., Berry, T., Verboven, P., Onwude, D., Schudel, S., … Rossi, R. M. (2021). Digital twins are coming: Will we need them in supply chains of fresh horticultural produce?. *Trends in Food Science & Technology*, *109*, 245–258.

7 Near-Field Microwave Sensing of Dual-Comb Patch for Non-Invasive Detection of Ammonia Gas in Stored Wet Grains

Abirami Karthikeyan and A. Rajesh

7.1 INTRODUCTION

In recent years, industrial development has grown enormously, which is influencing the development of the industrial Internet of Things (IoT). This development of the IoT system has an impact on various sensor designs, preferably non-invasive types of sensing. The release of ammonia gas (NH_3) during the warehouse stacks and the decay of grains due to microbial growth and protein breakdown. Higher ammonia levels not only specify spoilage but also pose health risks and affect grain quality. Therefore, rapid and real-time monitoring of ammonia gas release from stored grains is important for safeguarding food safety and preserving air quality. Early instruments for harmful gas detection are habitually expensive, time-consuming, and hardly used in real-time monitoring systems. Recent research has focused on developing advanced environmental sensors and discovering materials like carbon nanotubes, metal or metal oxide-type nanoparticles, and nanomaterials in 2D, graphene, and nanostructures in hybrid forms. This analysis shows the development of materials and skills for environmental monitoring applications [1].

Various fields like healthcare, chemical industries, agriculture, and medical applications deal with gas monitoring systems. This work speaks about gas sensor technologies covering various types, like magnetic sensors, thermal conductivity monitoring, catalytic, electrochemical, etc. This sensor is characterized by its principal advantages and limitations with future works. This literature work also discusses strongly emerging trends in gas sensor insights for engineers and research professionals [2]. The sensors were made of graphene [3], copper oxide nanomaterials [4], 2D MXene-based [5], and various strong conducting materials and deal with machine learning aspects [6].

Electronics have evolved to the next-generation paper-based material [7, 8] type of sensing, a flexible type of electronics offering a low-cost, one-use substitute to

DOI: 10.1201/9781003594093-7

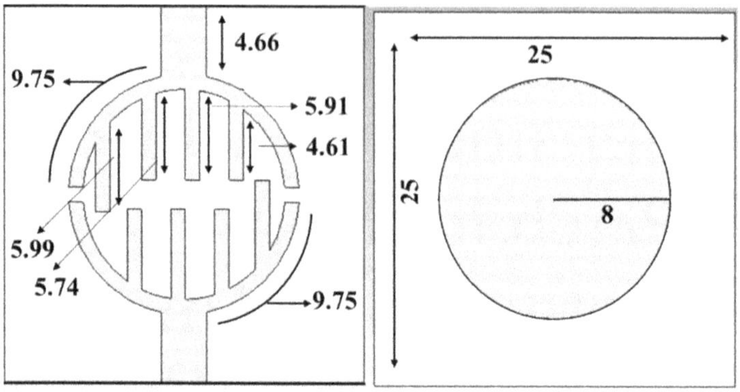

FIGURE 7.1 Schematic representation of the dual-comb patch sensor (top and bottom views) with dimensions in mm.

traditional plastic-made devices. However, their restricted concert has limited widespread acceptance. This design presents a good-performance, paper-made sensor for ammonia gas detection by using composite poly 3,4-ethylene dioxythiophene (PEDOT) and iron combinations. When compared to Kapton, this sensor demonstrates a 10-fold reduction in size, with good humidity control and consistent performance. This sensor is capable of applications including wearable masks for ammonia detection, nose filters, food decay detection, and wireless diapers for ammonia gas detection [9]. The dielectric constant of ammonia gas varies with its temperature [10–12]. In healthcare, hazard management, and environmental monitoring, the gas sensors play a vital role in various types of sensing, and microwave sensors play a good role in nanostructure sensing [13] [14], which is dependent on Interdigitated capacitors (IDC). This sensor operates in 1 to 6 GHz frequencies, is made of Roger's substrate, and has a sensitivity of 28 kHz/%O_2, in advanced methods of electromagnetic minimizing sources [15].

7.2 METHODOLOGY

Schematic top and bottom views of the dual-comb patch with its clear dimensions are shown in Figure 7.1. The dual-comb patch was designed with copper material of thickness 0.035 mm on the FR-4 substrate of length 25 × 25 mm with 1.6 mm thickness. This design carries a circular defective plane of radius 8 mm. The outer edge of the dual cone and this circular defective plane were the same, which improved this sensing system's Electromagnetic radiation (EM).

7.3 DUAL-COMB PATCH SENSOR DESIGN EVOLUTION

In Figure 7.2, the first step, the top plane design started with a circular ring with a center cut followed by rectangular slots merged into it, and in step 2, the extended center cut through the rectangular slots, and in the final step the parasitic block was joined with the origin slots to improve the electromagnetic stabilities.

In Figure 7.3, the evolution of the ground plane is shown. The first try with a plain copper ground plane and then with the complementary circular plane and as the final design circular patch was fixed due to its good resonance frequency. This resonance frequency has good sensitivity in the detection of ammonia gas.

Principles of Design Evolutions

Step 1

- A comb-like patch with a split ring structure is made with dual feed, and the bottom plane is made with a simple patch covering the entire ground plane.
- This structure has not produced a resonance frequency properly.

Step 2

- On considering the step 1 structure, the split between the two combs was made straight to create the capacitance effect between these two combs, and the bottom ground plane was also made by a complementary circular patch.
- This modified structure is good for sensing ammonia gas, but the sensitivity was low.

Step 3

- To address the sensitivity issue, the design was further optimized into two isolated patches that were joined to the original structure.
- The ground plane was made with a circular patch structure, which created an inductive effect and helped to improve the sensitivity of the sensor.

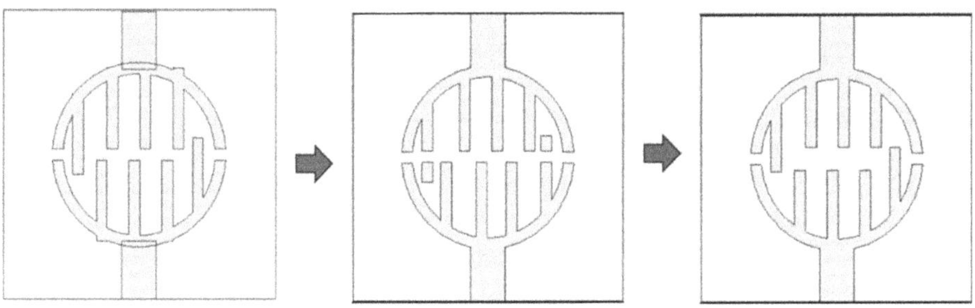

FIGURE 7.2 Design evolution of the top plane in dual-comb patch sensor.

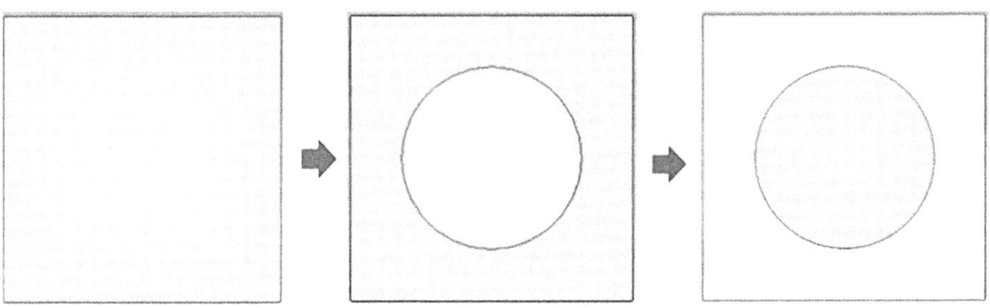

FIGURE 7.3 Design evolution of the bottom plane in the dual-comb patch sensor.

7.4 PRINCIPLES OF NEAR-FIELD DUAL-COMB SENSOR

The principle of dual-comb patch structure works based on a microwave principle through which this comb-type geometry makes more than one resonant track, with the dimensions of the patch wisely changed to resonance frequency propagation at 5.482 GHz. A feed line with this patch confirms the 50 ohms impedance matching for effective power transfer to this entire sensing system. The circular defected ground structure (DGS) creates LC (inductance and capacitance) effects, which improve the sensitivity of the sensor and refinement of the resonance frequencies. The substrate of FR4 has its own dielectric constant that impacts the wavelength of the signal and stimulates the resonance condition of 5.482 GHz.

The near-field dual-comb sensor detects ammonia gas by monitoring the shift in resonance frequency (Δf)

$$\Delta f = f_0\left(\sqrt{\frac{\varepsilon_r'}{\varepsilon_r}} - 1\right) \tag{7.1}$$

f_0 – initial resonance frequency,
ε_r – the relative permittivity of the drain grains,
ε_r' – relative permittivity of the wet grains.

7.5 MATERIAL PROPERTIES

The flame retardant-FR4 is made with epoxy lamination and has a 4.4 of the relative permittivity ε_r and ~0.02 of loss tangent, which is very low, because its constant dielectric nature makes it appropriate for high-frequency applications. The patch made by the conductive material of copper of 0.035 mm thickness has a good conductivity property of ~5.8×10^7 S/m, which has a very low resistive loss, confirming good electromagnetic wave propagation and minimal surface resistivity to the sensing system.It is composed of FR4 and copper materials that improve signal integrity and sensitivity,which are suitable for radio frequency (RF) applications.

7.6 RESULTS AND DISCUSSIONS

The sensor exhibits a resonance frequency of 5.374 GHz with a return loss of −42.334 dB. When the sensing unit is subjected to a Material Under Test (MUT) (ammonia gas), the dielectric property of this gas reacts with the sensor and exhibits the resonance frequency of 5.626 GHz with a return loss of −21.925 dB, which is of 50 MHz frequency shift with 20.409 dB variation in shift, which is shown in Figures 7.4 and 7.5.

The electrical field distribution of the top and bottom planes is shown in Figure 7.6. In the top plane, the current distribution is higher in the middle region, which is created by the facing of rectangular slots from each end. It creates L and C effects, making more sensitive regions. In the bottom plane, the outer ring of the circular patch had more current distribution. The surface current distribution of the top and bottom

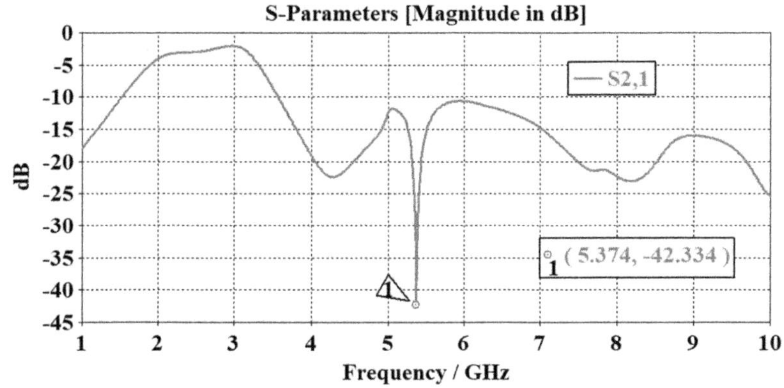

FIGURE 7.4 S_{21} resonance frequency of the dual-comb patch sensor.

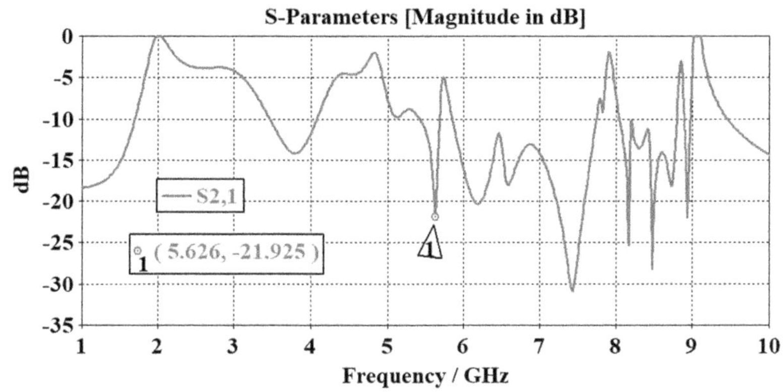

FIGURE 7.5 S_{21} resonance frequency of the dual-comb patch sensor with ammonia gas detected.

planes is shown in Figure 7.7 implicates surface current distribution in the patch of the sensor, where the arrowhead shows the current flow direction. From the observation, the surface current points to the middle area. When compared to other slots, corner slots have more current flow and points to the middle of the sensor. Hence, the sensitivity is higher in the center of the sensing unit. The top and bottom fabricated sensors are shown in Figure 7.8.

Table 7.1 shows the ideal resonance frequency of the sensor is at 5.374 GHz, with 42.133 dB return loss. When this sensor was subjected to the MUT, ammonia gas of relative permittivity 1.006 to 1.010 had a resonance frequency of 5.624 GHz with a return loss of 21.925 dB. Both frequencies were compared; it has a frequency shift of 50 MHz with 20.409 dB differences in return loss, demonstrating that the sensor has good sensitivity over ammonia gas detection with a 7.753 sensitivity.

Figure 7.9 shows the test experimental setup of ammonia gas kept in a glass chamber. This dual-comb patch sensor was kept inside the glass chamber and covered externally,

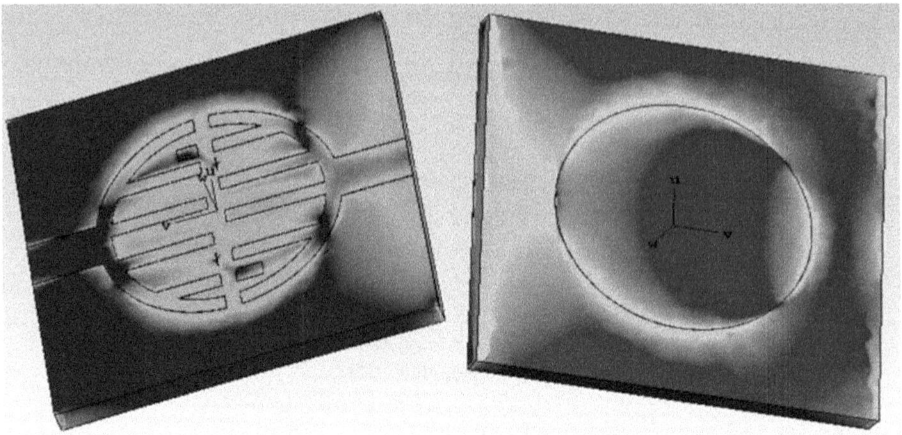

FIGURE 7.6 E-field distribution over the top and bottom planes of dual-comb patch sensor.

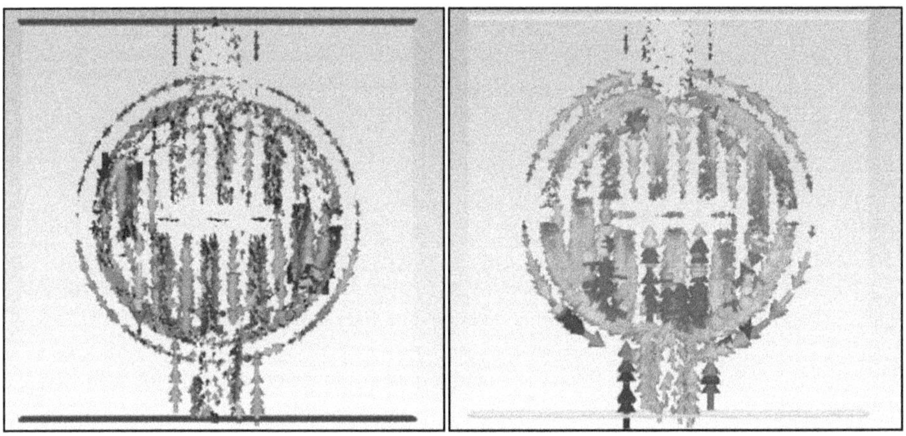

FIGURE 7.7 Surface current distribution over the top and bottom planes of dual-comb patch sensor.

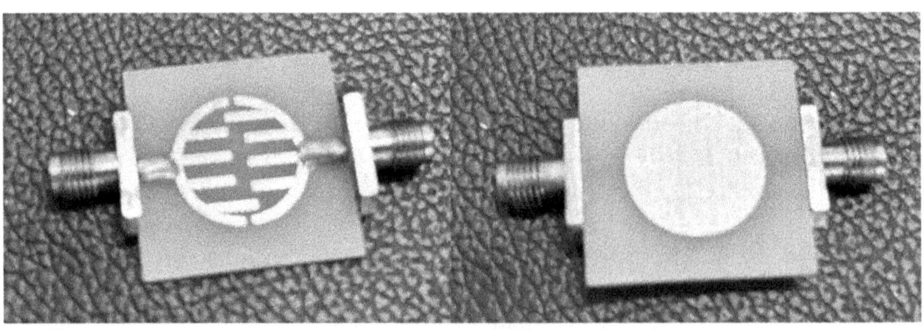

FIGURE 7.8 Fabricated: Top and bottom views of the dual-comb patch sensor.

TABLE 7.1
Comparison of Resonance Frequencies with and without MUT in the Dual-Comb Patch Sensor

Resonance Frequency of Sensor	Resonance Frequency of Sensor with MUT	Frequency Shift	Sensitivity
5.374 GHz in 42.133 dB	5.624 GHz in 21.925 dB	50 MHz in 20.409 dB	7.753

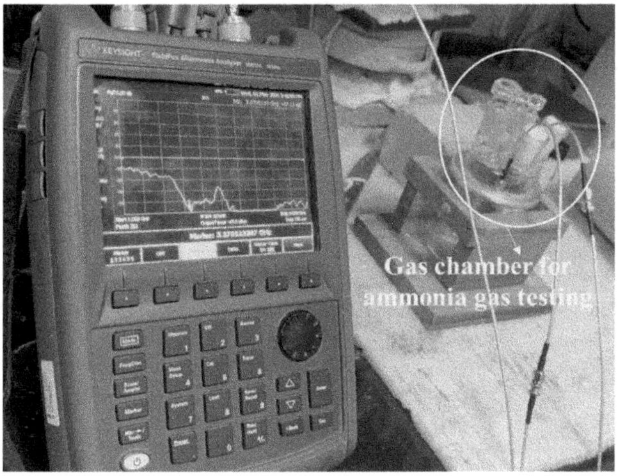

FIGURE 7.9 Dual-comb patch sensor is kept in the gas chamber.

where the probe from the sensor was connected with the Keysight Technologies vector network analyzer N9952A. The results were measured first when the glass chamber was empty, to calculate the sensor response alone. After the sensor measurement, ammonia gas was passed into the chamber. Then the valves were closed externally, and responses were measured for the ammonia gas effects on the sensor. These two measurements are shown in Table 7.1 with sensitivity analysis.

7.7 CONCLUSION

The dual-comb patch microwave sensor, exploiting a resonator made of copper on an FR4 substrate, validates accurate sensing at 5.482 GHz frequency, deploying the interaction of structural geometries and material properties to attain the most effective performance. The incorporation of the DGS of a circular patch improves electromagnetic field strength and gives extra inductive and capacitive effects that advance impedance matching and improve the sensitivity of the resonance frequency. This non-invasive ammonia detection compact design is extremely preferable, contributing improved sensitivity, good Q-factor, and real-time monitoring potential, making it a robust method for agricultural-based industrial applications. The sensor's sensitivity is degraded by environmental factors. On considering these limitations, future work should explore advanced materials to enhance frequency selectivity and sensitivity. Furthermore, real-time field testing and wireless systems need to be developed for smart monitoring systems.

REFERENCES

1. Dhall S, Mehta BR, Tyagi AK, Sood K. A review on environmental gas sensors: Materials and technologies. *Sensors Int.* 2021;2(May 2021):100116. Available from: https://doi.org/10.1016/j.sintl.2021.100116
2. Panda S, Mehlawat S, Dhariwal N, Sanger A, Kumar A. Comprehensive review on gas sensors: Unveiling recent developments and addressing challenges. *Mater Sci Eng B.* 2024;308(July):117616. Available from: https://doi.org/10.1016/j.mseb.2024.117616
3. Tian W, Liu X, Yu W. Research progress of gas sensor based on graphene and its derivatives: A review. *Appl Sci.* 2018;8(7)1118. Available from: https://doi.org/10.3390/app8071118
4. Steinhauer S. Gas sensors based on copper oxide nanomaterials: A review. *Chemosensors.* 2021;9(3):1–20.
5. Li Q, Li Y, Zeng W. Preparation and application of 2D MXene-based gas sensors: A review. *Chemosensors.* 2021;9(8)225.Available from: https://doi.org/10.3390/chemosensors9080225
6. Yaqoob U. Chemical gas sensors: Recent developments, challenges. *Sensors.* 2021;21(D):2877.
7. Maity A, Ghosh B. Fast response paper based visual color change gas sensor for efficient ammonia detection at room temperature. *Sci Rep.* 2018;8(1):1–10.
8. Maity A, Raychaudhuri AK, Ghosh B. High sensitivity NH3 gas sensor with electrical readout made on paper with perovskite halide as sensor material. *Sci Rep.* 2019;9(1): 1–10. Available from: http://doi.org/10.1038/s41598-019-43961-6
9. Fujita H, Hao M, Takeoka S, Miyahara Y, Goda T, Fujie T. Paper-based wearable ammonia gas sensor using organic–inorganic composite PEDOT:PSS with Iron(III) compounds. *Adv Mater Technol.* 2022;7(8). Available from: https://doi.org/10.1002/admt.202101486.
10. Constants TD. The dielectric constants of ammonia, phosphine and arsine. *Proc R Soc London Ser A, Contain Pap a Math Phys Character.* 1927;117(776):43–62.
11. Breitschwerdt KG, Schmidt W. Permittivity and dielectric relaxations in liquid-ammonia solutions. *Zeitschrift fur Naturforsch – Sect A J Phys Sci.* 1970;25(10):1467–1471.
12. Frederick G, Keye AJG. The dielectric constant of ammonia as a function of temperature and density. *Mod Views Electr.* 2013;V:18–29.
13. Ge L, Mu X, Tian G, Huang Q, Ahmed J, Hu Z. Current applications of gas sensor based on 2-D nanomaterial: A mini review. *Front Chem.* 2019;7(December):1–7.
14. Lun D, Xu K. Recent progress in gas sensor based on nanomaterials. *Micromachines.* 2022;13(6).Available from: https://doi.org/10.3389/fchem.2019.00839
15. Gugliandolo G, Naishadham K, Crupi G, Donato N. Design and characterization of a microwave transducer for gas sensing applications. *Chemosensors.* 2022;10(4):1–12.

8 Edge Computing in IoT-Driven Agriculture
A Transformative Enabler for Smart Farming

Abuzar Saeed, Naved Alam, and Aquib Mustafa

8.1 INTRODUCTION

The Internet of Things (IoT) has modernized inter-device communication with the world, enabling home appliances, industrial systems, etc., to store and process data. Nevertheless, traditional cloud-based techniques can struggle with latency, security, and resource demands, especially as IoT networks expand. Edge computing, by processing data closer to its source, tackles these hurdles by lowering the delays and improving data security. This chapter explores how edge computing enhances IoT capabilities, making faster, more efficient responses across sectors like healthcare, smart cities, and agriculture. By integrating edge computing with IoT, we can build sustainable, responsive systems that optimize energy, reduce waste, and bolster the developing needs of our interconnected world.

8.1.1 What Is IoT?

Internet of Things, or the IoT, as it stands today, is a network of tactile objects such as devices, instruments, vehicles, and other items that are equipped with sensors, electronic circuits, actuators, etc., to collect data and information. The veteran researchers Bahga and Medisetti, in their book, defined IoT as:

> A dynamic global network infrastructure with self-configuring capabilities based on standard and interoperable communication protocols where physical and virtual "things" have identities, physical attributes and virtual personalities and use intelligent interfaces, and are seamlessly integrated into the information network, often communicating data associated with users and their environments. [1]

8.1.2 Evolution of Internet

The Internet has undergone a vigorous evolution since its humble beginnings as a network of computers communicating with each other. In the period spanning from the 1960s to the 1970s, the technological progression in the field of Internet

DOI: 10.1201/9781003594093-8

flourished all over the world. This evolution started with the development of ARPANET (the first wide area packet switching network with distributed control, developed in 1969 by Advanced Research Projects Agency Network and served as pioneer of the Internet [2]), FTP, etc., which further laid the foundation and paved the paths for more development and evolution in various phases, which are discussed below:

- **Internet of Documents**: The term "Internet of Document" refers to the creation and sharing of text-based information as its sole purpose during its initial phases. It was only in the 1990s that the adoption of the World Wide Web enabled users to share information through web browsers.
- **Internet of Commerce**: Developed in the late 1990s to early 2000s, Internet of Commerce, commonly known as electronic commerce or e-commerce, is a widely used method for the sale of goods and services throughout the world digitally, which, at the same time, also enabled the digital payments systems in the forms of bank-to-bank transactions, through credit/debit cards, etc. Corporate giants, such as Amazon.co, Flipkart, and Ola, are all examples of the Internet of Commerce or E-Commerce.
- **Internet of Application**: After the evolution of Internet of Documents and Commerce, web-based applications and services also advanced in the late 2000s. It was only in the coming years that the development of dynamic, productive, and interactive applications such as social media platforms, cloud services, and productivity tools took place.
- **Internet of People**: After the massive success of the Internet of Application, "Internet of People" came into the picture in the 2010s. It was only in this decade that the massive following of social media took place, as well as the emergence of new social media platforms, leading to the transformation of these platforms. These platforms enabled people to connect and communicate with each other virtually, creating a global space or global civilization. Examples of these platforms are Facebook, Instagram, and Twitter.
- **Internet of Things**: The term "Internet of Things" represents a network of interconnected physical devices and circuits with the Internet. It is the most recent phase of evolution of the Internet. This technology allows us to gather and process real-world data in devices, actuators, sensors, etc. This also helps in enabling automation, data analysis, and more enhanced applications across different industries.

8.2 ARCHITECTURE OF IoT

The Architecture of IoT, as shown in Figure 8.1, encompasses the following layers:

- **Sensing layer**: It is the primary layer of any IoT device and consists of sensors and actuators, which are used for sensing its surroundings and collecting data, for example, temperature sensors and motion detectors. This layer

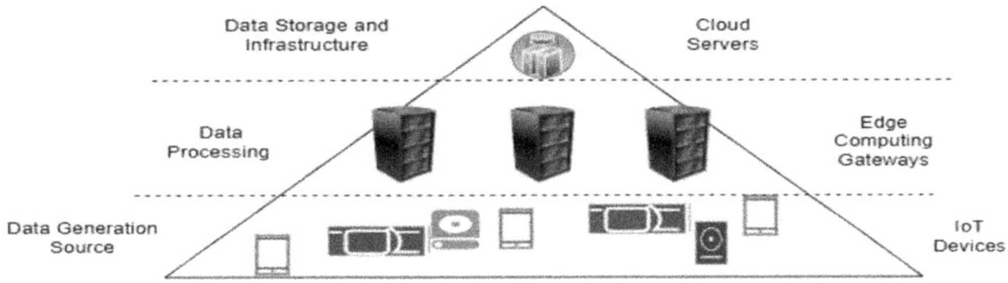

FIGURE 8.1 Structure of IoT system.

is also responsible for real-time data acquisition and acts as a foundation layer for the IoT system, which impacts accuracy, reliability, and efficiency of IoT applications.

- **Network layer**: This layer of the IoT consists of components responsible for connecting all devices and allowing them to be aware of their physical environment. It is only through this layer that the devices can exchange data with other connected devices, which in turn is essential for effective data management and processing for an IoT system and impacts energy consumption, scalability, and network efficiency.
- **Service layer**: Service layer, also known as data processing layer, is responsible for processing the collected data, performing analysis, extracting the information, and storing it. Cloud-based platforms and data centers are also included in this layer. It also enables intelligent decision-making through data analytics, reduces response time with edge/fog computing, and supports large-scale data management and automation.
- **Interface layer**: It is the layer in any IoT system that connects users and the IoT system, enabling interaction and control. It consists of software that includes a variety of user interfaces and Application Programming Interfaces (APIs), which enable users to interact with IoT systems. This layer is the innermost and final layer of an IoT system. This layer helps the IoT system by improving accessibility and usability and supports customization and automation for different applications.

8.3 EDGE COMPUTING

In this scenario, "Edge" refers to any network and computing resources positioned between data source and cloud data server/centers [3]. The term "edge computing" depicts the technology that allows computation on upstream data for IoT services and downstream data for cloud services to be done at the edge of the network. It is a computing model in which data is computed and processed nearer to the source of its production. This approach enhances response time and lowers the time taken for transferring the data and tackles computational storage by processing the data locally.

8.4 EDGE COMPUTING, CLOUD COMPUTING, AND FOG COMPUTING: A CLEAR COMPARISON

8.4.1 Edge Computing

- **Data Processing Location**: The processing of data in edge computing is done locally at or near the source of data generation, such as IoT devices, sensors, and local servers. This collocation enables real-time data processing with minimal latency.
- **Latency**: Many IoT applications, such as mobile gaming, warehouse logistics, remote health monitoring, linked cars, and industrial control systems, depend upon low-latency connectivity [4]. The data is handled locally in edge computing, which reduces the time for processing and response to data and is also known as reduced latency. This is a primary thing for applications that require immediate actions, such as autonomous vehicles or industrial automation.
- **Bandwidth Usage**: By the use of edge computing, the volume of data that needs to be transmitted over the network is also reduced, as the processing and screening of data is done locally. This technique also reduces the volume of data to be dispatched to the centralized servers, optimizing bandwidth and reducing network overcrowding.
- **Scalability**: By incorporating more edge nodes or devices, we can make it possible to expand edge computing. This enables multi-tier computing capabilities and can be tailored to diverse workloads without relying on a central data center.
- **Data Security**: By placing the sensitive data closer to its source, we can enhance the security of the data, which also minimizes the risk of breach during transmission. Localized processing also enables us to implement more tailored security measures at each edge node.
- **Use Cases**: The optimal use of edge computing is in applications that require low-latency and real-time responses, for example, smart cities, industrial automation, and healthcare monitoring.

8.4.2 Cloud Computing

- **Data Processing Location**: Central data centers or servers that are sometimes spatially far away from the data production sources come under the cloud computing systems.
- **Latency**: Higher latency is experienced in cloud computing in contrast to edge computing and fog computing due to distance, which can cause problems or glitches for applications requiring swift response from the servers.
- **Bandwidth Usage**: An enormous amount of data is transmitted to and from the central cloud server, resulting in higher bandwidth usage and network overcrowding in cloud computing.
- **Scalability**: Cloud computing grants various expandable assets through the digital environment in data centers, which are necessary for effective administration, computer power, storage, etc.

- **Data Security**: One major problem is the erroneous processing of real-world data sent from IoT devices to cloud computing through cloud-based IoT [5]. During the transmission and transfer of data, there is a risk of breach and exposure, in spite of all protective measures taken for data security at the data centers, including encryption, access controls, and monitoring.
- **Use Cases**: Extensive computing applications, large-scale data storage, and complex analytics such as big data processing, machine learning, and enterprise applications have major use of cloud computing.

8.4.3 Fog Computing

- **Data Processing Location**: In fog computing, data is processed at fog nodes. These fog nodes can be located at various locations like gateways, routers, base stations, local servers, or dedicated fog servers. Inherently, fog computing augments cloud computing by carrying out some processing tasks near the edge of the network, close to end-users [6].
- **Latency**: The fog nodes in fog computing result in intermediate processing, which primarily offers lower latency, which is often in the range of milliseconds to seconds, depending upon the distance between the data generation source and fog nodes and processing power.
- **Bandwidth Usage**: Due to intermediate data processing at fog nodes, fog computing offers larger bandwidth consumption as compared to edge computing, though it is still lower than that of cloud computing.
- **Scalability**: Properties like distributed architecture, local data processing, and dynamic resource allocations make fog computing a highly scalable technology. But due to some drawbacks like management complexity, network bandwidth, and interoperability, it becomes difficult to scale fog computing.
- **Data Security**: Due to the presence of multiple locations from edge devices to intermediate fog nodes, it becomes difficult to secure the data in fog computing as there is multiple-layer processing, and each layer in this distributed architecture needs to be secured for assurance of confidentiality, integrity, and availability of data.
- **Use Cases**: Characteristics of fog computing coupled with the advancement of IoTs and Wireless Sensor Network (WSNs) systems make it efficient for better sustainable development and smart living [7]. Fog computing has a major use in industries that require real-time data processing, low-latency, and decentralized architecture applications. It has significant advantages over conventional cloud computing, as it brings computation, storage, and networking closer to data sources.

8.5 EDGE COMPUTING AND IoT: A HARMONIOUS INTERACTION

The integration of edge computing with IoT devices enhances advanced data processing, decision-making, and productivity optimization by capitalizing on local computing resources. This unification of both technologies depicts a potent approach that augments the capabilities of both technologies. Edge computing boosts IoT applications by processing the data closer to the data generation source instead of relying solely on centralized cloud servers. This caters to several advantages over traditional cloud computing, particularly in areas requiring low-latency, real-time processing, bandwidth efficiency, and enhanced security. Some real-world examples where edge computing is more effective are as follows:

- **Tesla Autopilot** uses Edge artificial intelligence (AI) for real-time object detection and decision-making.
- **Smart Traffic Lights** adjust signal timing dynamically based on real-time vehicle density.
- **Siemens Edge Computing Solutions** reduce factory downtime by predicting maintenance needs.
- **John Deere Smart Tractors** use Edge AI to automate precision spraying and harvesting, etc.

The local processing and object isolation in edge computing address most of the privacy and security problems that are encountered by cloud and fog computing [8]. The required computing capability, storage space, and swift response times required by IoT applications are furnished by edge computing [9]. The implementation of edge computing transfigures the system to enable flexibility in adapting to changes in the resources of the production chain, which in turn fosters the development of IoT-based production. [10]. New applications that require synchronous processing power and the explosive growth of IoT systems and devices continue to propel comprehensive computing systems [11]. Now, let's explore how this integration unfolds and its key components:

- **Data Processing**: In lieu of sending all raw data to a centralized cloud, edge devices are designed with such processing power that they process the data locally and help them to analyze it locally too. This helps them to make judgments quickly and also helps in reducing latency and network congestion.
- **Decision-Making**: Use of edge computing with IoT devices also helps them in making decisions within a few milliseconds based on locally processed data. This technology makes it possible to react with real-time responses to environmental changes or operational conditions, improving the system's overall efficacy and efficiency.
- **Data Storage**: The edge device enables swift authorization and backup by storing the important data locally. This approach enables quick data recovery and analysis while diminishing the dependency on cloud storage.
- **Connectivity**: Setting up of link between IoT sensors and the cloud enables the edge devices to act as gateways. They expedite effortless communication between the local network and other systems, maintain connectivity, and process data.
- **Security**: Processing the data locally also helps in enhancing the security and preserving the integrity of the data. Storing and processing the sensitive data nearer to its source reduces the risk of data breach during transmission. IoT

ecosystems often use a vast number of constrained devices that are unable to execute the advanced encryption and decryption methods essential for data security. The edge computing platform is set up to encrypt and payload using AES-256 in such a case before sending it to the cloud platform. [12]

- **Low-Latency Applications**: Due to the generation of large amounts of data and long distances between clients and backend data centers, low-latency applications would make conventional centralized cloud computing insecure [13]. The functionality of the applications that require quick answers and responses, for example, augmented reality, industrial automation, and driverless cars, is enriched. These applications are also generously compensated by the integration of edge computing.

8.6 HURDLES AND ASPECTS

- **Hardware Limitations**: In contrast to centralized cloud servers, edge computing devices may have lower processing and storage capabilities due to several reasons, such as resource constraints, decentralization, cost efficiency, and connectivity limitations. These limitations result in hardware limitations. To overcome this, we have to use effective hardware accelerators (e.g., FPGAs, GPUs, TPUs for parallel processing and efficient computation) and efficient edge devices that utilize low-powered AI chips.
- **Software Development**: Designing and developing software for edge computing devices has certain challenges that have to be considered while designing. These challenges include interoperability, security and privacy, data management, and processing. Also, while designing this software, having enough knowledge of both technologies, edge computing and IoT, is essential. To overcome this, the use of containerization and virtualization, lightweight machine learning (ML) models, serverless edge computing, and cross-platform development can be beneficial and effective.
- **Security**: In any technological device, security is one of the important things that has to be considered. In edge devices, their security is the major hurdle due to many aspects: data privacy and protection, data vulnerability, authentication and access control, etc. This security issue can be solved through reinforcing the verification methods within edge computing and providing more feasibility studies for the validity of smart and sustainable development [14]. Also, the hardware-based security, end-to-end encryption, zero-trust architecture, and AI-driven anomaly detection models can be used to overcome these security hurdles.
- **Consumption of Power**: As many of the edge devices run on batteries, for conducting effective power management, various challenges arise: limited power resources, energy efficiency, communication overhead, scalability, etc. To overcome these challenges, we can use energy-efficient hardware (e.g., ARM-based processors, ultra-low-power MCUs), Dynamic Power Management (implement DVFS (Dynamic Voltage and Frequency Scaling) for adaptive power use), and Energy harvesting (utilize solar, RF, and piezoelectric energy harvesting for self-sustained IoT devices).

- **Network Connectivity**: For the effective integration of edge computing and IoT, it is necessary to have good network connectivity. Also, system performance and data consistency may be affected by network connectivity glitches and problems. This can be resolved by using adaptive networking, mesh networking, edge caching and preprocessing, and software-defined networking (SDN).

8.7 INTEGRATION STRATEGIES

8.7.1 Edge Gateways

Having a modular, user-friendly, Internet gateway is one of the most crucial components of telemetry prehabilitation via IoT [15]. The edge gateways are the component that serves as the bridge between IoT devices and cloud servers. They accumulate the data from several devices, translate data from different IoT protocols into a uniform format, and carry out preliminary data processing. Gateways reduce the network congestion and latency by transferring some processing work to the edge, enhancing system responsiveness. Among the various roles of edge gateways, some of them are as follows:

- Protocol Translation: It is the process of transforming data into the standard protocol.
- Data Aggregation: Combining the information and data from several devices into a sequence that is easy to manage.
- Data Filtering: For additional processing or transmission, data is selected; this process of selecting data is known as data filtering.
- Security: For safeguarding the data, devices, and networks in IoT systems, security is essential. It is done using encryption and authentication methods for protection.

8.7.2 IoT Platforms with Edge Capabilities

IoT platforms with integrated edge computing systems furnish a holistic approach for managing IoT installations. A single environment is offered for managing devices, gathering, processing, and analyzing data, in both Cloud and the Edge.

IoT solutions with edge capabilities offer several advantages, such as the following:

- Development Made Easier: Development is accelerated by pre-built edge computing components.
- Scalability: The capacity to manage growing volumes of data and IoT devices.
- Flexibility: The capacity to adjust to various deployment settings and use cases.
- Security: Integrated security measures to safeguard information and gadgets.

8.7.3 Hybrid Cloud Structures

Resultantly, cloud and the edge build a synergistic connection, a twinned architecture that works in tandem [16]. Flexible and efficient IoT solutions are produced in a hybrid cloud framework, which is the combination of the advantages of cloud computing and edge computing. In today's scenario, edge devices are deployed in the hub station of the network, so that data passes forward to the cloud server through the edge devices [17].

Data security, low expenses, and maximization of performance can be achieved by the organizations by prudently distributing workloads between edge and the cloud.

The key perks of hybrid cloud architectures include the following:

- For optimal resource implementation, hybrid cloud structures align workloads with the best environment (cloud or edge).
- The performance is also enhanced as hybrid cloud structures lower the latency for applications that require swift response.
- The dependability is also reduced on a single infrastructure as it provides both cloud and edge.
- Balancing the expenses of cloud and edge resources also enhances the cost.

8.8 BENEFITS OF INTEGRATION

- Improved Response Time: Processing the data locally in edge computing significantly reduces the latency and quickens the response time. The traffic flows can also be minimized to diminish the bandwidth requirements in IoT devices [19] [20]. As a result, decisions are made more quickly and efficiently.
- Enhanced Data Privacy: By processing sensitive data locally, the security and privacy are optimized, which lowers the chance of exposure during the data transfer.
- Increased Reliability: Since edge devices can continue to operate without a cloud connection, it results in smooth performance of operations even in service interruptions.
- Optimized Network Usage: Network bandwidth is also preserved by edge processing, by lowering the amount of data sent to the cloud server. This optimization results in reduced operating expenses and more effective network utilization.
- New Business Opportunities: For creative services and applications, edge computing creates emerging prospects. Making it possible to create advanced and complex systems that make use of site-based assessments and real-time data processing. Also, Edge-to-cloud platforms empower organizations with the agility to acknowledge business demands without delay, capture evolving market opportunities, and accelerate the introduction of new products to market [18].

8.9 CONCLUSION

The integration of IoT with edge computing presents a novel approach toward attaining sustainable development across various sectors. By presenting computation closer to data sources, edge computing improves the power of IoT systems with low-latency, real-time decision-making, and efficient data processing and storage. This integration not only lowers the environmental impact by optimizing energy usage and minimizing data transfer but also strengthens privacy and security, essential for delicate applications in healthcare, smart cities, and industrial automation. The hurdles in expansion, infrastructure costs, data management, and power consumption highlight areas demanding ongoing advancement. Addressing these hurdles with advanced hardware, efficient software development, and resilient security measures is crucial in maximizing the benefits of edge-enabled IoT systems. The uninterrupted progress of hybrid cloud architectures, edge gateways, and IoT platforms develops the basis of potential for scalable, resilient, and responsive systems. As edge computing and IoT continue to progress, their use cases promise to reinterpret industry standards, enabling sharper and more sustainable solutions that can adapt to future technological and environmental needs. This development spans a range of applications, from smart homes and cities to agriculture and healthcare.

REFERENCES

1. Bahga A, Medisetti V. *Internet of things: A hands-on approach*. 7th ed. Orient Blackswan Private Limited; 2022. p. 22.
2. Munn L. From the cloud to the edge. In: *Technical territories: Data, subjects, and spaces in infrastructural Asia*. University of Michigan Press; 2023. pp. 105–25. Available from: http://www.jstor.org/stable/10.3998/mpub.12584902.10
3. Shi W, Cao J, Zhang Q, Li Y, Xu L. Edge computing: vision and challenges. *IEEE Internet Things J*. 2016;3(5):637–46.
4. Anees T, Habib Q, Al-Shamayleh AS, Khalil W, Obaidat MA, Akhunzada A. The integration of WoT and edge computing: Issues and challenges. *Sustainability*. 2023;15(7):5983. doi:10.3390/su15075983
5. El-Sayed H, et al. Edge of things: The big picture on the integration of edge, IoT, and the cloud in a distributed computing environment. *IEEE Access*. 2018;6:1706–17. doi:10.1109/ACCESS.2017.2780087
6. Hosseinian-Far A, Ramachandran M, Slack CL. Emerging trends in cloud computing, big data, fog computing, IoT and smart living. In: Dastbaz M, Arabnia H, Akhgar B, editors. *Technology for smart futures*. Springer; 2018. doi:10.1007/978-3-319-60137-3_2
7. Nair S. Beyond the cloud-unravelling the benefits of edge computing in IoT. *Int J Comput Eng Technol*. 2023;14:91–7.
8. Al-Naime K, Al-Anbuky A, Mawston G. Internet of Things gateway edge for movement monitoring in a smart healthcare system. *Electronics*. 2023;12(16):3449. doi:10.3390/electronics12163449
9. Chen B, Wan J, Celesti A, Li D, Abbas H, Zhang Q. Edge computing in IoT-based manufacturing. *IEEE Commun Mag*. 2018;56(9):103–9. doi:10.1109/MCOM.2018.1701231
10. Datta SK, Bonnet C. An edge computing architecture integrating virtual IoT devices. In: *2017 IEEE 6th Global Conference on Consumer Electronics (GCCE)*; 2017; Nagoya, Japan. pp. 1–3. doi:10.1109/GCCE.2017.8229253
11. Ali O, Ishak MK, Bhatti MKL, Khan I, Kim K-I. A comprehensive review of Internet of Things: Technology stack, middlewares, and fog/edge computing interface. *Sensors*. 2022;22(3):995. doi:10.3390/s22030995
12. Kuchuk H, Malokhvii E. Integration of IoT with cloud, fog, and edge computing: A review. *Adv Inf Syst*. 2024;8(2):65–78.
13. Hundt R, Kornbluh K. *Building back better with distributed computing*. German Marshall Fund of the United States; 2021. Available from: http://www.jstor.org/stable/resrep28526

14. Yu W, Liang F, He X, Hatcher WG, Lu C, Lin J, et al. A survey on edge computing for the Internet of Things. *IEEE Access*. 2018;6:6900–19. doi:10.1109/ACCESS.2017.2778504
15. Kuchuk H, Malokhvii E. Integration of IoT with cloud, fog, and edge computing: A review. *Adv Inf Syst*. 2024;8(2). doi:10.20998/2522-9052.2024.2.08
16. Premsankar G, Di Francesco M, Taleb T. Edge computing for the Internet of Things: A case study. *IEEE Internet Things J*. 2018;5(2):1275–84. doi:10.1109/JIOT.2018.2805263
17. Pan J, McElhannon J. Future edge cloud and edge computing for Internet of Things applications. *IEEE Internet Things J*. 2018;5(1):439–49. doi:10.1109/JIOT.2017.2767608
18. Dave R, Seliya N, Siddiqui N. The benefits of edge computing in healthcare, smart cities, and IoT. arXiv preprint arXiv:2112.01250. 2021. doi:10.12691/jcsa-9-1-3
19. Kuchuk H, Malokhvii E. Integration of IoT with cloud, fog, and edge computing: A review. *Adv Inf Syst*. 2024;8(2):65–78.
20. Wang L, Wang Y. Supply chain financial service management system based on blockchain IoT data sharing and edge computing. *Alexandria Eng J*. 2022;61(1):147–58. doi:10.1016/j.aej.2021.11.019

9 Harnessing Artificial Intelligence in Animal Sciences

A Vital Frontier for Smart Agriculture

Ambreen Hamadani, Henna Hamadani, and Jerina Rugji

9.1 INTRODUCTION

9.1.1 Artificial Intelligence and Its Role in Animal Sciences

Artificial intelligence (AI) brings under its umbrella several tools and technologies. These technologies are used to solve complex problems. It is an amalgamation of many disciplines, which include data science, mechanistic modelling, engineering, and statistics, only to name a few [1]. Since its introduction back in the 1950s, numerous AI applications have been advanced or expanded due to enhancements in computer performance. Particularly, progress has been driven by the connections established between AI and other fields, such as biomedicine, along with the vast quantities of data generated in various areas, especially in healthcare [2]. The Fourth Industrial Revolution brought forth a variety of technologies that have fundamentally transformed most of our existing systems and have created newer and better business models and processes across multiple domains. Simultaneously, decision-making has become more decentralized, largely due to advancements in AI supported by Big Data Analytics (BDA). Notably, AI attempts to replicate intelligence and human cognitive skills through the use of machines or computers [3]. AI is said to increase by 20% the yield, operational efficiency and crop yields and also save water up to 30% [4]. As per the World Economic Forum, agricultural AI could grow from $1.7 billion (2023) to $4.7 billion by 2028.

For example, in precision agriculture, AI-driven analytics platforms process satellite imagery and IoT sensor data to optimize crop yield while minimizing resource use. In healthcare, AI-powered diagnostics, such as IBM Watson Health, assist doctors in making accurate and timely decisions. Similarly, in finance, algorithmic trading models leverage BDA to analyze market trends and execute trades in real time. Even in manufacturing, smart factories powered by AI and robotics, like those used by

 DOI: 10.1201/9781003594093-9

Siemens and Tesla, have enabled predictive maintenance and automation, significantly enhanced efficiency, and reduced costs. These advancements are possible due to the continuous evolution of AI, which is often categorized into three key subdivisions. Artificial Narrow Intelligence (ANI), also known as Weak AI, is designed to perform specific tasks efficiently, such as recommendation systems on Netflix or language translation by Google Translate. Artificial General Intelligence (AGI), or Strong AI, is the next limit wherein machines are capable of reasoning, learning, and making decisions like humans, though this level has not yet been fully realized. Super Intelligence, a theoretical stage, looks at AI surpassing human cognitive abilities, potentially changing our role in decision-making and problem-solving [3].

9.1.2 Big Data and Animal Sciences

The 21st century saw significant advancements in AI, particularly in ML, deep learning (DL), and neural networks (NNs). The rise of "Big Data" and augmentations in computing power helped create AI systems capable of performing tasks similar to human intelligence. These tools were primarily established to address simple problems that posed minimal challenges to human intelligence. Over time, their application expanded to uncover complex patterns that are noticeable to humans but difficult to articulate [5]. AI systems are already significantly influencing our daily lives, jobs, and social interactions. The market offers virtual personal assistants, recommendation systems, autonomous vehicles, surveillance technologies, crop prediction tools, smart grids, drones, banking and commerce solutions, and gene sequencing devices. An increasing number of multinational companies are shifting their business models to focus on data and predictive analytics to leverage the productivity advantages that AI provides. This transformation is driven by the pursuit of technological solutions to pressing global challenges such as climate change, economic development, security, and population dynamics.

New and cutting-edge developments in AI and sensor technologies have led to the rise of new opportunities in various sectors [6]. Leveraging AI tools in the field of animal science is rapidly expanding. AI models are being established for various applications in animal science, including health monitoring, breeding optimization, feed management, and behavior analysis. These tools help improve livestock management, enhance productivity, and ensure animal welfare, while also contributing to more sustainable farming practices [7]. Utilizing AI tools for monitoring animal welfare involves analyzing data from multiple sources to evaluate the health and ensure the well-being of animals. AI technologies, including machine learning (ML) and computer vision, can track behavioral patterns, identify anomalies, and monitor physiological indicators in real time. For example, AI systems can analyze videos and auto-detect signs of stress or discomfort, allowing for timely intervention. AI can analyze data from wearable sensors to monitor vital signs and activity levels, helping farmers and veterinarians make well-informed animal care decisions. Sensor technologies, including wearable devices and environmental monitoring systems, continuously collect diverse data points, such as body temperature, heart rate, movement patterns, feeding behaviors, and environmental conditions. With AI-driven algorithms, this information provides real-time, accurate insights into animal health and welfare. In addressing ongoing challenges like optimizing feeding behavior, these technologies can significantly enhance animal welfare, streamline supply chain operations, and boost productivity and profitability within the dairy livestock export industry [8].

9.1.3 Enhancing Animal Performance, Welfare, and Productivity Using AI

AI and ML are significantly influencing all aspects of animal welfare. This includes aspects such as monitoring and management of animal health, well-being, and farm economics. Systems and algorithms are now being developed that are capable of detecting early signs of illness, stress, or any kind of discomfort in animals. The technologies in vogue include advanced image recognition, behavioral analysis, and bioacoustics. For instance, AI-powered cameras and sensors can monitor livestock in real-time, identifying abnormalities in movement or behavior that may indicate health issues. This proactive approach allows for timely interventions, reducing suffering and improving overall animal welfare. This integration becomes fruitful due to the numerous benefits it offers to farmers. Farm automation also ensures animal welfare and productivity [9]. For example, in cattle, AI and robotic technologies are used for supervising various aspects of animal care, plus emotional well-being [10], milk production [11], herd management [12], feeding patterns [13], animal movement [14], and overall health [1]. With AI-powered sensors, it is possible to track the healthiness of livestock and detect stress or illness early. These innovations also help farmers optimize feeding schedules, minimize waste, detect diseases early on, and enhance the quality of their products. The use of AI in animal breeding and genetics ensures that the scientific selection of animals becomes possible on farms to increase the productivity of future generations [15, 16].

AI tools have also transfigured productivity and efficiency in animal-related industries. In agriculture, AI systems optimize feeding schedules, breeding programs, and environmental conditions to maximize productivity. ML algorithms analyze enormous amounts of data to predict optimal breeding times, enhancing reproductive success rates. Additionally, AI-driven automation in tasks such as milking, feeding, and cleaning reduces labor costs and increases operational efficiency. These advancements not only boost productivity but also ensure that animals are properly cared for. Some examples are given below:

- In the dairy industry, AI technologies have been employed to monitor the health and productivity of dairy cattle. Wearable sensors track vital indicators and movement levels, providing farmers with immediate data on each cow's health status. This data-driven approach helps in early disease detection, such as mastitis and lameness [17], allowing for prompt treatment and lowering the odds of severe health issues. Sensor systems for mastitis management in cows were optimized by researchers by developing algorithms tailored to four specific situations: severe clinical mastitis, non-severe mastitis, dry-off treatment, and herd-level monitoring [18]. By improving the sensitivity and specificity of these systems, farmers would be able to enhance udder health management and make timely interventions.

- In poultry, AI is transforming poultry farming to enhance animal welfare, performance, and productivity. Large datasets are processed using ML models to design optimal feed to supply birds with the required nutrients to promote health and growth. AI computer vision systems monitor birds around the clock to detect crowding or stress to promote welfare. AI climate control systems adjust automatically to control temperature, humidity, and ventilation to supply optimal living conditions. AI predictive analytics also allow farmers to forecast health issues to limit losses and increase farm efficiency in total. All of these contribute to enabling sustainable poultry farming with higher yields and improved welfare of animals.
- In the commercial sector, automated systems equipped with AI algorithms can automate the monitoring of the behavior and health of chickens, identifying signs of distress or disease, which is crucial to preventing sudden, sweeping outbreaks [19]. Systems are now being developed that not only improve biology in terms of mental and physical health and the capacity to express species-specific behavior but also provide profitable gains and are reasonable to implement [20].
- In wildlife conservation, AI technologies are used to track and protect endangered species. AI-powered drones and camera traps collect data on animal populations and their habitats, helping conservationists monitor species without disturbing them [21]. ML models analyze this data to identify trends and threats, enabling more effective conservation strategies.

9.2 AI FOR BREEDING

The accurate prediction of breeding values is essential in modern animal breeding, and AI has significantly advanced this process. By integrating genomic data and ML algorithms, researchers can now handle humongous datasets, leading to more precise and efficient predictions of breeding values [22]. Several ML algorithms like regularized regression, ensemble methods, and DL have been compared for their performance, showing promising results in identifying genetic variants associated with desirable traits and predicting the genetic merit of the animals [23]. In some cases, DL methods have been shown to have better accuracy than well-established techniques like genomic best linear unbiased prediction (GBLUP) [24]. This improves the accuracy of genetic predictions and accelerates breeding by reducing the need for extensive phenotypic evaluations and shortening the generation interval.

ML algorithms are especially valuable in dealing with high-dimensional datasets like Single Nucleotide Polymorphisms (SNPs), Copy Number Variations (CNVs), and gene expression data, which are common in animal breeding. Even in single-cell studies, DL methods are being seen as potential alternatives to conventional methods both in data preprocessing and downstream analysis. Artificial neural networks (ANNs) forecast genomic breeding values (GEBVs) across species like dairy and beef cattle, pigs, and broilers [25]. Ehret et al. (2015) used ANNs to combine metabolic, genomic, and milk data, achieving a correlation of 0.643 for subclinical ketosis risk prediction [26]. Support vector machines (SVMs) handle high-dimensional SNP data well, making them useful for classifying and predicting traits, particularly non-linear relationships, in breeding programs as well as breed classification based on the similar distribution of single-nucleotide polymorphisms per autosome [27]. Random forest (RF) models are effective in predicting complex traits and imputing missing genotypes in noisy genomic data [28]. Gradient Boosting Machine (GBMs), like XGBoost, are robust for genomic prediction tasks, improving GEBV accuracy by handling missing data and accounting for marker interactions [29]. DL models, such as convolutional and recurrent NNs, are adept at identifying patterns in large SNP datasets. Zeng et al. (2021) used DL to predict phenotypes in livestock by incorporating zygosity data [30]. Techniques such as gradient descent and grid search are used to optimize model performance.

9.3 IMAGE AND VIDEO ANALYSIS ON ANIMAL FARMS

The emergence of computer vision and image processing technologies in animals opens new avenues for improving the sector and leads to tremendous data generation with completely non-invasive and labor-free methods. Image data is being analyzed using various deep-learning algorithms. These include algorithms like Mask R-CNN – Mask Region-based Convolutional Neural Network, Faster R-CNN – Faster Region-based Convolutional Neural Network, YOLO (v3 and v4) – You Only Look Once, DeepLab v3 – (no official full form, but it's part of the DeepLab family, which focuses on semantic segmentation), U-Net – (no official acronym, but it's named U-Net due to its U-shaped architecture, commonly used for biomedical image segmentation), and others that have been used in animal science research [31]. These algorithms often rely on pre-trained models so that the training process can be optimized quickly while using fewer resources. Convolutional Neural Networks (CNNs) find their applications in animal farming for tasks like disease detection and animal behavior analysis [32]. Computer vision is now making possible round-the-clock animal monitoring on farms, including monitoring animal health, behavior, and welfare. It is also paving the way toward precision livestock farming through high-throughput phenotyping and data mining, as illustrated in Figure 9.1.

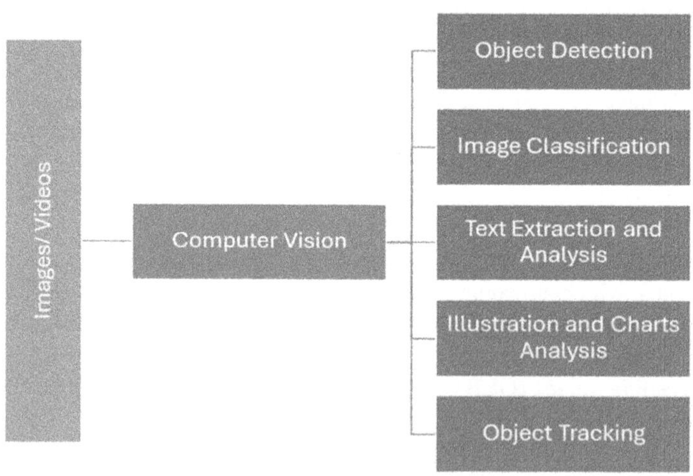

FIGURE 9.1 Applications of computer vision in animal sciences.

TABLE 9.1
Some Computer Vision Applications in Veterinary and Animal Sciences

Reference	Task	Species
[33]	Hip dysplasia classification, hip structures	Canine
	Fracture dating of long bones	Canine
	Fracture detection in long bones	Canine
[34]	Lameness detection in horse	Equine
[35]	Joint abnormalities classification	
	Normal vs. abnormal cardiac silhouette	Canine
[36]	Detection of cardiogenic pulmonary edema	Canine
[37]	Pulmonary nodules or masses detection	Canine
[38]	Detection of pulmonary coccidioidomycosis	Canine
[39]	Identification of common radiographic issues	Canine
[40]	Horse blink detection using deep nets	Equine
[41]	Thoracic and extra-thoracic abnormalities	Canine
[42]	Cardiomegaly classification (based on VHS)	Canine
[43]	Landmarks identification for VHS	Canine
[44]	Technical error detection	Canine + Feline
[45]	pododermatitis and neoplasia of the paw	Canine
[46]	Breed classification using deep learning	Ovine
[47]	detection of dairy cows in free stall barn	Bovine

Just as in humans, disease detection is an important aspect of animal health management, and AI algorithms are now used to analyze radiographs, CT scans, and MRI images. This is assisting veterinarians in diagnosing diseases with precision and helping in early, fast diagnosis. Some broad applications of computer vision in animals are given in Table 9.1. Computer vision technologies have also been successfully used for visual tracking and analysis, especially in wildlife. Systems for tracking and analyzing social interactions among animals, such as mouse models, have also been successful in animals.

9.4 AUTOMATION AND IoT TECHNOLOGIES

The AI and IoT tech integration in animal sciences is transforming traditional farming methods. From automating daily operations to generating actionable insights from data, smart farming solutions improve efficiency, sustainability, and productivity across various sectors in animal agriculture. The case studies highlighted demonstrate significant advancements in farm management, environmental control, and animal welfare, providing a blueprint for future innovations in this field. Some specific use cases concerning the dairy and poultry industry are given below:

- The use of IoT in dairy farms, such as in the Connected Cow project, demonstrates how sensor-equipped collars or wearables track data like movement, rumination, and health parameters. These devices transmit real-time data, allowing farmers to monitor cattle health, detect early signs of disease, and optimize breeding schedules. As a result, this enhances herd management and reduces labor and veterinary costs. Precision Livestock Farming (PLF) uses these IoT sensors to track individual animals, leading to data-driven decision-making [48]. Key benefits of this include early detection of illness, optimized feeding, reproduction cycles, and lower mortality rates due to preventive measures. Farms in the Netherlands have integrated IoT to collect vast amounts of data on feeding patterns, milk production, and animal health. Automated milking systems paired with sensors monitor the quantity and quality of milk produced. Data analytics tools are then used to predict trends and optimize feeding and care routines [49].

- The study of Arshad et al. (2024) analyzes the digitalization of AI in transforming sheep farming. The article cites innovations such as health monitoring via wearable devices, AI-based diagnosis of parasitism detection, and technologies such as virtual fencing and sky shepherding. The technologies enable better decision-making, improve the health of sheep, and improve farm productivity. The authors also briefly discuss the optimization of such systems to facilitate adoption by farmers [7].
- In poultry farms, IoT-based environmental sensors supervise temperature, humidity, air quality, and light inside barns [50]. One example is the Automated Poultry Management System, where environmental parameters are controlled via IoT, ensuring optimal living conditions [51]. This leads to better growth rates and reduces the risk of diseases spread by poor ventilation. AI technologies have been deployed to monitor climate conditions and predict environmental impacts on farm animals. In India, an AI-powered climate control system for poultry houses has been developed, which adjusts temperature and ventilation automatically based on bird behavior and weather patterns. This minimizes heat stress and maximizes productivity [52]. Internationally, smart climate control systems are also available for augmenting poultry production [53–55].
- Smart fencing equipped with IoT allows farmers to manage sheep grazing patterns remotely. Virtual fencing solutions, along with GPS tracking collars, monitor the location of animals and automatically herd them into designated areas. This technology has reduced the need for physical fences, saving on labor costs and resources, while improving pasture management. Marini et al. (2018) have developed a system designed to discourage sheep from crossing virtual boundaries using only auditory signals, addressing ethical concerns associated with electric stimuli [56]. This sound-based approach has shown effectiveness in guiding sheep within virtual confines [57, 58].
- Recently, research conducted in Australia utilized a neck collar that emitted sound signals, specifically adapting the eShepherd® device, originally designed for cattle, to sheep. While sheep learned to remain within designated zones, some boundary breaches were observed, likely due to the device's cattle-oriented design. Therefore, refining the algorithm specifically for sheep could enhance its efficacy. Furthermore, sheep with dense neck wool might respond differently to auditory cues compared to

breeds with less wool, and prolonged exposure to the signals may lead to reduced sensitivity, potentially necessitating a stronger audio stimulus [59]. Microsoft's Farmbeats project leverages AI and IoT to provide farmers with insights based on real-time data about soil, crops, and livestock.

9.5 FUTURE TRENDS AND ETHICAL CONSEQUENCES

AI and IoT's future in animal sciences is marked by transformative trends that promise to enhance productivity, sustainability, and animal welfare. PLF, for example, employs IoT devices and AI to monitor individual animals' health, behavior, and productivity in real time, enabling targeted breeding, feeding, and healthcare interventions [60]. Additionally, AI is revolutionizing animal health management, with predictive models that can detect early signs of illness and optimize treatment plans. Genomic advancements and AI-based breeding models allow for precision in selecting traits such as disease resistance, making animal production more sustainable and productive. Automated feeding and behavior monitoring systems further enhance efficiency and animal welfare, ensuring that animals receive the right nutrients and minimizing stress through early detection of discomfort [61]. Together, these advances promise a future where animal management is both resource-efficient and welfare-centered. AI applications in agriculture could generate approximately $100 billion in on-farm economic value by improving labor and input costs and increasing yields. Compared to traditional methods, AI can significantly enhance decision-making in various aspects of farm management, from feeding to health monitoring, leading to increased efficiency and productivity. AI helps farmers overcome various obstacles without reducing productivity, enabling them to enhance animal health and outcomes. However, integrating AI in animal sciences raises ethical considerations that must be navigated carefully. Continuous surveillance, while beneficial for monitoring, can encroach on farm privacy, requiring balanced approaches to avoid excessive intrusion and a need for data security. Data privacy for farm operators is also a concern, as the ownership and use of data need transparent governance to protect farmers' interests. Genetic manipulation poses ethical questions about long-term health impacts, as selective breeding might affect animals' lifespan and behavior, underscoring the importance of ethical breeding standards. Furthermore, the environmental impact of intensive AI-driven livestock production highlights the need for sustainable integration practices. Research opportunities abound in climate-resilient farming models, disease prediction, resource management, and food safety, where AI could play a pivotal role in addressing climate impacts and traceability and pave the way toward achieving global sustainability goals [62]. Establishing ethical frameworks and standards will ensure these technologies advance animal sciences responsibly, supporting both productivity and humane practices.

9.6 CONCLUSION

AI is revolutionizing animal sciences, enhancing productivity, health, and sustainability. Precision farming and predictive disease models are revolutionizing animal management. Responsible use requires ethical frameworks and sustainable methods, though. India's agri-tech industry is also going to witness huge growth, with AI and IoT going to be the drivers of it. Indian farmers will be able to reap maximum benefit from these technologies after overcoming infrastructure, finance, and education challenges, ensuring food security and sustainable growth. More research is going to be instrumental in finding a balance between global needs, animal welfare, and nature conservation, guiding agriculture toward sustainability. The starting points for this should include infrastructure strengthening, resource allocation in the sector, and skill upgradation in the sector. Customized solutions catering to specific problems of the Indian animal husbandry sector are much needed for an innovative and productive ecosystem in India. AI is revolutionizing animal sciences, making farming smarter and more efficient. But with this progress come important ethical questions that can't be ignored. Data ownership, surveillance, and privacy are of concern. This would prevent its use, misuse, and overuse. Constant surveillance, if unchecked, could infringe on farmers' privacy, turning farms into data-driven observation zones. In addition, AI systems generate vast amounts of farm data, from feeding patterns to genetic insights. However, without clear policies, there's a risk of corporations or third parties profiting from farmers' hard-earned insights in unethical ways. Transparency and fair governance would ensure that AI benefits farmers, not just big tech. AI and biotechnology together could raise questions about transgenic animals and to what extent genetic tampering could be allowed. AI can optimize resources and minimize waste, but if used without sustainability in mind, it could contribute to overproduction, excessive energy use, and ecological strain. Despite these challenges, AI has an incredible ability in disease prediction, resource management, and food safety.

REFERENCES

1. Ezanno P, Picault S, Beaunée G, Bailly X, Muñoz F, Duboz R, et al. Research perspectives on animal health in the era of artificial intelligence. *Veterinary Research*. 2021 December;52(1):40.
2. Hosny A, Parmar C, Quackenbush J, Schwartz LH, Aerts HJWL. Artificial intelligence in radiology. *Nature Reviews. Cancer*. 2018 August;18(8):500–10.
3. Thayyib PV, Mamilla R, Khan M, Fatima H, Asim M, Anwar I, et al. State-of-the-art of artificial intelligence and big data analytics reviews in five different domains: A bibliometric summary. *Sustainability*. 2023 February 22;15(5):4026.
4. Madrewar S, Khadkikar N, Suryawanshi O, Mulani J, Sagar A. Digital agriculture: impact of IoT and AI on Indian agribusiness. *International Journal of Applied Economics Accounting and Management (IJAEAM)*. 2024 August;2:327–50.
5. Esmaeily R, Razavi MA, Razavi SH. A step forward in food science, technology and industry using artificial intelligence. *Trends in Food Science and Technology*. 2024 January;143:104286.
6. Neethirajan S. Artificial intelligence and sensor innovations: Enhancing livestock welfare with a human-centric approach. *Human-Centric Intelligent Systems*. 2023 November 22;4(1):77–92.
7. Arshad MF, Burrai GP, Varcasia A, Sini MF, Ahmed F, Lai G, et al. The groundbreaking impact of digitalization and artificial intelligence in sheep farming. *Research in Veterinary Science*. 2024 April;170:105197.

8. Brault SA, Hannon SJ, Gow SP, Otto SJG, Booker CW, Morley PS. Calculation of antimicrobial use indicators in beef feedlots—Effects of choice of metric and standardized values. *Frontiers in Veterinary Science*. 2019 October 9;6:330.
9. Hamadani H, Khan A. Automation in livestock farming – A technological revolution. *International Journal of Advanced Research*. 2015;3(May):1335–44.
10. Dawkins MS. *The Science of Animal Welfare: Understanding What Animals Want*. Oxford University Press, USA; 2021.
11. Fuentes S, Gonzalez Viejo C, Cullen B, Tongson E, Chauhan SS, Dunshea FR. Artificial intelligence applied to a robotic dairy farm to model milk productivity and quality based on cow data and daily environmental parameters. *Sensors*. 2020 May 24;20(10):2975.
12. Mahato S, Neethirajan S. Integrating artificial intelligence in dairy farm management – Biometric facial recognition for cows [Internet]. 2024 [cited 2024 October 13]. Available from: https://www.preprints.org/manuscript/202401.2032/v1
13. Melak A, Aseged T, Shitaw T. The influence of artificial intelligence technology on the management of livestock farms. Tibaduiza DA, editor. *International Journal of Distributed Sensor Networks*. 2024 January 8;2024:1–12.
14. Hansen KD, Garcia-Ruiz F, Kazmi W, Bisgaard M, La Cour-Harbo A, Rasmussen J, et al. An autonomous robotic system for mapping weeds in fields. *IFAC Proceedings Volumes*. 2013 June;46(10):217–24.
15. Hamadani A, Ganai NA. Development of a multi-use decision support system for scientific management and breeding of sheep. *Scientific Reports*. 2022 November 12;12(1):19360.
16. Hamadani A, Ganai NA. Artificial intelligence algorithm comparison and ranking for weight prediction in sheep. *Scientific Reports*. 2023 August 15;13(1):13242.
17. Haladjian J, Haug J, Nüske S, Bruegge B. A wearable sensor system for lameness detection in dairy cattle. *MTI*. 2018 May 15;2(2):27.
18. Hogeveen H, Klaas IC, Dalen G, Honig H, Zecconi A, Kelton DF, et al. Novel ways to use sensor data to improve mastitis management. *Journal of Dairy Science*. 2021 October;104(10):11317–32.
19. Jacobs L, Blatchford RA, De Jong IC, Erasmus MA, Levengood M, Newberry RC, et al. Enhancing their quality of life: Environmental enrichment for poultry. *Poultry Science*. 2023 January;102(1):102233.
20. van de Weerd HA, Day JE. A review of environmental enrichment for pigs housed in intensive housing systems. *Applied Animal Behaviour Science*. 2009;116(1):1–20.
21. Samiappan S, Krishnan BS, Dehart D, Jones LR, Elmore JA, Evans KO, et al. Aerial wildlife image repository for animal monitoring with drones in the age of artificial intelligence. *Database*. 2024 July 23;2024:baae070.
22. Feng W, Gao P, Wang X. AI breeder: Genomic predictions for crop breeding. *New Crops*. 2024 December;1:100010.
23. Hamadani A, Ganai NA, Mudasir S, Shanaz S, Alam S, Hussain I. Comparison of artificial intelligence algorithms and their ranking for the prediction of genetic merit in sheep. *Scientific Reports*. 2022 November 4;12(1):18726.
24. Pedrosa VB, Chen SY, Gloria LS, Doucette JS, Boerman JP, Rosa GJM, et al. Machine learning methods for genomic prediction of cow behavioral traits measured by automatic milking systems in North American Holstein cattle. *Journal of Dairy Science*. 2024 July;107(7):4758–71.
25. Chafai N, Hayah I, Houaga I, Badaoui B. A review of machine learning models applied to genomic prediction in animal breeding. *Frontiers in Genetics*. 2023 September 6;14:1150596.

26. Ehret A, Hochstuhl D, Krattenmacher N, Tetens J, Klein MS, Gronwald W, et al. Short communication: Use of genomic and metabolic information as well as milk performance records for prediction of subclinical ketosis risk via artificial neural networks. *Journal of Dairy Science*. 2015 January;98(1):322–9.
27. Kasarda R, Moravčíková N, Mészáros G, Simčič M, Zaborski D. Classification of cattle breeds based on the random forest approach. *Livestock Science*. 2023 January;267:105143.
28. Petrazzini BO, Naya H, Lopez-Bello F, Vazquez G, Spangenberg L. Evaluation of different approaches for missing data imputation on features associated to genomic data. *BioData Mining*. 2021 December;14(1):44.
29. Erfanian N, Heydari AA, Feriz AM, Iañez P, Derakhshani A, Ghasemigol M, et al. Deep learning applications in single-cell genomics and transcriptomics data analysis. *Biomedicine & Pharmacotherapy*. 2023 September;165:115077.
30. Zeng S, Mao Z, Ren Y, Wang D, Xu D, Joshi T. G2PDeep: a web-based deep-learning framework for quantitative phenotype prediction and discovery of genomic markers. *Nucleic Acids Research*. 2021 July 2;49(W1):W228–36.
31. Borges Oliveira DA, Ribeiro Pereira LG, Bresolin T, Pontes Ferreira RE, Reboucas Dorea JR. A review of deep learning algorithms for computer vision systems in livestock. *Livestock Science*. 2021 November;253:104700.
32. Li G, Huang Y, Chen Z, Chesser GD, Purswell JL, Linhoss J, et al. Practices and applications of convolutional neural network-based computer vision systems in animal farming: A review. *Sensors*. 2021 February 21;21(4):1492.
33. McEvoy FJ, Amigo JM. Using machine learning to classify image features from canine pelvic radiographs: Evaluation of partial least squares discriminant analysis and artificial neural network models. *Veterinary Radiology & Ultrasound*. 2013 March;54(2):122–6.
34. Feuser AK, Gesell-May S, Müller T, May A. Artificial intelligence for lameness detection in horses-A preliminary study. *Animals (Basel)*. 2022 October 17;12(20):2804.
35. Shim H, Lee J, Choi S, Kim J, Jeong J, Cho C, et al. Deep learning-based diagnosis of stifle joint diseases in dogs. *Veterinary Radiology & Ultrasound*. 2023 January;64(1):113–22.
36. Yoon Y, Hwang T, Lee H. Prediction of radiographic abnormalities by the use of bag-of-features and convolutional neural networks. *The Veterinary Journal*. 2018 July;237:43–8.
37. Müller TR, Solano M, Tsunemi MH. Accuracy of artificial intelligence software for the detection of confirmed pleural effusion in thoracic radiographs in dogs. *Veterinary Radiology & Ultrasound*. 2022 September;63(5):573–9.
38. Pomerantz LK, Solano M, Kalosa-Kenyon E. Performance of a commercially available artificial intelligence software for the detection of confirmed pulmonary nodules and masses in canine thoracic radiography. *Veterinary Radiology & Ultrasound*. 2023 September;64(5):881–9.
39. Ott J, Bruyette D, Arbuckle C, Balsz D, Hecht S, Shubitz L, et al. Detecting pulmonary coccidioidomycosis with deep convolutional neural networks. *Machine Learning with Applications*. 2021 September;5:100040.
40. Dimitrova S, Orchard E, Paudel B, McBride S, Hemmings A, Akanyeti O. Automatic Horse Blink Detection Using Computer Vision and Deep Nets. In: Panoutsos G, Mahfouf M, Mihaylova LS, editors. *Advances in Computational Intelligence Systems [Internet]*. Springer Nature Switzerland; 2024 [cited 2024 October 22]. pp. 436–47. (Advances in Intelligent Systems and Computing; vol. 1454). Available from: https://link.springer.com/10.1007/978-3-031-55568-8_37
41. Banzato T, Wodzinski M, Tauceri F, Donà C, Scavazza F, Müller H, et al. An AI-based algorithm for the automatic classification of thoracic radiographs in cats. *Frontiers in Veterinary Science*. 2021 October 15;8:731936.

42. Li S, Wang Z, Visser LC, Wisner ER, Cheng H. Pilot study: Application of artificial intelligence for detecting left atrial enlargement on canine thoracic radiographs. *Veterinary Radiology & Ultrasound*. 2020 November;61(6):611–8.
43. Burti S, Longhin Osti V, Zotti A, Banzato T. Use of deep learning to detect cardiomegaly on thoracic radiographs in dogs. *The Veterinary Journal*. 2020 August;262:105505.
44. Tahghighi P, Norena N, Ukwatta E, Appleby RB, Komeili A. Automatic classification of symmetry of hemithoraces in canine and feline radiographs. *Journal of Medical Imaging*. 2023 July 25 [cited 2024 October 22];10(04). Available from: https://www.spiedigitallibrary.org/journals/journal-of-medical-imaging/volume-10/issue-04.04.004/Automatic-classification-of-symmetry-of-hemithoraces-in-canine-and-feline/10.1117/1.JMI.10.4.04.004.full
45. Smith A, Carroll PW, Aravamuthan S, Walleser E, Lin H, Anklam K, et al. Computer vision model for the detection of canine pododermatitis and neoplasia of the paw. *Veterinary Dermatology*. 2023 December 6;35:13221.
46. Hamadani A. Artificial intelligence on farms: Sheep breed classification using computer vision. *Indian Journal of Animal Production and Management*. 2024 October;40(4):260–8.
47. Tassinari P, Bovo M, Benni S, Franzoni S, Poggi M, Mammi LME, et al. A computer vision approach based on deep learning for the detection of dairy cows in free stall barn. *Computers and Electronics in Agriculture*. 2021 March;182:106030.
48. Taneja M, Jalodia N, Malone P, Byabazaire J, Davy A, Olariu C. Connected cows: Utilizing fog and cloud analytics toward data-driven decisions for smart dairy farming. *IEEE Internet of Things Magazine*. 2019;2(4):32–7.
49. Wang X, Demirci A, Graves RE, Puri VM. Conventional and Emerging Clean-in-Place Methods for the Milking Systems. In: *Raw Milk [Internet]*. Elsevier; 2019 [cited 2024 October 22]. pp. 91–115. Available from: https://linkinghub.elsevier.com/retrieve/pii/B9780128105306000055
50. Sitaram KA, Ankush KR, Anant KN, Raghunath BR. IoT based Smart Management of Poultry Farm and Electricity Generation. In: *2018 IEEE International Conference on Computational Intelligence and Computing Research (ICCIC) [Internet]*. Madurai, India: IEEE; 2018 [cited 2024 October 22]. pp. 1–4. Available from: https://ieeexplore.ieee.org/document/8782308/
51. Ramteke B, Dongre S. IoT Based Smart Automated Poultry Farm Management System. In: *2022 10th International Conference on Emerging Trends in Engineering and Technology - Signal and Information Processing (ICETET-SIP-22)*, Nagpur, India. 2022. pp. 1–4.
52. Bose R, Roy S, Mondal H. A novel algorithmic electric power saver strategies for real-time smart poultry farming. e-Prime – Advances in electrical engineering, *Electronics and Energy*. 2022;2:100053.
53. Fang C, Zhang T, Zheng H, Huang J, Cuan K. Pose estimation and behavior classification of broiler chickens based on deep neural networks. *Computers and Electronics in Agriculture*. 2021;180:105863.
54. Raj AAG, Jayanthi JG. IoT-based Real-time Poultry Monitoring and Health Status Identification. In: *2018 11th International Symposium on Mechatronics and its Applications (ISMA)*, Sharjah, United Arab Emirates. IEEE; 2018. pp. 1–7.
55. Ribeiro R, Casanova D, Teixeira M, Wirth A, Gomes HM, Borges AP, et al. Generating action plans for poultry management using artificial neural networks. *Computers and Electronics in Agriculture*. 2019;161:131–40.
56. Marini D, Meuleman M, Belson S, Rodenburg T, Llewellyn R, Lee C. Developing an ethically acceptable virtual fencing system for sheep. *Animals*. 2018 February 27;8(3):33.

57. Campbell DLM, Marini D, Lea JM, Keshavarzi H, Dyall TR, Lee C. The application of virtual fencing technology effectively herds cattle and sheep. *Animal Production Science*. 2021 March 30;61(13):1393–402.
58. Kleanthous N, Hussain A, Sneddon J, Khan W, Khan B, Aung Z, et al. Towards a virtual fencing system: Training domestic sheep using audio stimuli. *Animals*. 2022 October 25;12(21):2920.
59. Campbell DL, Belson S, Lea JM, Ouzman J, Lee C, Kalinowski T, et al. Automated virtual fencing can effectively contain sheep: Field trials and prospects. *Animals*. 2023;13(4):619.
60. Chelotti JO, Martinez-Rau LS, Ferrero M, Vignolo LD, Galli JR, Planisich AM, et al. Livestock feeding behaviour: A review on automated systems for ruminant monitoring. *Biosystems Engineering*. 2024 October;246:150–77.
61. Zhang L, Guo W, Lv C, Guo M, Yang M, Fu Q, et al. Advancements in artificial intelligence technology for improving animal welfare: Current applications and research progress. *Animal Research and One Health*. 2023 December 29;2:44.
62. Vinuesa R, Azizpour H, Leite I, Balaam M, Dignum V, Domisch S, et al. The role of artificial intelligence in achieving the sustainable development goals. *Nature Communications*. 2020 January 13;11(1):233.

10 The Role of IoT in Promoting Sustainability in Green Agriculture

Khan Jabroot, Himanshu, Deepika Saxena, and Rahmatullah Kazimi

10.1 INTRODUCTION

Agriculture plays a pivotal role in many developing countries as currency exchange. This serves as the main source of employment and earnings for marginal farmers. Additionally, the world is battling the problems of COVID-19 in every sector. The agriculture sector is one of these sectors that faces various scarcities in labor supply, transportation, and risks linked to agricultural markets. The Internet has expanded rapidly across the globe over the past two decades, creating tremendous benefits for organizations and people. Agriculture has contributed substantially to the development of human civilizations that settled centuries ago and remains one of the essential sectors. Most agricultural operations rely on traditional techniques of farming techniques [1]. With the world population expected to hit 9.6 billion by 2050, there is a need to increase food production by 70% to feed the people. Furthermore, it has been estimated that about 33% of the food produced globally is wasted, which amounts to around 1.3 billion tons annually. The Internet of Things has emerged to be a great, promising concept by offering innovative technologies that improve users' experiences and capabilities by transforming their work environments. IoT offers diverse solutions across various sectors, including healthcare, retail, traffic management, security, smart homes, smart cities, and agriculture. In particular, its application in agriculture is promising since this sector requires constant monitoring and control. Agriculture 4.0 increases crop yield but reduces input costs, labor, and environmental pollution. So, Agriculture 4.0 is the most relevant to this century, in which food demand is increasing more rapidly.

Machine learning (ML) is a technology of AI that facilitates automatic data reading without human intervention. It has a greater impact on the global economy's status, especially ML which has made a paramount difference in international trade [2]. Communication barriers in agriculture have been greatly reduced due to advancements in multilingual support, banking, FinTech, and improved crop, soil, and livestock management. Different ML terminologies, such as Bayesian models and deep learning, are introduced to agriculture through artificial intelligence (AI). Agriculture practices that contribute to the fulfillment of present-day and future

DOI: 10.1201/9781003594093-10

food and nutrition of the population without having detrimental effects on the ecosystem, society, and the earth are referred to as sustainable agriculture. The concept of sustainable agriculture has gained exponential attention in the modern world with the increase in technological approaches [3]. In the ancient era, farming was practiced using sticks, sickles, gathering crops, and animal husbandry. Concerning smart farming, more efficient techniques are used to observe their farms with the help of smartphones and control devices. Genetically modified seeds are now cultivated, which prevents infestation and diseases. Unfortunately, many areas of the Earth's surface are unsuitable for agriculture due to limitations like soil quality, topography, temperature, and climate, and even the most viable cultivable regions often lack uniformity. Additionally, farming areas are increasingly fragmented by political boundaries, economic factors, and the spread of urbanization, putting constant pressure on the availability of arable land. Recently, the overall land dedicated to food production has decreased. In conventional farming methods, farmers often make regular visits to their fields throughout the crop's growth cycle to closely monitor crop conditions. Modern sensor and communication technologies provide an accurate field overview, enabling farmers to monitor activities remotely [4]. The current chapter aims to provide recent IoT applications for sustainable farming. Due to the increasing food requirements from a growing human population on one hand and shrinking arable land on the other hand, it has become imperative to produce crops with more efficiency and less area. Everyone needs to understand food security within the framework of sustainable agriculture. The emergence of new technologies aimed at boosting crop yields is crucial for attracting innovative young individuals.

The importance of emerging technologies, especially the Internet of Things (IoT), receives major focus in this chapter because it demonstrates how they transform agriculture into an intelligent, sustainable effective system that addresses future requirements. This research explores current business problems in the field and discusses prospective development areas for meaningful findings that benefit engineers and researchers. To maximize crop yields through available land resources, farmers should implement IoT-based sensors with communication technologies during farmland management. Future investigation in this field has the potential to build on existing knowledge by conducting deeper analyses regarding other IoT architecture components that enable smart farming technology. Holds a strong position among revolutionary agricultural development through precision farming systems because they allow enhanced agricultural sustainability and productivity. Modern agricultural transformation occurs through the integration of Industry 4.0 technologies, which consist of IoT and AI, along with remote sensing and Geographical Information System (GIS). These technologies surpass basic tools because they introduce an innovative approach to securing better results in agricultural resource management. Occupying the field and crop needs requires high-resolution data obtained from satellites, along with unmanned aerial vehicles (UAVs), combined with digital analysis through AI. The approach enhances efficiency through customized practice implementation, which suits individual land features, so yields get maximized with reduced waste production and environmental consequences.

10.2 INTERNET OF THINGS (IoT) IN AGRICULTURE

The Internet of Things (IoT) is a relatively new technology through which devices are connected to remotes to optimize smart farming practices. The IoT has started to shape almost all fields of activity, including health, trade, communications, power, and agriculture, thus improving the output across all sectors. The currently implemented applications of IoT present information on the effects this concept and current practice still have. As technology advances, IoT is expected to add value to different farming activities, such as the use of communication networks, data acquisition, smart devices, sensors, mobile technologies, cloud-based decision intelligence, decision-making tools, and automation of agricultural activities, and is illustrated in Figure 10.1. Sensors and tools assist in assessing the weather and establishing yield levels among farmers. Smart water management is also a part of IoT because in areas like water harvesting, controlling flow rates, assessing the water requirements of crops, programming the process of watering plants or fields, and improving ways of saving water more efficiently in comparison with the previous methods [5]. Sensors, gates, and cloud incorporation allow water status and supply to be monitored and can be customarily altered to best serve the needs of the soil and plants. It is not easy for the farmers to try to address issues of nutrient deficiencies, pests, and diseases individually at the plant level.

Nevertheless, IoT technology is very helpful and profitable, which has brought the meaning of contemporary agriculture to a new level. There has been extensive development in the domain of IoT technologies, which has injected profound change in the functioning of the farming sector, primarily dealing with the improvement of communication frameworks. Some of this progress includes making smart devices connect, obtaining data from devices at a remote place and sensors through mobile and Internet networking, performing intelligent analysis from the cloud, making interface systems with smart devices, and deciding and automating agricultural tasks. These capabilities have helped revolutionize the agriculture industry by enabling a better use of resources, influencing climatic conditions, and enhancing productivity. Scientists have provided various techniques, structures, and tools to monitor and share crop information through the development of physical forms suitable for specific crop and field types. Many manufacturers provide communication devices, various types of sensors, robots, heavy equipment, and drones to collect and disseminate information. These technologies are governed by food and agriculture organizations as well as government agencies that set standards regarding their usage, keeping in mind the safety of food as well as the earth. The easy-to-use and low-cost tracking system provides information on old school farming practices, equipment, animals, pests, diseases, and so on, collected from various sources for the promotion of sustainable agriculture [6].

10.2.1 Role of IoT in Promoting Sustainable Agriculture

The Internet of Things (IoT) has the potential to revolutionize sustainable agriculture by enabling activities associated with agriculture to be monitored and controlled with precision, minimizing depletion of resources, and minimizing environmental degradation. Soil sensors, weather base stations, and GPS-fitted equipment can easily provide real-time information on parameters of interest like moisture content of the soil, nutrient status in the crops, the health of the crops, and weather conditions, and are presented in [Figure 10.2]. As such, technological innovation will place the farmers in the position of making the right dispensation of water as well as chemicals, fertilizers, and pesticides at the right time and place. All these improvements in the practice will lead to the prevention of wastage, only more efficient use of resources,

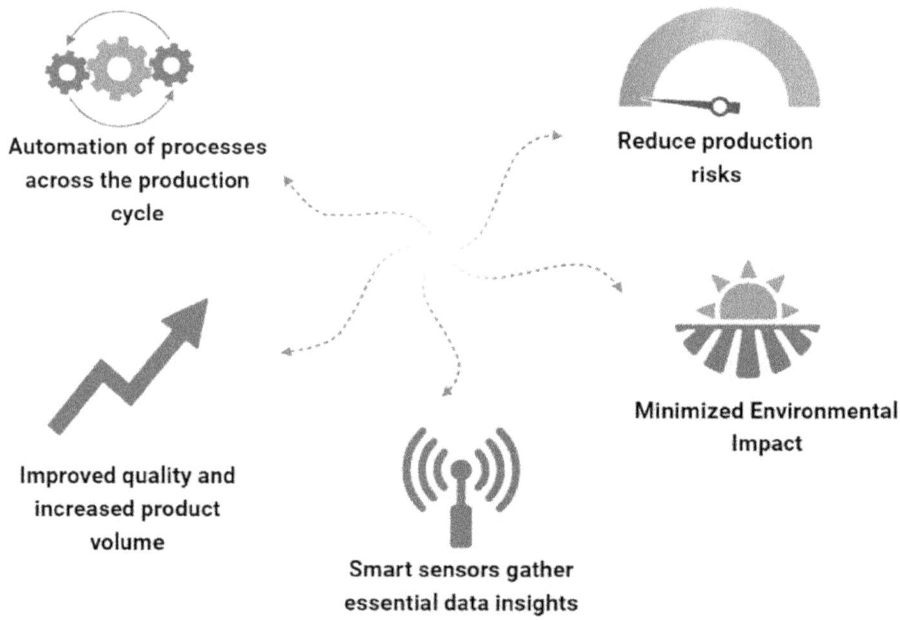

FIGURE 10.1 Benefits of using IoT in agriculture.

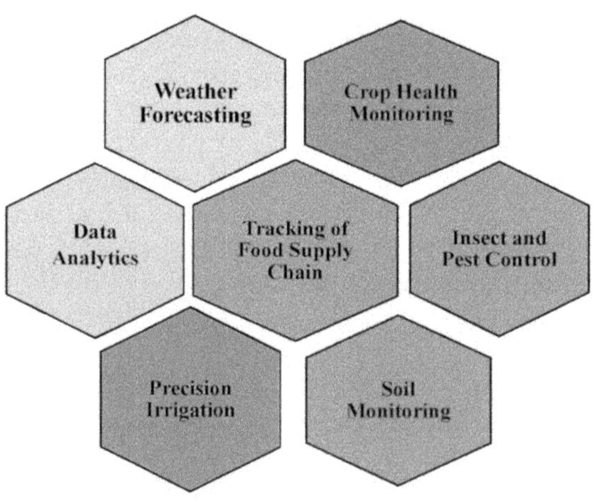

FIGURE 10.2 Role of IoT in sustainable agriculture.

and safeguard against the excess application of chemicals that do not support the ecological balance of the soil and its biosystem.

By employing IoT technologies, smart irrigation scheduling, which is soil moisture and weather-dependent, is deployed to mitigate water waste through over-irrigation. This also facilitates early detection of the crop because of the imaging sensors mounted on the drones, which allows for timely intervention so that the yield losses are minimized. Supply chain management is enhanced due to the IoT by tracking the farm produce during the whole process and at what cost, thus maintaining the quality, minimizing wastage, and improving transport. Therefore, the Internet of Things (IoT) stimulates productivity and economic returns as well as the promotion of suitable and healthy farming practices, restoration of the environment, diminishing of harmful gases associated with farming, as well as building a prosperous and healthy agriculture sector [7].

10.3 APPLICATION OF IoT IN SUSTAINABLE AGRICULTURE

Modern agriculture needs the Internet of Things (IoT) to operate because technology enables accurate data-based farming methods. Throughout their activities, farmers enhance irrigation as well as fertilizer control and pest regulation through IoT device data collection, including weather stations, soil moisture sensors, and GPS-equipped equipment. The implementation of IoT devices allows farmers to decrease environmental effects, as well as raise production output and lower resource consumption. Serious agricultural practices boost both surveillance systems in food production and distribution systems and reduce food insecurity. The integration of IoT technologies brings support for next-generation, environmentally friendly farming systems that will become essential for future food production operations.

10.3.1 Soil Mapping and Plant Monitoring

Global Positioning System (GPS) satellites send signals that enable GPS receivers to determine their location in real-time, continuously providing positional data while in motion. This accurate location information allows farmers to identify specific field data points, such as pest infestations, soil types, weeds, and other obstacles. The system aids in pinpointing various areas within a field, enabling the targeted application of inputs like seeds, fertilizers, herbicides, pesticides, and water. Using GPS position and field-specific data, the nutrient status of the field is approximated through soil analysis, and based on this nutrient deficiency at different growth phases of the crop, critical management decisions are made. Soil fertility status is affected by factors such as topography, type and texture, cropping pattern, and use of fertilizers, while irrigation and some others may also influence the status of soil fertility. Physical, chemical and biological properties of the soil should be measured most efficiently in determining whether a certain field is good for growing a particular type of crop [8]. Currently, there is an availability of many sensors and instruments through which farmers can learn about soil characteristics like water retention capacity, types of soil, and how quickly it can absorb water, among others; farmers then find out ways to prevent deterioration of soils through erosion, alkalization, acidification, salinization, and pollution.

Drought is another stress factor under which plants experience reduced growth and production of crops. A variety of remote sensing methods that can provide soil moisture data often help monitor agricultural drought in remote areas. Soil type, nutrient status of the soil, water supply, fertilizers, and pests are major factors that determine the yield and quality of rice. The use of IoT helps to control crops and offers soil nutrition and property updates through the developed mobile application. The system components include electrical conductivity (EC), temperature sensors with a T-Beam microcontroller, and the Internet of Things (IoT) capabilities, as well as the estimated EC value close to the calibration solution. Through plant monitoring through IoT Agricultural Data Collection Online (ADCON)-based station, use of sensors, and mobile devices (smartphones and tablets), farmers can monitor soil and ambient parameters like leaf wetness, air and soil temperature, and humidity, and thus increase grape productivity and crop quality from seeding to harvesting [9].

10.3.2 Sensing Technologies

The methods include photoelectric sensors, electromagnetic sensors, conductivity measurements, and ultrasound for measuring the soil texture and structure, nutrients, vegetation, humidity and vapor, air temperature, and much more. Remote sensing data are capable of identifying varieties of crops, pests and weeds, stress in soil and plant health, and drought conditions. As part of the plant health assessment, the following parameters are considered: moisture of the ground, nutrients, light exposure, humidity, amount of rainfall, and the color of the leaves. To control the conditions, their temperature and light intensity the plants are controlled, and water and energy are saved by using micro irrigation systems. There are so many sensors that are used for various purposes. With the help of all these parameters, if it goes beyond some limited range, then the sensors detect these changes and send a signal to the microcontroller to take some steps until the proper range of the parameter is restored. Most sensing technologies employ one or more sensors to would measure temperature, humidity, soil patterns, airflow, position, carbon dioxide, pressure, light, and moisture levels. Thus, the peculiar features of these sensors include such important characteristics as reliability, memory, portability, durability, spreads, and computational capacity to provide the necessary agricultural application. Currently, available wireless sensors are an important input for gathering crop conditions data and for informative purposes. These sensors function autonomously and can be incorporated with high-tech machinery and agricultural equipment, apart from other large equipment, based on the particular application needs [10].

The performance of IoT technologies for agriculture will improve only when experts conduct comprehensive evaluations of different sensor types, including communication standards, alongside their appropriate usage scenarios. Soil moisture and temperature sensors and humidity sensors function best under different agricultural situations since they possess distinctive advantages. Soil moisture sensors determine optimal water distribution through their ability to show current soil water amounts, while temperature sensors aid in crop health evaluation. The selection of sensors depends on three main elements: crop types, present soil conditions, along with local environmental conditions. The process of choosing communication protocols between Zigbee, LoRaWAN, and NB-IoT remains essential. Protocols have varying suitability for agricultural environments because they offer specific capabilities regarding distance coverage and power usage, together with data transfer performance. LoRaWAN operates better for long-distance communications that consume low power, but Zigbee functions efficiently within short distances for fast data exchange. Farmers can improve their resource management while minimizing costs through informed decisions based on a comprehensive evaluation of sensors, along with communication methods, which lead to better agricultural sustainability. The application of this method produces both precise and effective farm-specific agricultural practices. The primary types of sensors, along with their respective operating methods and purposes, are outlined in Table 10.1.

10.3.3 Variable Rate Technology and Grid Soil Sampling

Emerging technologies, including variable rate technology and related practices like grid soil sampling, have the potential to enhance soil fertility management and evaluate the spatial distribution of nutrients and yields. The world's crop production is an important factor in addressing the food scarcity in the world, forecasting weather changes, and challenging changes in climate conditions. ML makes human thinking easier by reducing future decision-making problems into an optimally solvable problem. In this case, the modality applied to the method of bringing an understanding of training the machine and reasoning to the said decision is used to predict the future.

TABLE 10.1
Types of Sensors and Their Uses

Sensors	Applications	References
Acoustic sensors	Fruit harvesting, seed variety classification, and pest monitoring and detection	[11]
Airflow sensors	Evaluating the structure, moisture content, and air permeability of soil in a stationary or mobile setup	[12]
Eddy covariance sensors	Monitoring CO_2, water vapor, methane, and other gas transfers. Evaluating trace gas fluxes and surface atmospheric conditions in various agricultural environments	[13]
Electrochemical sensors	Evaluating soil nutrient content and pH levels	[14]
Field-Programmable Gate Array (FPGA) based sensor	Monitoring the humidity, watering, and transpiration of plants in real-time	[15]
Light Detection and Ranging (LIDAR)	Crop yield prediction, land mapping, soil categorization, 3D farm modelling, and soil loss and erosion monitoring	[16]
Telematics sensors	Evaluating the position, routes travelled, and activity related to farm and machine operations	[17]
Ultrasonic ranging sensors	Crop canopy monitoring, tank monitoring, spray distance measurement, uniform spray coverage, and object detection	[18]

Yield forecasts are considered to be one of the biggest issues in this sphere. Aspects like rainfall, temperature, and other climatic conditions, in addition to the use of pesticides, affect yield. This is particularly so when passing judgment on managing agricultural risk because prior crop yield is central to the process. Thus, satellite images afford relatively up-to-date information at a regional scale due to their overall coverage and frequency of repetition. However, basic ratio- and depth-based vegetation indices such as Normalized Difference Vegetation Index (NDVI) are used for the quantification of the health of vegetation, vegetation density, yield-vegetation productivity, and the Leaf Area Index (LAI), and for comparison with photosynthetically active radiation (PAR) of green plants across the spectrum [19].

10.3.4 Crop Management

The agriculture sector is part of the world economy; worldwide crop production is a critical factor in solving the global food deficit, forecasting climatic variations, and combating changes in climate conditions. ML makes human thinking easier by reducing the probability of future decision-making into an easily solvable problem. In this case, the approach used for training the machine and arriving at the given decision is commonly used for predicting the future. Predictions of crop yields are one of the biggest challenges within agriculture. Climate factors such as rainfall, temperature, and others, and the use of pesticides influence yield. This is especially important when making judgments about managing agricultural risk since previous crop yield information is vital to the process. Satellite images provide relatively current data at the regional level because of their spatial extent and temporal return rate. Spectral properties of crops are generally connected with biomass and yield trials, specifically spectral reflectance properties of vegetation, or vegetation indices that include red and near-infrared bands associated with green leaf area. However, simple ratio- and depth-based vegetation indices, such as the NDVI, are preferred for the analysis of the relative health of vegetation, growth of vegetation, crop yields, and LAI across the spectrum and comparison with PAR of green plants [20].

10.3.5 Site-Specific Nutrient Management

Fertilizer is a natural or chemically compounded substance possessing the properties for the growth of plants and the fertility of the soil. Lack of nutrients and the use of high concentrations of fertilizer are also undesirable for the soil, plants, and the environment. In smart agriculture, the site-specific soil nutrient fertilization in an inefficient way would only provide an estimate of the amount of nutrients needed; hence, the rate at which nutrients can pollute the soil and environment is highly controlled. Regarding the on-site fertilizer measurements, they vary in the type of soil, crop type, required yield, cation exchange capacity, exchange efficiency, and fertilizer type utilized by the smart IoT approach to the spatial distribution of the nutrient. From satellite data, NDVI data were used to capture nutrient status, crop and vegetation activity, plant population and density, as well as soil nutrients. Among the emergent technologies that have come close to supporting IoT-based smart fertilization are GPS/geo-mapping, Variable Rate Technology (VRT) and auto-steering, and autonomous vehicles. Others are fertigation, where water-soluble fertilizer is used in the amendments, and chemigation, where pesticides are used in the control of soil-borne diseases [21].

10.4 IoT FOR CROP HEALTH AND YIELD OPTIMIZATION

The Internet of Things provides real-time monitoring data about plant growth environments, thereby revolutionizing crop health assessment along with production quality enhancement. The filtering of temperature data, together with moisture measurements, as well as pest detection and soil nutrient reports by smart sensors, provides early sickness or stress recognition. Agricultural outcomes improve when farmers use IoT-based interventions such as specialized watering techniques and bug control methods. The examination process achieves greater definition through satellite imagery and drone operations because they deliver detailed information about fields. The precise utilization of resources enabled by IoT-powered automation reduces expenses and creates favorable environmental outcomes. Agribusinesses benefit from IoT-driven solutions, which lead to sustainability through enhanced farms while improving crop quality, combined with increased harvest production.

10.4.1 Crop Pest and Disease Management

According to the Food and Agriculture Organization (FAO), the annual crop yield loss globally due to pests and diseases is only 20–40%, with pests and diseases being controlled through pesticides and other agrochemicals. In their great percentage, they are poisonous to both man and animals and therefore contaminate the structures of the environment. The IoT-enabled devices include robots, wireless sensors, and drones to locate and manage crop rivals by monitoring, modelling, and disease prediction; overall efficiency is thereby improved beyond the pest elimination practice. In the case of disease and pest management using the IoT, the approach involves detection and image processing. The remote sensing imagery and field sensors are used to capture data, including plant health and pest incidence, in each field for the full growing period. IoT technologies include automated traps that capture, count, and describe the species and varieties of insects, and in addition, share information with the cloud for computing. Due to recent advances in robotic engineering, an agricultural robot with multispectral image sensing devices and scheduled spraying nozzles is used for pest identification [22].

The IoT devices are used for early detection of pests and diseases in fruit and vegetable crops, as shown in Table 10.2.

10.4.2 Yield Monitor

A yield monitor includes sensors and products such as a memory device, a computer, and an interface, where the components are connected and communicate with each other. These sensors continually monitor yield by evaluating the volume of grain flow. The mass flow sensor works under the mechanism of utilizing

TABLE 10.2
Use of IoT in Disease Detection in Fruits and Vegetables

Crop	Method used	Disease/Pest Detection	Outcome	Reference
Onion	Machine learning	Detects and investigates the life cycle of thrips	Diseases can be detected with 96.3% accuracy	[23]
Potato	ResNet-50	Potato blight	Obtains a 97% accuracy rate	[24]
Tomato	GoogLeNet, SquezeNet, and ResNet-50	Tomato leaf disease	99.87%	[25]
Apple	deep learning algorithms	Apple scab	92%	[26]
Mango	TFLite model	Powdery Mildew, Black Sooty Mold, Cutting Weevil, Die Back, Anthracnose	100%	[27]
Maize	Decision support system (DSS)	Bacterial stalk rot	99.65%	[27]
Papaya	convolutional neural network (CNN)	Anthracnose, Bacterial spot, Curl, Ring spot	-	[28]
Chilli	YOLO model	Thrips	98.5%	[29]
Guava	Convolutional Neural Network	Wilt	93.82% and 95.17%	[30]
Litchi	YOLO model	Downy Blight	89.58%	[31]

microwave energy beams and assessing the reflected energy after an impact. Yield monitors have GPS receivers that produce yield maps, given yield data from specific locations. Installed on a harvester, the yield monitor pairs with a mobile application that provides real-time results of the harvesting or post-harvest information to a web application. From the app, yield maps can be created and shared with agronomists at high resolution, while other farm management data can be exported for further analysis [32]. In horticultural crops, the study defined fruit growth element as the most important factor in quantifying yield quantity and yield quality during the crop progression period. Described uses of color images help in tracking fruit conditions to predict their ripening and make decisions on harvesting and marketing. Real-time information on producing food crops can also be obtained, and large-area observation is made possible through satellite imagery. The crop yield estimation system is made up of both software and hardware such as a yield estimation application that is installed on any Android device. This is an application that uses mathematical algorithms and is connected by Bluetooth that estimates the yield of crops. Furthermore, satellite data can give spectral signatures that will give effective estimates close to real yields. A combination of ML and satellite data assimilation has been accomplished in predicting the maize yield under different environmental conditions within the crop models [33].

10.4.3 Drones and Imaging Technologies for Crop Monitoring

Drone solution technology and imaging are now indispensable tools in crop monitoring for the timely provision of information to farmers. Drones are capable of shooting images across wavelengths that are favorable for the interpretation of plant health indices such as chlorophyll content, canopy cover, and water stress and supply this information to the appropriate roles. These views allow farmers to identify diseased plants, pest outbreaks, high and low nutrient content early and reduce regular all-encompassing chemical spray. Thermal and hyperspectral imaging technologies add to monitoring, as temperature and slight spectral differences in plant composition are otherwise undetectable. Farmers who decide to conduct flights regularly are allowed to determine which areas of their fields need attention, using the observations to improve yields. It also finds application in sustainability, since with these technologies, the targeted application of various resources is possible, where it is most needed. In general, drones and imaging technologies mean an effective and easily scalable solution to modern agriculture that encourages increased production and better stewardship of the world's resources [34].

10.5 ENERGY EFFICIENCY AND RESOURCE MANAGEMENT

Through IoT implementations, farmers gain exact automated farming controls that enhance management and energy efficiency in green agriculture operations. Fabricated sensors with network capabilities monitor environmental elements as well as water usage, together with soil attributes and power consumption patterns in real time. This data enables farmers to cut down unnecessary resource expenditure by adjusting schedules for irrigation and lighting, along with equipment operations. Devices under the Internet of Things approach utilize solar power as a renewable energy solution that promotes sustainability practices. The integration of Internet of Things creates sustainable farming systems through high-productivity management solutions, which promote climate-smart agriculture and sustainable food production by maintaining resource efficiency. It also lowers operating expenses.

10.5.1 Renewable Energy Integration with IoT

This system improves crop production quality and quantity as the environment is controlled manually in greenhouse farming conditions. The utilization of IoT for the creation of a smart greenhouse includes mounting various kinds of sensors and motors to know and control the climate conditions necessary for the plants. This new upcoming agricultural system will allow the operations of some electrical appliances to be done mechanically, such as opening and closing of windows, control of the cooling and heating system, among other things, and also the turning on/off of the light bulbs. The irrigation methods that are being used at the moment are slightly sophisticated and include watering at certain times; hence, minimal control is needed. However, at the same time, it is as much guesswork as anything else and can be extremely profligate in terms of water and energy use as well. Nevertheless, certain field parts may be over- or under-irrigated even with modern techniques of irrigation [35]. Besides, while farmers would turn off their irrigation systems when rain is expected, at other times, they do not bother to look at the sky to know when to stop/turn off the irrigation systems. All of this can be thought of by smart irrigation management systems by employing precision farming methods along with IoT devices that keep a check on moisture content for soil, humidity, and temperature almost at every part of the field. Based on this data, the control system can set practical irrigation regimes. If you're looking to get a better outcome and water and energy efficiency, then this device gives the precise amount of water where and when needed to make that happen for the plant. The use of renewable energy in the smart grid has much to gain from the massive advances being witnessed by IoT and embedded systems in improving the generation, transmission, usage, and storage of energy. It also optimizes the utilization of power in assembly lines by minimizing line loss and maximizing the optimization of means of production through harmonized yield about usage in the real-time sense [36].

10.5.2 Efficient Use of Farm Machinery and Automation

The use of farm machinery and automation is changing agriculture by minimizing the use of resources, reducing the number of workers, and maximizing production. Most machinery used in modern farming has a global positioning system and improved Internet of Things sensors, which make planting, administration of fertilizers and food, and other essential ingredients with high precision where and when they are required. These lesser applications preserve fuel use, fertilizer, and pesticide and therefore decrease costs while preserving the environment. Self-driving tractors and robotic pickers are helpful on the farm in the sense that they reduce the amount of time a farmer needs to take to repeat tedious processes and execute them properly. Intelligent systems offer actual time information concerning the status of the equipment and help in choosing optimal maintenance policies that reduce outages. In addition to productivity improvement, automated machinery entails cost advantage by utilizing energy and reducing wastage. It also has the potential to make farming more efficient, environmentally friendly, and profitable as agriculture automation continues to improve [37].

10.6 IoT IN SUPPLY CHAIN AND POST-HARVEST MANAGEMENT

Real-time condition tracking will be another major benefit of post-harvest integration with big data and IoT from IoT environmental sensors installed in the stored produce, vehicles, and warehouses. Common conditions affecting the quality of vegetables include temperature, humidity, and carbon dioxide, all of which can be monitored in real time. For instance, products like fruits and vegetables require a special temperature range to avoid spoiling. IoT technology does this because the technology is capable of continually measuring the environmental conditions in the storage space, and should the value of the measured parameter stay outside the optimal value, the system can send alarm notifications or physically realign the storage space for the right preservation of produce [7].

IoT technologies are reshaping post-harvest management by introducing the concept of prediction for equipment maintenance. One can also place IoT sensors and storage equipment, such as refrigeration units, with a view to constantly monitoring their performance and identifying initial signs of failure. However, by closely observing the data gathered by thousands of sensors over time, big data algorithms can predict with a very high probability at what time the equipment is likely to fail. Such a predictive feature enables the stakeholders to plan for the repair or replacement of the components so that a failure toward the end of the harvesting period is avoided because this would lead to additional losses after harvesting losses. This approach is effective in not only preventing incidents of spoilage occasioned by the breakdown of equipment but also in making sure that the produce maintains the right quality and freshness through storage and transportation [38].

10.7 CHALLENGES AND BARRIERS TO ADOPTING IoT

At present, the World Health Organization (WHO) employs the term 'food insecure' to denote the condition affecting more than 800 million people globally. However, the global population is continuously increasing, and people need good-quality food, so the prospects of improving the production of food and cash crops through horticulture cannot be denied. Technology has a variety of heterogeneous parameters that determine its implementation. Applying technology to agriculture has led to increased precision, reduced time, and increased efficiency. Smart farming can increase the productivity of food crops; yet, there are setbacks in the use of these technologies. The attributes that are making the useful arable land continue to diminish and the appropriateness of certain crops in given locations are geographic and ecological. Fluctuations in weather increase calamities, including drought, depletion of groundwater, and soil erosion, and hence have a negative effect on crops. The failure to exercise necessary precautions when applying the chemicals has adverse effects on the environment. Agricultural production has a number of issues in developing countries, including crop choice deficiency, soil analysis deficiency, ineffective irrigation, poor weather forecasts, and numerous others. Although the influence of technological sophistication has brought about numerous beneficial changes both economically and in terms of quality, about half of the population of the developing nations is directly or indirectly engaged in agricultural production [21]. According to the current prediction, future agriculture is expected to incorporate AI and big data services. As a result, the enterprise will consist of numerous interconnected systems, and farm machinery, as well as management systems, will cover the entire spectrum from sowing to forecasting of production. Some of the major technologies and methods, in this regard, are given priority to have sustainable agriculture in the future.

10.7.1 Cost of Technology and Lack of Financial Resources

Current technologies reduce the need for manual labor and operate at a high speed with great precision. As a result, it is expected that machines may eventually take over many tasks traditionally performed by humans. However, this transition is unlikely, as numerous countries have faced poverty where the agricultural sector relies heavily on human labor. Moreover, adopting these devices and technologies often involves significant financial investment, making it challenging for farmers to afford alternatives to conventional tools. Financial institutions could offer sufficient loans to farmers in cases where they do not achieve the expected yield, potentially due to unforeseen events such as droughts, floods, pest infestations, or diseases affecting their crops [39].

10.7.2 Telecommunications Infrastructure

Agriculture is predominantly a rural business, and farming activities take place more on fertile soils than on polluted soils. However, mobile phone and tablet usage and telecommunication infrastructure do not allow effective transfer of data in some instances. The right use of information requires a strong Internet connection that is the basis of smart farming. Moreover, operating various control systems, including those about fertilizers, pesticides, seed volume, and improved performance, is possible only when connected to high-quality Internet. Thus, while mobile phones have enhanced the availability of the mobile Internet to rural producers, signal quality and data downloads remain poor [40].

10.7.3 Communication

The success of the Internet of Things (IoT) in agriculture largely hinges on the connectivity between devices. While many telecom operators provide connectivity services, they account for only a small fraction of the overall sustainable farming ecosystem. Cellular operators are offering new services that will be focused on growers, and the market access has also been developed in most rural areas. Cellular technology can only be said to be effective depending on the service providers being able to prove the benefits of using it, including flexibility, portability, as well as the ability to afford two-way communication. In developing countries, and for intending farmers who want to increase production, mobile services and smartphone technology features show a promising future. The technology selected as low-power wide area or LPWA is forecast to have a larger and more profound influence on smart farming, as this has improved functionalities, efficient coverage, low power consumption, and low cost. Successful offerings by the cellular operators on the Internet of Things market can be highly lucrative, giving smart agriculture services in conjunction with LPWA [41].

10.8 FUTURE TRENDS OF IoT IN SUSTAINABLE AGRICULTURE

IoT solutions for smart farming leverage cloud-based computing platforms and scalability to manage and store vast amounts of environmental data collected from sensors. ML can be employed to analyze these large datasets, thereby improving the management capabilities of smart farming. For instance, processing big data can provide insights into crop growth and development, optimize resource usage (such as energy, fertilizers, water, and pesticides), and enhance both crop quality and productivity. Consequently, integrating single-board computers (SBCs) and UAVs with IoT-based sensors is highly recommended for effective crop monitoring in both indoor (greenhouse) and outdoor (field) environments. Computer vision represents a crucial application of IoT in precision agriculture. When integrated with IoT technologies, computer vision facilitates the monitoring and management of crops, evaluates plant health, and supports data-driven decision-making to boost agricultural productivity. A key benefit of employing IoT-based devices equipped with computer imaging is their ability to reduce the reliance on human labor. This enables farmers to effectively manage large areas of land. These sensors can provide insights into which finite natural resources to utilize and the optimal timing for their use. Additionally, they facilitate crop monitoring and soil quality assessment for farmers.

These sensors can provide guidance on which finite natural resources to utilize and the appropriate timing for their application. They also enable farmers to monitor crops and assess soil quality. By leveraging information such as soil quality and weather conditions, farmers can adjust their production strategies accordingly. Additionally, these IoT-based smart sensors and devices can help minimize crop losses and maximize yields from the same plot of land. A variety of network connections, including both wired and wireless options, have been examined for communication between IoT-based devices. It is widely recognized that two common wired networks, such as Controller Area Network (CAN) and Ethernet, are frequently utilized in indoor agriculture, such as greenhouse environments, due to their physical components being less vulnerable to climatic factors. In contrast, wireless connections like Wi-Fi, ZigBee, and LoRa are employed in both greenhouse and field settings. Wi-Fi presents several significant challenges, including high power consumption and limited signal range. In contrast, ZigBee and LoRa are recognized as energy-efficient protocols suitable for communication within wireless networks. Consequently, enhancing algorithms to reduce power consumption during data transfer remains a necessary change.

10.9 CONCLUSION

The effectiveness and applicability of Internet of Things (IoT) solutions in farming address the aforementioned key challenges experienced by the agricultural sector, as well as promote sustainable agriculture. Internet of Things technology provides farmers with an opportunity to acquire firsthand information and trends in several aspects of their business. It integrates responsibilities such as keeping livestock and water, the health of the soil, and the condition of the crop, among others. This approach, which is underlined by data, makes it possible to inform decisions that would enhance resource productivity, gain greater yields from resources, and, consequently, reduce the negative impact on the environment. Problems that must be addressed to achieve IoT's applicability in the field of agriculture are associated with data security, as well as the problem of accessibility for those farmers who live in relatively remote areas. Constant investigation, development, and collaboration among end-users and multiple organizations must undertake crucial steps to optimize the utilization of IoT in agriculture and build a food system that is much stronger, more effective, and more sustainable in the future.

REFERENCES

1. Agarwal, P., Swami, S., & Malhotra, S. K. (2024). Artificial intelligence adoption in the post-COVID-19 new-normal and role of smart technologies in transforming business: a review. *Journal of Science and Technology Policy Management*, *15*(3), 506–529.
2. Batra, I., Sharma, C., Malik, A., Sharma, S., Kaswan, M. S., & Garza-Reyes, J. A. (2024). Industrial revolution and smart farming: a critical analysis of research components in Industry 4.0. *The TQM Journal*, *37*(6), 1497–1525.
3. Zhang, J., & Drury, M. (2024). Sustainable agriculture in the EU and China: A comparative critical policy analysis approach. *Environmental Science & Policy*, *157*, 103789.
4. Fuentes-Peñailillo, F., Gutter, K., Vega, R., & Silva, G. C. (2024). Transformative technologies in digital agriculture: Leveraging Internet of Things, remote sensing, and artificial intelligence for smart crop management. *Journal of Sensor and Actuator Networks*, *13*(4), 39.
5. Salam, A. (2024). Internet of Things in Agricultural Innovation and Security. In Abdul Salam (eds) *Internet of Things for Sustainable Community Development: Wireless Communications, Sensing, and Systems* (pp. 71–112). Springer International Publishing.
6. Askerbekov, D., Garza-Reyes, J. A., Ghatak, R. R., Joshi, R., Kandasamy, J., & de Mattos Nascimento, D. L. (2024). Embracing drones and the Internet of drones systems in manufacturing–An exploration of obstacles. *Technology in Society*, *78*, 102648.
7. Kumar, S., & Sharma, S. C. (2024). Intelligent transportation storage condition assessment system for fruits and vegetables supply chain using internet of things enabled sensor network. *International Journal of System Assurance Engineering and Management*, *16*, 1–15.
8. Swathy, R., Geethalakshmi, V., Pazhanivelan, S., Kannan, P., Annamalai, S., & Hwang, S. (2024). Real-time nitrogen monitoring and management to augment N use efficiency and ecosystem sustainability–A review. *Journal of Hazardous Materials Advances*, *16*, 100466.
9. Vekariya, P., Kumari, P., Ghetiya, R., & Kumar, M. S. (2024). Empowering Agriculture: IoT for Sustainable Farming Practices. In Potluri, S., Satpathy, S., Basa, S. S., Zuorro, A., (eds) *AI in Agriculture for Sustainable and Economic Management* (pp. 119–140). CRC Press.
10. Savaliya, M., & Raj, M. (2024). Advancement in Multisensor Remote Sensing Studies for Assessing Crop Health. In Datta, D., Kundu, A., Patel, N. R., *Vegetation Dynamics and Crop Stress* (pp. 93–112). Academic Press.
11. Vardhan, V. H., Lavanya, M., & Sowmyavani, M. (2024). Application of Machine Learning Approaches in Crop Management. In Malik, B. B., Mukherjee, G., Kar, R., Shaw, K. S., Mukherjee, A., (eds) *Green Industrial Applications of Artificial Intelligence and Internet of Things* (pp. 199–210). Bentham Science Publishers.
12. Soussi, A., Zero, E., Sacile, R., Trinchero, D., & Fossa, M. (2024). Smart sensors and smart data for precision agriculture: a review. *Sensors*, *24*(8), 2647.
13. Swain, C. K., Chatterjee, D., Nayak, A. K., Mohapatra, K. K., Sahoo, R., Pradhan, A., & Singh, N. R. (2024). Greenhouse Gas and Energy Flux Measurements with Eddy Covariance Technique Under Lowland Rice Ecology. In: Pathak, H., Chatterjee, D., Saha, S., Das, B. (eds) *Climate Change Impacts on Soil-Plant-Atmosphere Continuum* (pp. 631–666). Springer Nature Singapore.
14. Ameer, S., Ibrahim, H., Kulsoom, F. N. U., Ameer, G., & Sher, M. (2024). Real-time detection and measurements of nitrogen, phosphorous & potassium from soil samples: A comprehensive review. *Journal of Soils and Sediments*, *24*(7), 2565–2583.
15. Palarimath, S., Maran, P., Balakumar, C., & Sujatha, T. (2024, April). Exploring Sensor-based Smart Farming Technologies in the Internet of Things (IoT). In *2024 International Conference on Computing and Data Science (ICCDS)*, Chennai, India, (pp. 1–6). IEEE.

16. Vinod Chandra, S. S., Hareendran, A., & Albaaji, G. F. (2024). Precision farming for sustainability: An agricultural intelligence model. *Computers and Electronics in Agriculture*, *226*, 109386.
17. Nannar, S., Sindhuchao, S., Chaiyaphan, C., & Ransikarbum, K. (2024). Optimization of the sustainable food supply chain with an integrative data envelopment analysis approach. *International Journal of Management Science and Engineering Management*, *19*(4), 278–293.
18. Taseer, A., & Han, X. (2024). Advancements in variable rate spraying for precise spray requirements in precision agriculture using Unmanned aerial spraying Systems: A review. *Computers and Electronics in Agriculture*, *219*, 108841.
19. Parida, P. K., Somasundaram, E., Krishnan, R., Radhamani, S., Sivakumar, U., Parameswari, E., & Gangai Selvi, R. (2024). Unmanned aerial vehicle-measured multispectral vegetation indices for predicting LAI, SPAD chlorophyll, and yield of maize. *Agriculture; Basel*, *14*(7), p. 1110.
20. Liu, X., Du, R., Xiang, Y., Chen, J., Zhang, F., Shi, H., & Wang, X. (2024). Estimating winter canola aboveground biomass from hyperspectral images using narrowband spectra-texture features and machine learning. *Plants*, *13*(21), 2978.
21. Getahun, S., Kefale, H., & Gelaye, Y. (2024). Application of precision agriculture technologies for sustainable crop production and environmental sustainability: A systematic review. *The Scientific World Journal*, *2024*(1), 2126734.
22. Nyakuri, J. P., Nkundineza, C., Gatera, O., & Nkurikiyeyezu, K. (2024). State-of-the-art deep learning algorithms for internet of things-based detection of crop pests and diseases: A comprehensive review. *IEEE Access*, *12*, 169824–169849.
23. Banerjee, S., Sarda, K., Khandelwal, S., Gajarushi, A., Gawande, S., Velmurugan, R., & Baghini, M. S. (2024, January). IoT-based Sensing System for Thrips Pest and Disease Management in Onion Crop. In *2024 IEEE Applied Sensing Conference (APSCON)*, Goa, India, (pp. 1–4). IEEE.
24. Lata, M. A., Khatun, T., Sultana, M., Badhan, I. A., Ahmed, T., & Hasan, M. R. (2024). Development of an IoT-based smart potato leaf disease monitoring and controlling system with image processing. *TELKOMNIKA (Telecommunication Computing Electronics and Control)*, *22*(6), 1502–1510.
25. Arafat, I. S., Aswath, S., & Nishath, S. H. (2024). Automatic early detection of tomato leaf disease using IoT and deep learning.
26. Ahmed, I., & Yadav, P. K. (2024). Predicting apple plant diseases in orchards using machine learning and deep learning algorithms. *SN Computer Science*, *5*(6), 700.
27. Al-Otaibi, S., Khan, R., Ali, J., & Ahmed, A. (2024). Artificial intelligence and Internet of Things-enabled decision support system for the prediction of bacterial stalk root disease in maize crop. *Computational Intelligence*, *40*(1), e12632.
28. Mustofa, S., Ahad, M. T., Emon, Y. R., & Sarker, A. (2024). BDPapayaLeaf: A dataset of Papaya leaf for disease detection, classification, and analysis. *Data in Brief*, *57*, 110910.
29. Manoj, R., Girish, S., Pushpa, B. R., Rani, N. S., & Sangamesha, D. (2024, July). Early-Stage Disease Prediction in Chilli Plant Using YOLO Models. In *2024 Second International Conference on Advances in Information Technology (ICAIT)*, Chikkamagaluru, Karnataka, India, (Vol. 1, pp. 1–9). IEEE.
30. Mirjat, R. M., Mahar, S. A., Siddiqui, M. I. M., Mahar, J. A. M., & Magsi, A. M. (2024). A framework for guava wilt disease segmentation using k-means clustering and neural network techniques. *VAWKUM Transactions on Computer Sciences*, *12*(1), 76–93.
31. Lu, J., Wu, Z., Lan, Y., Deng, X., & Huang, J. (2024). Study on the prediction model of litchi downy blight damage based on IoT and hyperspectral data fusion. *IEEE Internet of Things Journal*, *11*(16), 27184–27200.

32. Hitimana, E., Kuradusenge, M., Sinayobye, O. J., Ufitinema, C., Mukamugema, J., Murangira, T., & Ngabonziza, J. (2024). Revolutionizing coffee farming: A mobile App with GPS-enabled reporting for rapid and accurate on-site detection of coffee leaf diseases using integrated deep learning. *Software*, *3*(2), 146–168.
33. Kumar, V., Sharma, K. V., Kedam, N., Patel, A., Kate, T. R., & Rathnayake, U. (2024). A comprehensive review of smart and sustainable agriculture using IoT technologies. *Smart Agricultural Technology*, *8*, 100487.
34. Sharma, V., Honkavaara, E., Hayden, M., & Kant, S. (2024). UAV remote sensing phenotyping of wheat collection for response to water stress and yield prediction using machine learning. *Plant Stress*, *12*, 100464.
35. Ghiasi, M., Wang, Z., Mehrandezh, M., & Paranjape, R. (2023). A systematic review of optimal and practical methods in design, construction, control, energy management, and operation of smart greenhouses. *IEEE Access*, *12*, 2830–2853.
36. Kataray, T., Nitesh, B., Yarram, B., Sinha, S., Cuce, E., Shaik, S., & Roy, A. (2023). Integration of smart grid with renewable energy sources: Opportunities and challenges–A comprehensive review. *Sustainable Energy Technologies and Assessments*, *58*, 103363.
37. Soori, M., Arezoo, B., & Dastres, R. (2023). Machine learning and artificial intelligence in CNC machine tools, a review. *Sustainable Manufacturing and Service Economics*, *2*, 100009.
38. Schmidt, D., Butturi, M. A., & Sellitto, M. A. (2023). Opportunities of digital transformation in post-harvest activities: A single case study of an engineering solutions provider. *AgriEngineering*, *5*(3), 1226–1242.
39. Wang, W., Guo, Q., Yang, Z., Jiang, Y., & Xu, J. (2023). A state-of-the-art review on robotic milling of complex parts with high efficiency and precision. *Robotics and Computer-Integrated Manufacturing*, *79*, 102436.
40. Onyeneke, R. U., Ankrah, D. A., Atta-Ankomah, R., Agyarko, F. F., Onyeneke, C. J., & Nejad, J. G. (2023). Information and communication technologies and agricultural production: new evidence from Africa. *Applied Sciences*, *13*(6), 3918.
41. Majumdar, P., Mitra, S., Bhattacharya, D., & Bhushan, B. (2024). Enhancing sustainable 5G powered agriculture 4.0: Summary of low power connectivity, internet of UAV things, AI solutions and research trends. *Multimedia Tools and Applications*, *84*, 1–45.

11 Data Security in IoT-Enabled Smart Farming Monitoring Frameworks

Sofia Singla, Navdeep Singh Sodhi, Isha Batra, Arun Malik, and Ghadah Naif Alwakid

11.1 INTRODUCTION

Agriculture has undergone a great enhancement across the entire globe. Moreover, it has now converted into smart agriculture [1]. Farmers are utilizing the systematic processes of crop rotations as well as other multiple modifications in the conventional farming practices with the integration of Information and Communication Technology (ICT) in agriculture [2]. Smart farming or smart agriculture is defined as handling all of the farming practices by using modern and enhanced technologies like Internet of Things (IoT), machine learning (ML), deep learning, blockchain, and many more. It basically helps in showing improvement in the quality as well as quantity of products, along with optimization of human labor. By considering these types of technologies, farmers are basically utilizing data analytics processes, robotics or robotic process automation, location-based practices, connectivity using models like cellular model, LoRa model, software, sensors for the measurement of temperature, humidity, light, water, and soil [3].

Integration of modern technologies and agriculture has decreased numerous challenges for farmers. Data are now gathered by using smart sensors and other practices. Those data are further recorded in a digitized manner using technologies [1]. These kinds of practices help in ensuring more generalized work plan management. It has also reduced the requirement of manual labor, improved work culture and quality in crop productivity, and increased sustainable farming. There are three different factors of sustainable agriculture, that is, society, energy, and environment. Sustainable farming refers to serving all basic requirements of society without showing any of the negotiation between environmental factors and energy, like controlling deficiencies in yields, water harvesting, recycling, etc. The seven best practices of sustainable agriculture are integration of pest management practices, crop diversity and crop rotation, renewable energy sources, water management, soil management, organic farming, and agroforestry, as shown in Figure 11.1.

DOI: 10.1201/9781003594093-11

Modern technologies have completely reshaped the entire farming process. It is not wrong to state that smart farming can be defined as a novel concept of modernized farm management [4]. It helps to handle all of the agricultural practices like pest management, soil management, fertilizer management, water irrigation management, and crop health monitoring by using automated software embedded with smart physical objects. As shown in Figure 11.2, smart farming architecture provides the crop yield to the farmer by undergoing all processes, utilizing technologies or physical objects embedded with the Internet. Moreover, the smart farming concept is also effectively applicable to small farms rather than just to large companies with the help of enhanced technologies, i.e. IoT, ML, deep learning, blockchain, and many more. Despite this, deploying IoT-driven security mechanisms is still a challenging constraint in smart farming. In this, a trade-off is also seen due to cloud computing techniques, as it enhances communication time along with a reduction in the requirement of bandwidth. Researchers keep discuss the frameworks based on the better and improved analysis rate rather than considering the concerned security approaches [5]. So, it is also seen that smart farming is required to ensure cloud security without showing any major issues toward the overall performance of the system. Further, this book chapter identifies major research gaps in the context of trade-offs between data security and accurate monitoring in smart farming. Smart systems based on technologies like the IoT process a huge volume of real-time data and apply secure encryption mechanisms over the data [6]. It generally causes a delay in the overall processing of data. It has addressed how to handle data sharing agreements, privacy, and ownership among the different parties.

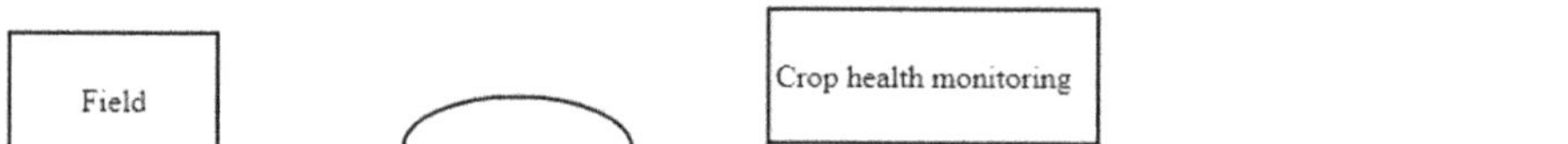

FIGURE 11.1 Sustainable agriculture.

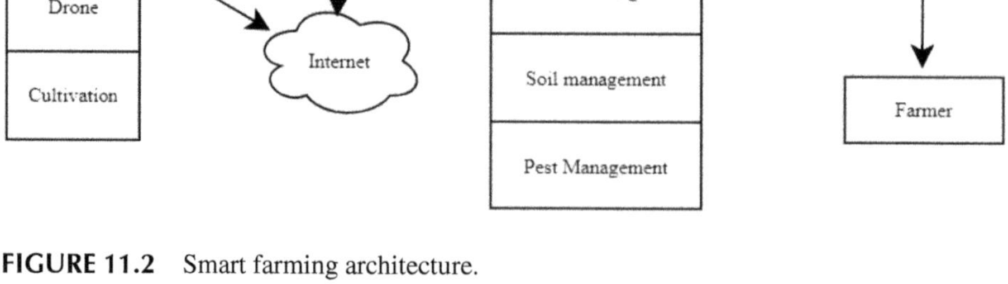

FIGURE 11.2 Smart farming architecture.

11.2 OBJECTIVES OF THE CHAPTER

The major objectives of this chapter are as follows:

- To review the existing frameworks for performing smart farming using enhanced technologies like the IoT, ML, deep learning, and artificial intelligence (AI)
- To compare all reviewed frameworks based on the frameworks, agricultural monitoring type, advantages, disadvantages, and outcomes
- To analyze all previously compared frameworks based on their performance in terms of accuracy
- To find a research gap among all of the existing frameworks for future research directions

11.3 RELATED WORK

This section reviews existing frameworks based on the type of agricultural monitoring, along with their advantages, disadvantages, and outcomes. Further, it includes a review related to the accuracy related to the implementation of the Intrusion Detection System in smart farming. Crop monitoring is really required to scrutinize in order to ensure the exponential growth in the farming economy. IoT technology has totally converted agriculture into smart agriculture. Earlier, traditional agriculture methodologies were utilized to gather the data and communicate with the numerous sources regarding the requirement for farmland. However, those methodologies were not able to provide the best possible solution for the betterment of agriculture and farmers. In this context, researchers had proposed a framework for smart farming based on IoT technology. It is completely focused on plant disease detection as well as water irrigation. Further, it also integrates ML algorithms for the automated identification of pests and the required level of water on the farmland. It is found that this respective framework was able to provide an overall accuracy of 84%[7]. But this accuracy is seen to be very low, so it is not wrong to say that this framework was not able to provide the best performance to farmers.

Later on, a novel framework was proposed by researchers using 6G-enabled Internet of Things (6G-IoT) and artificial intelligence (AI) for an autonomous irrigation system [8]. This framework was dependent on the prediction algorithm for climate change and unpredictable rainfall patterns. The main advantage of this framework is that it not only analyzes the environmental conditions but also measures the soil's moisture content. The accuracy found by this framework was 86.34% [8]. The major drawback is that this framework hasn't discussed the security of data related to environmental conditions. Researchers had proposed a sustainable and smart framework by integrating two different technologies, that is, deep learning and IoT [9]. This framework is primarily a disease recognition model that gathers data from multiple smart sources. The performance of the framework is developed in terms of accuracy (98.8%), precision (98.2%), recall (97.4%), and F-score (97.7%) [9].

Further, a framework was proposed by researchers by integrating camera calibration and geometric ranging [10]. Researchers had proposed the YOLOv8 framework in order to range and identify the range of outdoor cattle. Overall, the framework is able to achieve 95% accuracy. Moreover, it is also evaluated that accuracy seems to be facing a very small loss in accuracy, that is, 0.31%, with the variation in height [10]. A novel framework using an Unmanned Aerial Vehicle sensing system is proposed to monitor as well as identify the health of crops [11]. This study has introduced new and smart techniques to reduce the cost as well as improve the actual quality and yield of the crop. The results have been evaluated in terms of accuracy, precision, recall, and F-1 score [11]. Three different types of datasets have been utilized, and the respective framework is able to provide the best and most efficient results over each utilized dataset. However, it is found that a very small sample size is used for the entire study. Reliable and cost-effective decision-making systems are required to be introduced in the smart agriculture sector. Later on, an IoT framework based on a Convolutional Neural Network (CNN) model was proposed for the control of strategic crop growth, including beans, tomatoes, peppers, and cucumbers. IoT helped in collecting the data from numerous sensors. The overall performance of the framework is about 97.56% in terms of accuracy [12]. Further, a framework for the improvement of crop productivity using IoT technology has been proposed. This framework is integrated with an AI model. It gathers the data from the remote sensors for precise prediction of the growth of crops in smart farming. IoT technology is utilized to gather as well as process the data for smart farming. ML frameworks are utilized for précised prediction as well as monitoring the productivity of the crops. Further, blockchain technology helps in ensuring the end-to-end data security collected from the remote sensors in smart farming. It achieved the highest accuracy of 97%, with reduced cost and data security [13].

In this respect only, researchers have proposed a hybrid model for the agriculture sector based on the IoT and ML techniques. It is seen that the model is achieving an overall 97% of the accuracy while monitoring soil type [14]. However, it is also found that this model is unable to work with reliable and efficient data augmentation technologies. Rather than crop monitoring, water level monitoring, and other studies, food security is one of the greatest areas to address when discussing smart agriculture. Advanced technologies like IoT, ML, deep learning, and blockchain are utilized by the farmers and researchers to ensure best practices and services for the entire population. In this context, researchers had proposed a framework based on the pest management system and tomato cultivation considering IoT technology [15]. This framework integrates the moisture sensors and camera module for the collection of data from numerous sources, like field soil and tomato plants. Findings of this respective study showed that this proposed framework can successfully achieve an overall accuracy of 94% [15]. It is also found that this framework is not able to work for the large-scale sector. No doubt, automatic systems have been implemented along with classification as well as detection of other characteristics in the framework, but it is lacking in its rationale to provide a hundred percent of the best performance. Maintaining water quality is also considered one of the most critical factors in smart shrimp farming. Recently, researchers have proposed a framework integrating technologies like ML algorithms, web applications, cloud storage, IoT, and physical devices based on the microcontroller system that results in a six-layer model [16]. These six layers comprise physical layer, network layer, cloud layer, service layer,

ML layer, and application layer. It is found that this framework is able to provide real-time data visualization on shrimp farming with 97.84% accuracy [16]. In order to ensure more improvement in the accuracy of detection and monitoring of agricultural health and related strategies, researchers have recently proposed multiple frameworks. The proposed framework discussed the methodology to ensure reliable data analysis and communication using Isolation Forest [17]. It had proved the effective identification of any kind of data anomaly in the aspects of humidity and temperature. It is found that modern farming is able to achieve better results in terms of efficiency, productivity, and other related factors [17]. Further, integration of Industry 4.0 with smart farming can introduce automation, data-driven decision-making strategies, and digitization among the traditional sectors. Researchers had discussed the actual research patterns followed in the previous research regarding smart farming and industrial revolution by using the Latent Dirichlet Allocation (LDA) methodology [18]. It is found that this revolution can show actual adoption of smart agricultural practices.

Rather than just discussing the smart farming frameworks in order to monitor numerous characteristics like crop productivity, disease, water irrigation level, soil moisture, pest management, and farm health, it is very important to review and analyze the security of the framework. Numerous vulnerabilities related to data security are introduced with the enhancement of technologies in the agriculture sector. In this regard, a concise review has been conducted based on the security of smart farming frameworks using secondary data sources. The IoT has also integrated numerous cryptographic frameworks for data security in each sector. Discussing smart farming, Logical Security Framework for Offering Security in Internet of Things (LLSFIoT) can be considered as one of the most effective frameworks. It has involved three different phases, including data security, authentication, and registration, for enhanced solutions to all data breaches [19]. Researchers had given a commendable proposal regarding the security of frameworks by developing intrusion detection systems. A novel IoT-based intelligent agriculture framework was developed for secure data sharing among the sources [20]. It basically includes the authentication as well as key management scheme for the smart and secure transmission of data using IoT-enabled frameworks. It also includes the smart contract systems to only execute the legitimate transmission. This verification is performed with the help of blockchain technology. This framework has shown an overall accuracy of 98.33% as compared to other baseline algorithms [20]. Moreover, in the view of smart agriculture, researchers have proposed an Intrusion Detection System IDS framework in the combination of Gaussian Mixture Model (GMM) and Multilayer Perceptron (MLP) for enhanced security. This study utilized the CIC-DDoS2019 dataset and was able to achieve 99.99% accuracy [21]. However, it had not considered characteristics like memory consumption, real-time network traffic patterns, and time consumption in the detection of new attacks.

Recently, a framework named Enhanced Black Widow Optimization with Hybrid Deep Learning Enabled Intrusion Detection (EBWO-HDLID) has been proposed to detect as well as capture the specific attacks or threats for increased reliability and security of smart agriculture. This framework is able to provide satisfactory results by achieving 98.81% of accuracy as performance [22]. It is also seen that an enhanced and hybrid framework using a truncated back-propagation mechanism has been proposed in order to manage the large data stream. In this, an attack scenario was considered so that the actual performance of the framework can be evaluated. It is found that overall, 99.82% accuracy has been achieved [23]. Further, an enhanced and secure framework named Downsized Kernel Partial Least Squares (DKPLS) was developed to identify cyber-attacks. An overall accuracy of 99.92%has been found by this proposed framework [24]. Recently, a novel framework has been proposed by researchers regarding the security of IoT-enabled devices utilizing Generative Artificial Intelligence (GenAI) [25]. This study can be found as one of the most progressive among all reviewed frameworks due to the illustration of real-time use cases for precise optimization of cybersecurity strategies, data privacy mechanisms, and related issues like energy consumption. It is found that GenAI can be able to ensure improved IoT security and reduced threats of cyber-attacks [25].

11.4 COMPARATIVE ANALYSIS

This analysis shows a comparison among the existing frameworks regarding the smart farming utilized for agricultural monitoring, along with its respective advantages, disadvantages, and outcomes, as shown in Table 11.1. Moreover, an accuracy analysis in the context of data security given by existing frameworks is also discussed, as shown in Table 11.2. This concise security analysis is discussed to have a clear understanding of the recent studies on the security of data communicated among multiple sources.

11.5 RESULTS AND DISCUSSION

Two different comparisons have shown, based on the accuracy related to agricultural monitoring and data security in Tables 11.1 and 11.2. In this context only, Figure 11.3 shows the actual curve of accuracy achieved by existing frameworks for agricultural monitoring. This graphical result is only found based on the recently proposed frameworks in 2023 and 2024. It is seen that the accuracy percentage was very low, that is, 84% at the start of the year 2023. Later on, it showed improved results with an increase of nearly 2% in the accuracy. However, accuracy had increased exponentially with the percentages of 98.80%, 95%, 97.56%, and 97.56%. These higher levels of accuracy percentages are still unable to provide the best framework to farmers in the context of smart farming due to multiple drawbacks, as discussed in the comparative analysis. These frameworks are seen to be facing consistent improvement and deterioration in the year 2024. At the end, it is found that smart farming frameworks are able to provide the best accuracy percentages in the context of agricultural monitoring, but still, data security is completely compromised by researchers in each existing framework.

Discussing the accuracy in the context of data security from the existing framework, it is found that researchers are progressively proposing frameworks for the security of agricultural data from the year 2023. In the year 2023, the accuracy percentage has reached 99.99, but analysis shows that there was a major issue with the considered dataset, as shown in Figure 11.4. The size of the dataset can't actually be

TABLE 11.1
Comparison Among Existing Frameworks Based on Agricultural Monitoring, Advantages, Disadvantages, and Outcomes

Ref.	Year	Framework	Agricultural Monitoring	Advantages	Disadvantages	Outcome
[7]	2023	Smart farming using IoT and machine learning	Disease detection and water irrigation monitoring	The framework integrates the machine learning algorithms for the automated identification of pests and the required level of water on the farmland.	Collected data is directly sent for further categorization. No intermediate processing over the data is considered in order to remove outliers from the data or clean the data.	The respective framework was able to provide an overall accuracy of 84%.
[8]	2023	6G-IoT (6G-enabled Internet of Things) and AI (Artificial Intelligence) framework	Autonomous irrigation system	It has analyzed the environmental conditions along with the prediction of rainfall patterns and other climatic changes.	This framework hadn't discussed the security of data related to environmental conditions.	The respective framework was able to provide an overall accuracy of 86.34%.
[9]	2023	Customized CNN (Convolutional Neural Network) framework	Crop health	Smart sensors are utilized to track the actual crop health by assessing the soil moisture level, temperature, and humidity.	Researchers haven't discussed gathering real-time data from multiple angles, i.e., remote sites in the field.	Results are evaluated in terms of accuracy (98.8%), precision (98.2%), recall (97.4%), and F-score (97.7%).
[10]	2023	YOLOv8	Range of outdoor cattle from the fields	This study can be introduced as cost-effective agriculture based on smart technologies, along with better range management of cattle.	No dynamic ground monitoring system is seen in the framework.	This framework is able to achieve 95% accuracy. Moreover, it is also evaluated that accuracy seems to be facing a very small loss in accuracy, i.e., 0.31%, with the variation in height.

(Continued)

TABLE 11.1 (continued)

Ref.	Year	Framework	Agricultural Monitoring	Advantages	Disadvantages	Outcome
[12]	2023	Farming adaptation framework based on IoT integrated with a CNN (Convolutional Neural Network) deep learning model	Monitoring of strategic crops' growth	The major advantage of this framework is that it helps in generalizing information regarding the right amount of temperature, light, water, and fertilizer.	No data security model was implemented over the data collected from the numerous real-time smart sources.	The respective framework was able to provide an overall accuracy of 97.56%.
[13]	2023	Smart use based on IoT, machine learning, and blockchain	Crop productivity	IoT, machine learning, and blockchain technology have integrated together to provide precise and secure data collection, along with the evaluation of latency, throughput, execution time, and gas consumption.	The framework established a safe, profitable, and sustainable environment for agriculture, but still, security over stored data is not effectively considered. It had only discussed end-to-end communication security but not about the storage data at the server level.	Framework had achieved the highest accuracy at 97% with reduced cost and end-to-end data security.
[14]	2024	IoT-enabled hybrid model based on deep learning and machine learning	Soil type monitoring	Decision-making and a cost-effective, smart hybrid framework are introduced in the modern farming sector.	While monitoring the soil type, other related environmental conditions like cyber-attack over data and, privacy of data have not been considered.	This model is able to achieve an overall accuracy of 97% while monitoring soil type.
[15]	2024	IoT-based model using machine learning and deep learning methods	Pest management and tomato cultivation	IoT integration in the framework helped in precise transmitting and capturing of data from numerous sources.	Data security is essential to consider while connecting to the internet and cloud sources. This is not considered by this framework.	Findings of this respective study showed that this proposed framework can successfully achieve an overall accuracy of 94%.
[16]	2024	IoT-based monitoring in the integration of a machine learning model	Shrimp farm monitoring	The proposed framework also worked as per farmers' feedback so that further smart automation in terms of disease detection can be ensured.	Real-time insights for farm condition, timely intervention, and proactive management are really missing in the framework.	This framework is able to provide real-time data visualization on shrimp farming, along with 97.84% accuracy.

TABLE 11.2
Comparison of the Accuracy of Existing Frameworks in the Context of Data Security

Ref.	Year	Framework	Accuracy (%)	Findings
[20]	2023	Secure data sharing based on IoT and a smart contract mechanism	98.33	Performance has not been evaluated in terms of latency and scalability while dealing with real-time data.
[21]	2023	GMLP-IDS	99.99	The framework had not considered characteristics like memory consumption, real-time network traffic patterns, and time consumption in the detection of new attacks.
[22]	2024	Enhanced Black Widow Optimization with a Hybrid Deep Learning Enabled Intrusion Detection Framework	98.81	Performance has not been evaluated in terms of latency while dealing with real-time data.
[23]	2024	Time-Based Intrusion Detection System Framework	99.82	The entire performance is based on a considered attack scenario, and this framework can only be utilized for those attacks that will show similar signatures in the future.
[24]	2024	Downsized Kernel Partial Least Squares framework	99.92	This framework is basically proposed to handle cyber-attacks in the future.

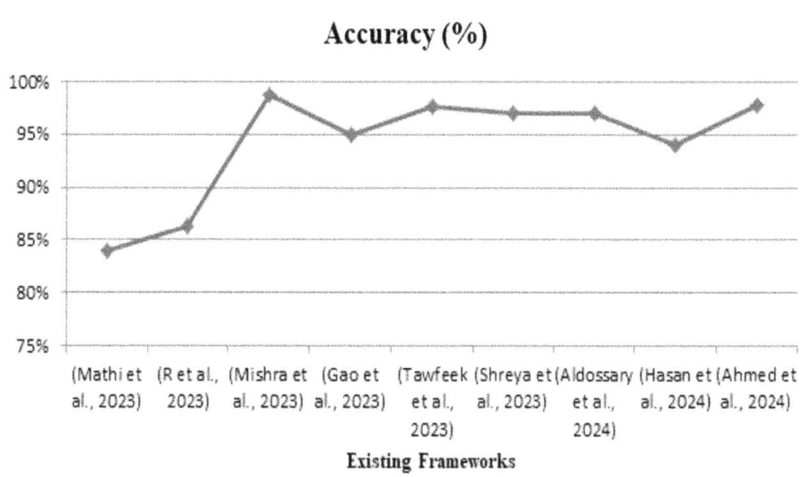

FIGURE 11.3 Curve of accuracy of existing frameworks for agricultural monitoring.

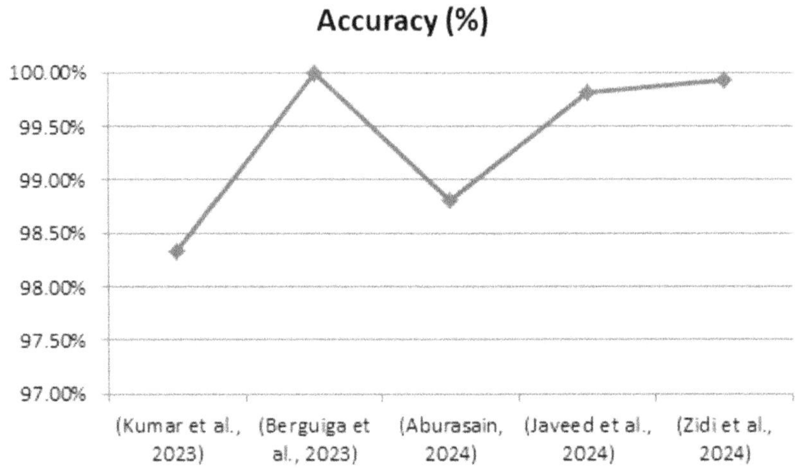

FIGURE 11.4 Curve of accuracy of recent frameworks for agricultural monitoring.

considered on a large scale. In the year 2024, accuracy was achieved with an effective percentage. Moreover, this framework can be considered the best for data security as researchers have discussed a framework for the cyber-attacks and threats that can occur in the future.

11.6 CONCLUSION AND FUTURE DIRECTIONS

Exponential and revolutionary improvement in agriculture using IoT technology has drawn great attention among researchers across the globe. Nowadays, researchers, cybersecurity professionals, and developers are working simultaneously to provide novel and secure agriculture frameworks with enhanced performance in all aspects. In this book chapter, it is concluded that existing smart agricultural frameworks are working effectively and ensuring improved accuracy while providing results to farmers, but integration of smart farming with data security along the specific agricultural monitoring types is completely missing. Moreover, if any of the existing frameworks have achieved a better accuracy percentage, then it is found that researchers had utilized an insufficient dataset, that is, the size of the dataset is small. Hence, it is very crucial to maintain agricultural data security while conducting any kind of monitoring in the field in the future.

REFERENCES

1. Sushanth, G., & Sujatha, S. (2018, March). IOT based smart agriculture system. In *2018 International Conference on Wireless Communications, Signal Processing and Networking (WiSPNET)* (pp. 1–4). IEEE.
2. Walter, A., Finger, R., Huber, R., & Buchmann, N. (2017). Smart farming is key to developing sustainable agriculture. *Proceedings of the National Academy of Sciences, 114*(24), 6148–6150.

3. Prathibha, S. R., Hongal, A., & Jyothi, M. P. (2017, March). IoT based monitoring system in smart agriculture. In *2017 International Conference on Recent Advances in Electronics and Communication Technology (ICRAECT)* (pp. 81–84). IEEE.
4. Patil, K. A., & Kale, N. R. (2016, December). A model for smart agriculture using IoT. In *2016 International Conference on Global Trends in Signal Processing, Information Computing and Communication (ICGTSPICC)* (pp. 543–545). IEEE.
5. Krisnawijaya, N. N. K., Tekinerdogan, B., Catal, C., van der Tol, R., & Herdiyeni, Y. (2025). Implementing FAIR principles in data management systems: A multi-case study in precision farming. *Computers and Electronics in Agriculture*, *230*, 109855.
6. Manogaran, G., Alazab, M., Muhammad, K., & De Albuquerque, V. H. C. (2021). Smart sensing based functional control for reducing uncertainties in agricultural farm data analysis. *IEEE Sensors Journal*, *21*(16), 17469–17478.
7. Mathi, S., Akshaya, R., & Sreejith, K. (2023). An internet of things-based efficient solution for smart farming. *Procedia Computer Science*, *218*, 2806–2819.
8. Sitharthan, R., Rajesh, M., Vimal, S., Kumar, S., Yuvaraj, S., Kumar, A., … Vengatesan, K. (2023). A novel autonomous irrigation system for smart agriculture using AI and 6G enabled IoT network. *Microprocessors and Microsystems*, *101*, 104905.
9. Mishra, S., Volety, D. R., Bohra, N., Alfarhood, S., & Safran, M. (2023). A smart and sustainable framework for millet crop monitoring equipped with disease detection using enhanced predictive intelligence. *Alexandria Engineering Journal*, *83*, 298–306.
10. Gao, T., Li, M., Xue, L., Bao, J., Lian, H., Li, T., & Shi, Y. (2023). Height-variable monocular vision ranging technology for smart agriculture. *IEEE Access*, *11*, 92847–92856.
11. Eladl, S. G., Haikal, A. Y., Saafan, M. M., & ZainEldin, H. Y. (2024). A proposed plant classification framework for smart agricultural applications using UAV images and artificial intelligence techniques. *Alexandria Engineering Journal*, *109*, 466–481.
12. Tawfeek, M. A., Yanes, N., Jamel, L., Aldehim, G., & Mahmood, M. A. (2023). Adaptive deep learning model to enhance smart greenhouse agriculture. *CMC-Computer Mater & Continua*, *77*(2), 2545–2564.
13. Shreya, S., Chatterjee, K., & Singh, A. (2023). BFSF: A secure IoT based framework for smart farming using blockchain. *Sustainable Computing: Informatics and Systems*, *40*, 100917.
14. Aldossary, M., Alharbi, H. A., & Hassan, C. A. U. (2024). Internet of Things (IoT)-enabled machine learning models for efficient monitoring of smart agriculture. *IEEE Access*, *12*, 75718–75734.
15. Hasan, M. R., Rahman, M. M., Shahriar, F., Khan, M. S. I., Uddin, K. M. M., & Hasan, M. M. (2024). Smart farming: Leveraging IoT and deep learning for sustainable tomato cultivation and pest management. *Crop Design*, *3*(4), 100079.
16. Ahmed, F., Bijoy, M. H. I., Hemal, H. R., & Noori, S. R. H. (2024). Smart aquaculture analytics: Enhancing shrimp farming in Bangladesh through real-time IoT monitoring and predictive machine learning analysis. *Heliyon*, *10*(17).
17. Alumfareh, M. F., Humayun, M., Ahmad, Z., & Khan, A. (2024). An intelligent LoRaWAN-based IoT device for monitoring and control solutions in smart farming through anomaly detection integrated with unsupervised machine learning. *IEEE Access*, *12*, 119072–119086.
18. Batra, I., Sharma, C., Malik, A., Sharma, S., Kaswan, M. S., & Garza-Reyes, J. A. (2024). Industrial revolution and smart farming: a critical analysis of research components in Industry 4.0. *The TQM Journal*, *37*(6), 1497–1525..
19. Batra, I., Hamatta, H. S., Malik, A., Baz, M., Albogamy, F. R., Goyal, V., & Alshamrani, S. S. (2021). LLSFIoT: Lightweight logical security framework for internet of things. *Wireless Communications and Mobile Computing*, *2021*(1), 8526206.

20. Kumar, R., Kumar, P., Aljuhani, A., Islam, A. N., Jolfaei, A., & Garg, S. (2022). Deep learning and smart contract-assisted secure data sharing for IoT-based intelligent agriculture. *IEEE Intelligent Systems*, *38*(4), 42–51.
21. Berguiga, A., Harchay, A., Massaoudi, A., Ayed, M. B., & Belmabrouk, H. (2023). GMLP-IDS: A novel deep learning-based intrusion detection system for smart agriculture. *Computers, Materials & Continua*, *77*(1).
22. Aburasain, R. Y. (2024). Enhanced black widow optimization with hybrid deep learning enabled intrusion detection in Internet of Things-based smart farming. *IEEE Access*, *12*, 16621–16631.
23. Javeed, D., Gao, T., Saeed, M. S., & Kumar, P. (2023). An intrusion detection system for edge-envisioned smart agriculture in extreme environment. *IEEE Internet of Things Journal*, *11*(16), 26866–26876.
24. Zidi, K., Abdellafou, K. B., Aljuhani, A., Taouali, O., & Harkat, M. F. (2024). Novel intrusion detection system based on a downsized kernel method for cybersecurity in smart agriculture. *Engineering Applications of Artificial Intelligence*, *133*, 108579.
25. Khan, A., Jhanjhi, N., Abdulhabeb, G. A. A., Ray, S. K., Ghazanfar, M. A., & Humayun, M. (2025). Securing IoT devices using generative AI techniques. In *Reshaping CyberSecurity With Generative AI Techniques* (pp. 219–264). IGI Global.

12 DeepPest ID

Enhancing Agricultural Pest Classification with Vision Transformers and CNNs

Dhiviya Bharathi Duraimanickam, Ajantha Vijayakumar, Joseph Abraham Sundar Koilraj, Muthaiah Rajappa, and Elakkiya Rajasekar

12.1 INTRODUCTION

Agriculture stands as a cornerstone of global food security and economic stability, yet it faces persistent challenges from pest infestations that can devastate crops and compromise yields. The Food and Agriculture Organization (FAO) estimates that pests destroy up to 40% of global crop production annually, resulting in losses exceeding $220 billion [1]. This significant threat to food security is further intensified by climate change, which has altered pest distributions and behavior patterns, making traditional identification methods increasingly inadequate. The imperative for effective pest management has never been more critical, particularly as the global population continues to grow, necessitating a 60% increase in food production by 2050 [2]. Traditional pest identification methods, which often rely on manual inspection and expert knowledge, are time-consuming, labor-intensive, and frequently subject to human error. Furthermore, the increasing push for sustainable agriculture and reduced pesticide use demands more precise and timely pest identification to enable targeted interventions. Recent advances in artificial intelligence (AI), particularly in deep learning and computer vision, have opened new possibilities for automated pest identification and classification. These technologies offer the potential for rapid, accurate, and scalable pest monitoring systems that can revolutionize integrated pest management (IPM) strategies. Deep learning models, with their ability to learn complex patterns from large datasets, present a promising solution for real-time pest identification, enabling farmers to implement timely and precise control measures while minimizing unnecessary pesticide application [3].

The integration of deep learning in agricultural pest management aligns with the broader trend toward smart agriculture and precision farming. As farming systems become increasingly digitized and data-driven, automated pest classification systems serve as crucial components in the technological ecosystem that supports sustainable agricultural practices. This digital transformation in pest management not only

DOI: 10.1201/9781003594093-12

addresses immediate crop protection needs but also contributes to broader goals of environmental sustainability and food security through optimized resource utilization and reduced chemical inputs.

Key Points of the Research

- Investigates the comparative effectiveness of AlexNet, ResNet50, VGG16, and Vision Transformer (ViT) in agricultural pest classification.
- Emphasizes the importance of data preprocessing, including augmentation and normalization, to enhance model performance.
- Aims to bridge the gap between theoretical deep learning capabilities and practical agricultural applications.
- Seeks to contribute to more efficient and sustainable farming practices.
- Addresses food security challenges through improved automated pest classification solutions.

12.2 LITERATURE SURVEY

Recent advancements in deep learning have significantly improved agricultural pest classification, offering automated solutions for early detection and control. Various studies have explored different architectures, datasets, and techniques to enhance classification accuracy and robustness. This section synthesizes key contributions in the field, focusing on deep learning methodologies, dataset challenges, and specialized techniques.

12.2.1 Deep Learning Architectures for Pest Classification

Deep learning models, particularly convolutional neural networks (CNNs) and ensemble methods, have demonstrated strong performance in pest classification. A comprehensive review by Kasinathan et al. [4] summarized modern machine learning and deep learning techniques for insect classification and detection, highlighting both their potential and challenges such as dataset diversity, environmental variability, and computational costs. Deep learning models, particularly Convolutional Neural Networks (CNNs) and ensemble methods, have demonstrated strong performance in pest classification. However, challenges such as dataset diversity, environmental variability, and computational costs persist. CNN-based approaches have been widely adopted, with Tetila et al. [5] leveraging transfer learning to classify 13 types of soybean pests using Unmanned Aerial Vehicle (UAV) imagery. While achieving high accuracy, their model required a large and diverse dataset. Similarly, Thenmozhi et al. [6] applied CNNs for crop pest classification, emphasizing the need for more robust datasets. Alves et al. [7] used ResNet for cotton pest classification, demonstrating the effectiveness of deep residual learning, though its applicability was limited by specific environmental conditions. To further improve accuracy, ensemble methods have been explored. Khanramaki et al. [8] proposed an ensemble CNN model for citrus pest classification, achieving 99.04% accuracy with data augmentation. However, its reliance on large datasets raises concerns about real-world applicability. Ayan et al. [9] introduced a genetic algorithm-based weighted ensemble, enhancing model robustness but increasing computational complexity, which limits real-time deployment. Additionally, Kasinathan et al. [10] combined multiple classifiers with image processing techniques to improve feature extraction, though computational efficiency remains a challenge.

12.2.2 Specialized Techniques and Applications

Beyond traditional CNNs, researchers have investigated hybrid deep learning models and alternative pest detection techniques. Ullah et al. [11] developed DeepPestNet, integrating CNNs with recurrent neural networks (RNNs) to extract spatial and temporal features. While this improved classification performance, high computational demands hinder real-time applications. Mallick et al. [12] applied transfer learning for pest and disease detection in mung beans, demonstrating adaptability to various plant conditions but facing challenges related to image quality and dataset variability. Alternative approaches, such as color model-based detection and image enhancement techniques, offer potential improvements. Jothiaruna et al. [13] utilized multi-resolution channels and region-growing techniques for disease detection, but their direct application to pest classification remains underexplored. Sundar et al. [14] introduced super-resolution image reconstruction to enhance low-resolution pest images, though its integration with deep learning-based classification requires further investigation. These studies highlight the potential of advanced image processing methods, but challenges in computational efficiency and real-time implementation persist. Similarly, [15] proposed a segmentation method that combines Advanced Comprehensive Color Feature (ACCF) extraction with region growing to enhance disease spot detection, demonstrating robustness to cluttered backgrounds and illumination variations.

The literature underscores the effectiveness of CNN-based architectures like ResNet, AlexNet, and VGG16 in pest classification [5–7], while also highlighting their reliance on large and diverse datasets. Ensemble methods improve robustness [8, 9] but often demand significant computational resources, making real-time applications difficult. Hybrid models such as DeepPestNet [11] introduce spatial and temporal feature extraction, though at the cost of increased training time. Additionally, alternative techniques like color model-based detection [13], super-resolution imaging [14], and disease spot segmentation using ACCF with region growing [15] present promising enhancements, yet their direct applicability to pest classification remains an open research area. Building on these findings, this study evaluates AlexNet, ResNet50, VGG16, and ViT for automated pest classification. A structured preprocessing pipeline, including data augmentation and normalization, improves model generalization and reduces error rates. While CNNs achieve strong performance, the ViT outperforms them, achieving an accuracy of 88%. These insights reinforce the importance of integrating deep learning with efficient data preprocessing to enhance pest classification accuracy. Ultimately, this research contributes to more sustainable agriculture and improved food security through precise and timely pest management.

12.3 MATERIALS AND METHODS

The workflow of the Crop Pest and Disease dataset involves several stages, as illustrated in Figure 12.1. Initially, images undergo data preprocessing, which includes resizing, visualization, augmentation, and normalization to prepare standardized inputs. These processed images are then fed into four parallel deep learning models – ViT, VGG16, ResNet50, and AlexNet – for feature extraction and analysis. A performance evaluation step follows, where the outputs of the models are compared to identify the most effective one. The best-performing model is selected for the final classification, which determines whether a disease is present and identifies its type.

12.3.1 System Requirements

The experimental setup was implemented on a high-performance computing system equipped with an Intel® Xeon® Silver 4210R CPU @ 2.40GHz (dual processors, 20 cores, 40 threads total) and 128 GB DDR4 RAM, ensuring efficient handling of large-scale deep learning computations. The system ran MATLAB R2023b for model experimentation and analysis, alongside a Python-based deep learning environment leveraging Torchvision and the Transformers library. The study employed four state-of-the-art deep learning architectures: AlexNet, VGG16, ResNet50, and

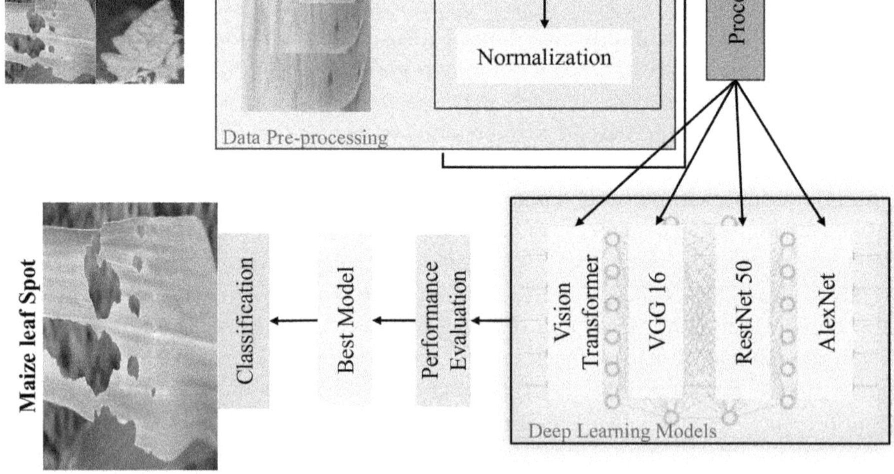

FIGURE 12.1 Workflow of end-to-end deep learning pipeline for automated pest classification.

ViT. For classification tasks, SequenceClassifierOutput was used to handle output representations effectively. The models were trained under a supervised learning paradigm, leveraging standard weight initialization techniques and adaptive optimization strategies. The training process was conducted for 100 epochs, using a mini-batch size of 32 and an optimized learning rate of 2e-5, ensuring stable convergence.

12.3.2 Dataset for Crop Pest and Disease Detection

The use of AI in the agricultural sector has gained significant traction in recent years, with the primary aim of enhancing crop yields, managing pest and disease control, and minimizing costs. In developing countries, the agricultural industry faces serious challenges, including pest and disease outbreaks, a knowledge gap between farmers and modern technology, and insufficient storage facilities. To contribute to addressing these issues, this study presents crop pest and disease datasets collected from local farms in Ghana. The dataset comprises two parts: the raw image collection, which includes 24,881 images distributed across 22 classes (6,549 for cashew, 7,508 for cassava, 5,389 for maize, and 5,435 for tomato) [16]. The crop-wise class distribution of the dataset is shown in Table 12.1.

TABLE 12.1
Crop-Wise Class Distribution of the Dataset

Crop	Disease/Pest	Count
Tomato	Septoria Leaf Spot	5,300
	Leaf Blight	2,700
	Leaf Beetle	1,800
	Verticillium Wilt	1,600
	Leaf Curl	1,200
Cassava	Bacterial Blight	5,200
	Brown Spot	3,000
	Mosaic	2,500
	Green Mite	2,300
	Healthy	2,400
Cashew	Anthracnose	3,500
	Red Rust	3,200
	Leaf Miner	2,900
	Gummosis	1,000
	Healthy	2,800
Maize	Leaf Spot	2,600
	Leaf Blight	2,200
	Streak Virus	2,100
	Grasshopper	1,400
	Fall Armyworm	900
	Healthy	800

12.3.2.1 Data Pre-Processing

Before applying preprocessing steps, it is essential to inspect the dataset for corrupted or unreadable images. These images should be removed as they can lead to errors during the data loading and training phases.

12.3.2.2 Image Resizing

To standardize image dimensions for uniform input into deep learning models, all images are resized to a fixed resolution (e.g., 224 × 224 pixels) to ensure compatibility with the input requirements of models such as AlexNet, ResNet50, VGG16, and ViT. Resizing images helps to reduce computational complexity and speeds up training time.

12.3.2.3 Data Visualization

To gain insights into the dataset by observing the distribution and variety of images across classes, visual exploration helps identify class imbalances, the quality of images, and potential outliers. This step ensures that the dataset is well-represented before applying preprocessing techniques. Figure 12.2 illustrates the distribution of classes in the crop pest and disease dataset, providing an overview of how images are distributed across different categories. Additionally, Figure 12.3 presents sample images from various classes in the crop pest and disease dataset, offering a visual representation of the dataset's diversity.

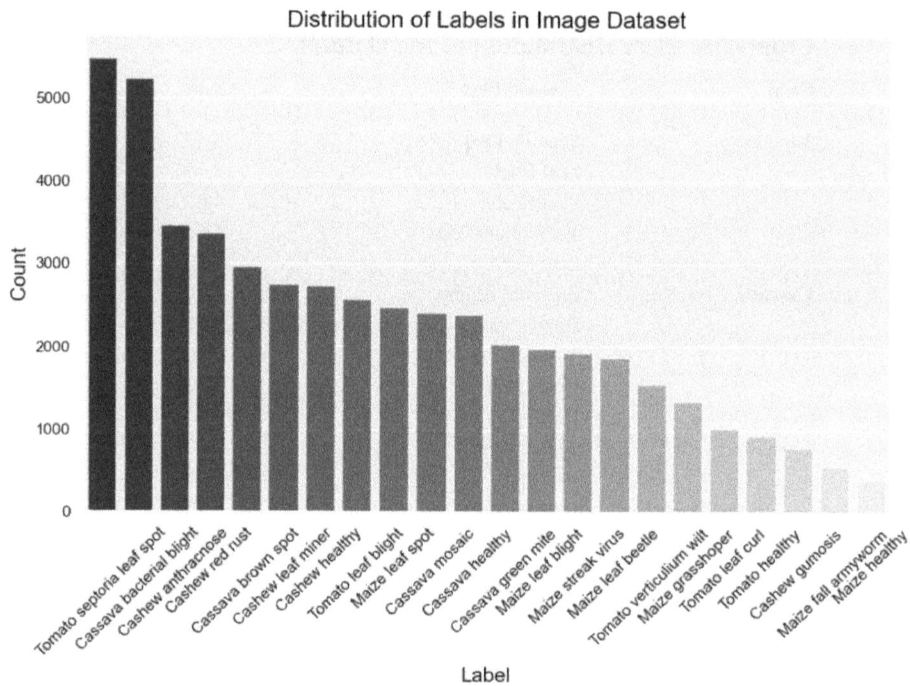

FIGURE 12.2 Distribution of classes in the crop pest and disease dataset.

FIGURE 12.3 Sample images from various classes in the crop pest and disease dataset.

12.3.2.4 Data Augmentation

To introduce variability, reduce overfitting, and enhance generalization, several data augmentation techniques were applied to the training images. These techniques included random rotations within ±30 degrees to ensure orientation invariance, horizontal flipping with a probability of 0.5 to improve robustness against left-right variations, and random scaling between 0.8x and 1.2x to simulate changes in object size. Additionally, random cropping was applied within 80–100% of the original image size to introduce spatial variability, while brightness adjustments of ±20% were made to adapt to varying lighting conditions. These augmentations enriched the training data, enabling the models to learn more robust and invariant features.

12.3.2.5 Normalization

To standardize pixel intensity values and stabilize training, all images were first normalized using min-max scaling, which rescaled the pixel values to the range [0, 1]. This ensured a uniform input representation across the dataset. Additionally, each image channel underwent mean and standard deviation normalization, where the dataset-specific mean (μ) and standard deviation (σ) were used for normalization. This preprocessing step, commonly employed in deep learning-based image classification, helped maintain a consistent data distribution, thereby improving both training efficiency and convergence stability.

12.3.3 Comparative Analysis of Deep Learning Architectures for Image Classification

12.3.3.1 Experimental Setup and Architecture Selection

This comprehensive study evaluated four distinct deep learning architectures, each representing a different evolutionary stage in computer vision. Figure 12.4 provides a visual representation, while Table 12.1 presents a detailed analysis of these architectures. The experiments were designed to ensure consistent evaluation parameters across the traditional architectures, while also leveraging modern frameworks for the transformer-based models.

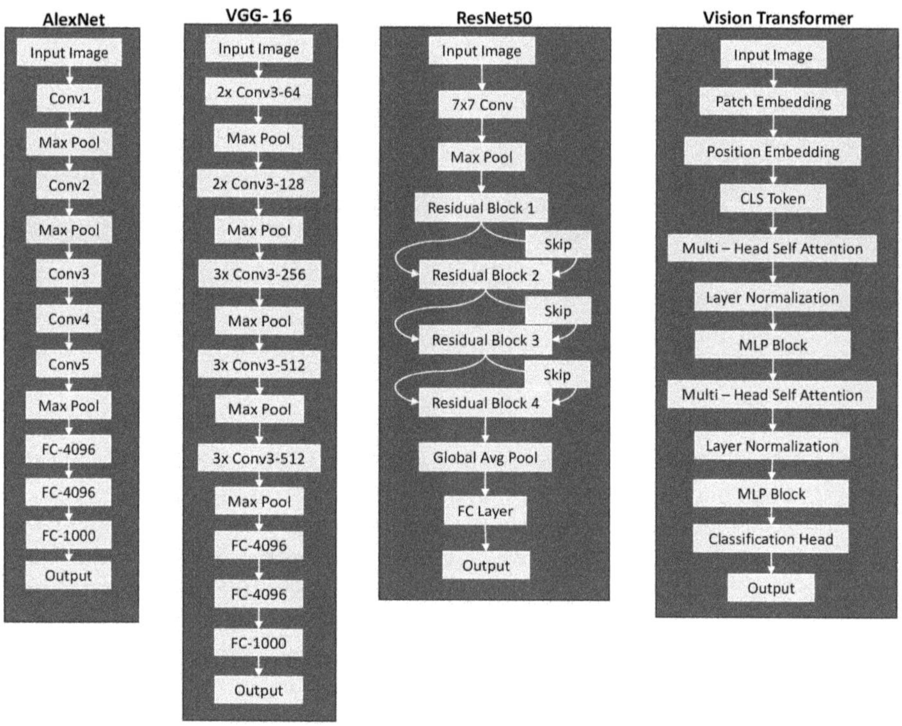

FIGURE 12.4 Comparative visual representation of AlexNet, VGG16, ResNet50, and Vision Transformer architectures.

12.3.3.2 Alex Net

Figure 12.5 illustrates the model training accuracy and loss curves for the AlexNet classification model, which follows a traditional deep-learning training process. The model was applied to a 22-class pest classification dataset. Standard hyperparameters were used in the training setup, including a categorical cross-entropy loss function, Adam optimizer with default values (β1 = 0.9, β2 = 0.999, ε = 1e-8), a batch size of 64, and a learning rate of 0.001. The training was carried out on a single GPU, spanning 100 epochs with 275 iterations per epoch, totaling 27,500 iterations.

The training process unfolded in three distinct phases: an initial sluggish start, followed by a phase of substantial improvement around 5,000 iterations, and finally, a phase of progressive stabilization. This progression is reflected in the accuracy curve, which shows an increase from an initial accuracy of approximately 10% to a final validation accuracy of 52.56%. The stability of the training is further confirmed by the loss curve, which starts at a high value and steadily converges throughout the training process [17].

12.3.3.3 VGG16

Figure 12.6 presents the model training accuracy and loss curves for the VGG16 classification model, a deeper CNN than AlexNet, applied to the same 22-class pest classification task. The training used standard hyperparameters: a batch size of 64, learning rate of 0.001, Adam optimizer with default parameters (β1 = 0.9, β2 = 0.999, ε = 1e-8), and categorical cross-entropy loss function. Training was conducted on a single GPU, consisting of 27,400 iterations across 100 epochs, with 274 iterations per epoch.

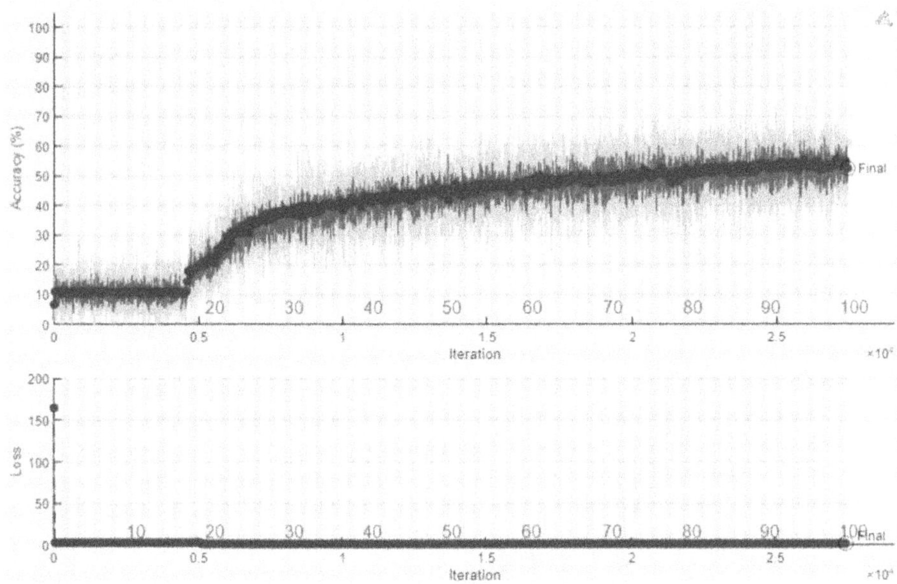

FIGURE 12.5 Training accuracy and loss curves of AlexNet classification model.

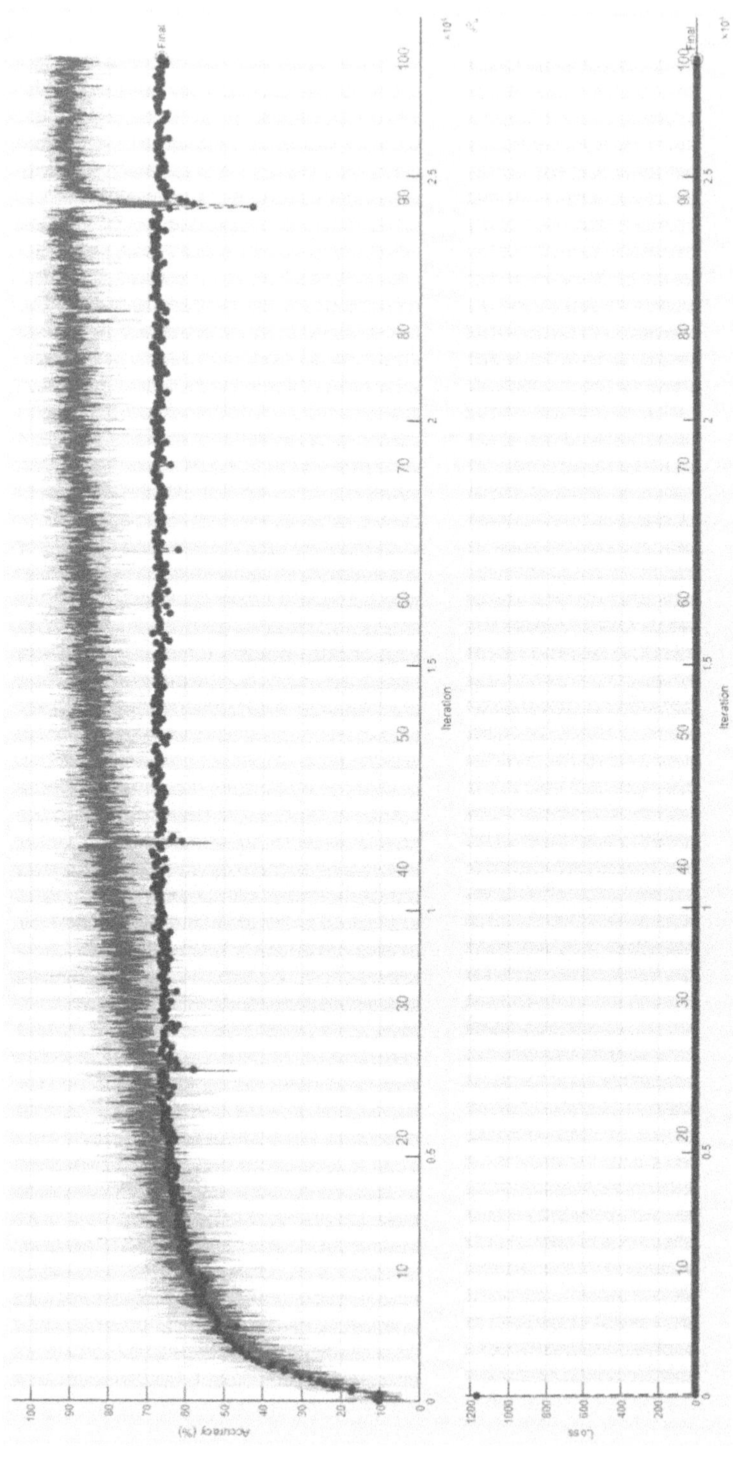

FIGURE 12.6 Training accuracy and loss curves of VGG16 classification model.

Compared to AlexNet's performance, VGG16 achieved a higher validation accuracy of 66.55%, showing a significant improvement from its initial accuracy. The accuracy curve exhibits distinct phases, with rapid learning up to around 10,000 iterations, followed by more gradual improvements with minor oscillations, suggesting slight overfitting. This is evident from the steady gap between training accuracy (blue line) and validation accuracy (dotted line) [18].

12.3.3.4 Res Net50

Figure 12.7 illustrates the model training accuracy and loss curves for the ResNet50 Classification Model. ResNet50, a deep residual network incorporating skip connections, was employed for the 22-class pest classification task, achieving a validation accuracy of 81%, surpassing AlexNet (52.56%) and VGG16 (66.55%). The training was conducted using standard hyperparameters, including a batch size of 64, learning rate of 0.001, Adam optimizer with default parameters ($\beta 1 = 0.9$, $\beta 2 = 0.999$, $\varepsilon = 1e\text{-}8$), and categorical cross-entropy loss function. The training graph demonstrates ResNet50's effective learning capability, with a steep initial learning curve up to 4,000 iterations, followed by gradual improvement until convergence. However, the model exhibits overfitting, as indicated by the training accuracy (solid blue line) exceeding 90%, while validation accuracy (dotted line) stabilizes around 81%.

The loss curve further confirms this behavior, with training loss smoothly converging toward zero, while validation loss remains slightly higher, suggesting overfitting. The superior performance of ResNet50 compared to AlexNet and VGG16 can be attributed to its residual learning framework, which enables effective training of deeper networks by mitigating vanishing gradient issues [19].

12.3.3.5 Vision Transformer

The ViT, applied to the 22-class pest classification task, achieved the highest validation accuracy of 86%, outperforming AlexNet (52.56%), VGG16 (66.55%), and ResNet50 (81%). As illustrated in Figure 12.8, ViT's unique design segments input images into fixed-size patches, which are then linearly embedded along with position embeddings. These patches are processed through a transformer encoder, composed of Multi-Head Self-Attention (MSA) blocks and multilayer perceptron (MLP) layers with normalization. Each transformer block implements self-attention mechanisms, allowing the model to dynamically assign importance to different image patches, followed by MLP layers for feature extraction and classification [20].

Key Advantages of ViT Over CNN Architectures

- Global Feature Learning: Unlike CNNs, which extract features hierarchically, ViT attends to global relationships between patches immediately using self-attention.
- Position-Aware Learning: Explicit position embeddings help maintain spatial relationships without requiring convolution operations.
- Flexible Receptive Field: Self-attention dynamically adapts the receptive field, unlike the fixed kernels used in CNNs.

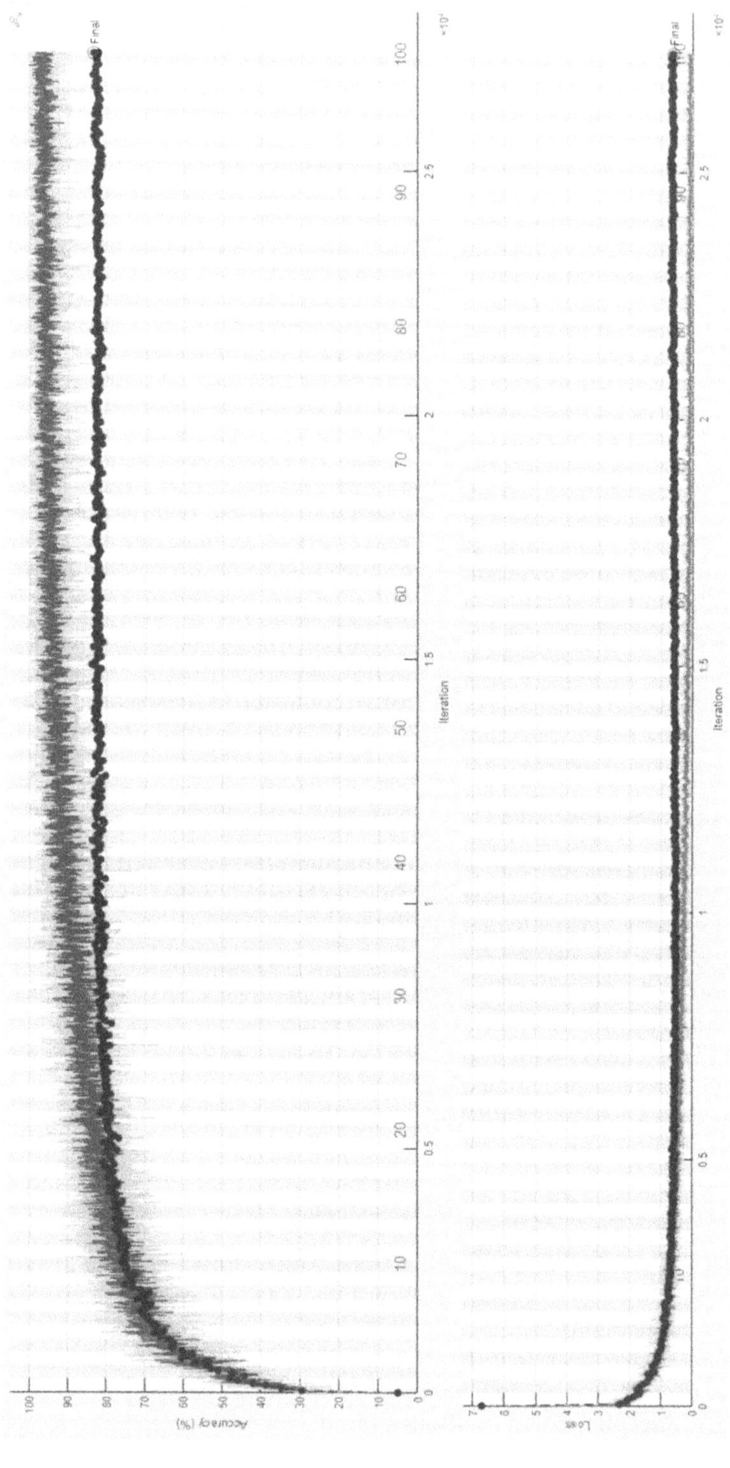

FIGURE 12.7 Training accuracy and loss curves of ResNet50 classification model.

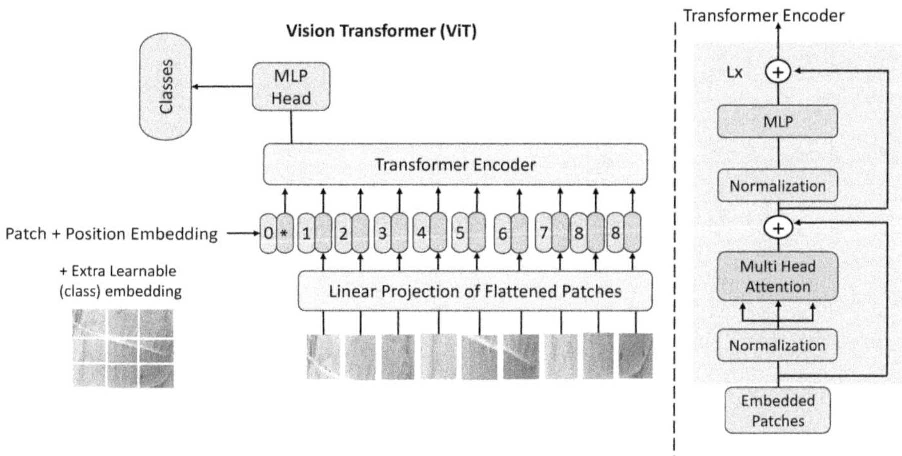

FIGURE 12.8 Vision Transformer (ViT) Architecture with transformer encoder.

The training configuration for ViT closely matched that of CNN models, but the results demonstrated several advantages:

- Faster convergence to higher accuracy levels.
- More stable training, with reduced fluctuation in validation metrics.
- Better generalization, indicated by a smaller gap between training and validation accuracy.
- Lower final loss values, suggesting higher confidence in predictions [21].

12.4 RESULTS AND DISCUSSIONS

The ViT model's classification report revealed a macro-average precision of 0.88, a recall of 0.84, and an F1-score of 0.84, demonstrating its strong performance. The confusion matrix further emphasized the model's effectiveness across various classes while also identifying challenges in distinguishing similar pest types. For instance, high precision and recall were achieved for the "Maize beetle leaf" class, as illustrated in Figure 12.9, which depicts an accurate prediction matching the actual target.

Beyond classification metrics, the confusion matrix, Figure 12.10, provides a comprehensive overview of the model's predictions, highlighting areas where the ViT model excelled and where it struggled to classify specific pest types accurately. This matrix helps identify common misclassifications and potential areas for improvement, further detailed in Figure 12.11. Additionally, Table 12.2 presents a detailed comparative analysis of the model architectures used, while Table 12.3 highlights key differences between the four deep learning architectures considered for pest classification.

The ViT model's 86% accuracy underscores its effectiveness in capturing complex pest features using its self-attention mechanism, which plays a crucial role in distinguishing subtle differences between pest classes. This strong performance

TABLE 12.2
Detailed Comparative Analysis of Model Architectures for Performance Evaluation

AlexNet	VGG16	ResNet50	Vision Transformer
1. 5 convolutional layers	1. 13 convolutional layers	1. Initial 7 × 7 conv, 64 filters	1. Splits image into fixed-size patches (e.g., 16 × 16)
2. 3 max pooling layers	2. 5 max pooling layers	2. 3 × 3 max pooling	2. Flattens and linearly embeds patches
3. 3 fully connected layers	3. 3 fully connected layers	3. 16 residual blocks arranged in 4 stages	3. Adds position embeddings
4. Variable filter sizes (11 × 11, 5 × 5, 3 × 3)	4. All conv layers use 3 × 3 filters	4. Global average pooling and Final fully connected layer	
Key Components:	**Key Components**:	**Key Components**:	**Key Components**:
1. Large initial filter (11 × 11)	1. Fixed filter size (3 × 3)	1. Bottleneck blocks with 1 × 1, 3 × 3, 1 × 1 convolutions	1. Transformer encoder blocks: 12
2. Channel sizes: 96, 256, 384, 384, 256	2. Channel progression: 64 → 128 → 256 → 512	2. Channel sizes: 64, 128, 256, 512	2. Self-attention heads: 12
3. ReLU activation	3. Max pooling after each block	3. Skip connections	3. MLP size: 3072
4. Local response normalization	4. Large FC layers (4096 neurons)	4. Batch normalization after each conv	4. Layer normalization
Distinctive Features:	**Distinctive Features**:	**Distinctive Features**:	**Distinctive Features**:
1. First successful deep CNN	1. Uniform architecture	1. Identity shortcuts	1. No convolution layers
2. Simple architecture	2. Simple design	2. Deep architecture (50 layers)	2. Global self-attention
3. Large filters	3. Deep but regular structure	3. Bottleneck design	3. Position embeddings
4. Dropout for regularization	4. High parameter count	4. Residual learning	4. CLS token for classification

Classification Report:

	precision	recall	f1-score	support
Cashew anthracnose	0.94	0.91	0.92	66
Cashew gumosis	1.00	1.00	1.00	18
Cashew healthy	0.96	1.00	0.98	50
Cashew leaf miner	0.93	0.93	0.93	56
Cashew red rust	1.00	1.00	1.00	58
Cassava bacterial blight	0.86	0.93	0.89	81
Cassava brown spot	0.90	0.85	0.87	52
Cassava green mite	0.96	0.96	0.96	45
Cassava healthy	1.00	0.98	0.99	51
Cassava mosaic	1.00	0.98	0.99	47
Maize armyworm fall	1.00	1.00	1.00	6
Maize beetle leaf	1.00	1.00	1.00	38
Maize blight leaf	0.82	0.79	0.81	34
Maize grasshoper	1.00	1.00	1.00	21
Maize healthy	0.78	0.88	0.82	8
Maize leaf spot	0.69	0.79	0.74	39
Maize streak virus	0.92	0.80	0.85	44
Tomato blight leaf	0.73	0.23	0.35	47
Tomato curl leaf	0.86	0.38	0.52	16
Tomato healthy	0.69	0.92	0.79	12
Tomato leaf septoria spot	0.63	0.95	0.76	99
Tomato verticulium wilt	0.60	0.30	0.40	30
accuracy			0.86	918
macro avg	0.88	0.84	0.84	918
weighted avg	0.87	0.86	0.85	918

FIGURE 12.9 Sample prediction result for "maize beetle leaf."

TABLE 12.3
Comparative Analysis of Key Differences of Deep Pest Classification Architectures

S. No.	Description	AlexNet	VGG16	ResNet50	Vision Transformer
1.	Depth Complexity	8 layers sequential	16 layers sequential	50 layers with shortcuts	12 transformer blocks
2.	Feature Extraction	Basic hierarchical	Deep hierarchical	Residual learning	Self-attention based
3.	Parameter Efficiency	~60M parameters	~138M parameters	~23M parameters	~86M parameters
4.	Modern Features	Basic CNN operations	Uniform filters	Skip connections	Attention mechanisms
5.	Advantages	GPU training breakthroughs	Simplicity, strong performance	Training deep networks	Strong on large datasets
6.	Drawbacks	Outdated by deeper models	Computationally expensive	High memory use	Outdated by deeper models

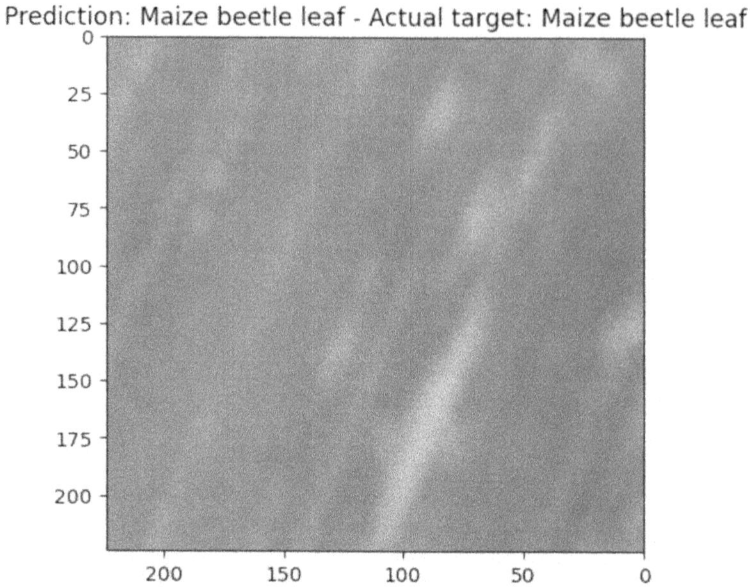

FIGURE 12.10 Classification report for 22-class pest classification task.

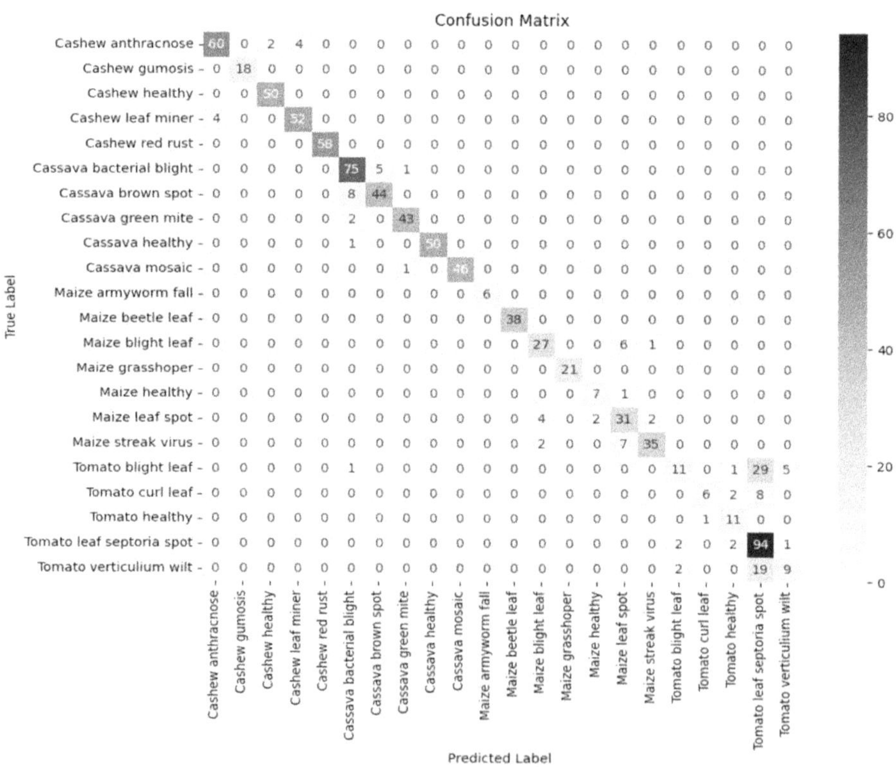

FIGURE 12.11 Confusion matrix for 22-class pest classification task.

suggests that the model's ability to capture long-range dependencies and leverage global context makes it particularly well-suited for pest classification, where fine-grained details are essential for accurate identification.

12.5 CONCLUSION AND FUTURE DIRECTIONS

The comparative analysis of ViT against other models demonstrated its strong performance in handling diverse pest classification tasks, achieving balanced macro-average metrics and excelling in precision and recall for key categories. However, distinguishing between visually similar pest types remains a challenge, highlighting the need for further improvements. Future research could focus on enhancing ViT through domain-specific pretraining or hybrid approaches that integrate the strengths of CNNs and transformers. Additionally, expanding the dataset with more diverse and high-resolution images, applying advanced data augmentation techniques, and leveraging ensemble learning strategies could further enhance model accuracy and generalization. These advancements will contribute to more reliable and practical applications in agricultural pest management, ensuring comprehensive and effective solutions.

REFERENCES

1. Skendžić, Sandra, Zovko, Monika, Živković, Ivana Pajač, Lešić, Vinko, and Lemić, Darija. "The impact of climate change on agricultural insect pests." *Insects* 12, no. 5 (2021): 440.
2. Yadav, Ravindra. "An innovative method: identifying pests through artificial neural networks and image processing." *International Journal of Intelligent Systems and Applications in Engineering* 12, no. 4 (2024): 366–372. https://ijisae.org/index.php/IJISAE/article/view/6223
3. Tiwari, Awanindra Kumar. "Insect pests in agriculture identifying and overcoming challenges through IPM." *Archives of Current Research International* 24, no. 3 (2024): 124–130.
4. Kasinathan, Thenmozhi, Singaraju, Dakshayani, and Uyyala, Srinivasulu Reddy. "Insect classification and detection in field crops using modern machine learning techniques." *Information Processing in Agriculture* 8, no. 3 (2021): 446–457.
5. Tetila, Everton Castelão, Machado, Bruno Brandoli, Astolfi, Gilberto, de Souza Belete, Nícolas Alessandro, Amorim, Willian Paraguassu, Roel, Antonia Railda, and Pistori, Hemerson. "Detection and classification of soybean pests using deep learning with UAV images." *Computers and Electronics in Agriculture* 179 (2020): 105836.
6. Thenmozhi, K., and Srinivasulu Reddy, U. "Crop pest classification based on deep convolutional neural network and transfer learning." *Computers and Electronics in Agriculture* 164 (2019): 104906.
7. Alves, Adão Nunes, Souza, Witenberg S. R., and Borges, Díbio Leandro. "Cotton pests classification in field-based images using deep residual networks." *Computers and Electronics in Agriculture* 174 (2020): 105488.
8. Khanramaki, Morteza, Asli-Ardeh, Ezzatollah Askari, and Kozegar, Ehsan. "Citrus pests classification using an ensemble of deep learning models." *Computers and Electronics in Agriculture* 186 (2021): 106192.
9. Ayan, Enes, Erbay, Hasan, and Varçın, Fatih. "Crop pest classification with a genetic algorithm-based weighted ensemble of deep convolutional neural networks." *Computers and Electronics in Agriculture* 179 (2020): 105809.

10. Kasinathan, Thenmozhi, and Uyyala, Srinivasulu Reddy. "Machine learning ensemble with image processing for pest identification and classification in field crops." *Neural Computing and Applications* 33, no. 13 (2021): 7491–7504.
11. Ullah, Naeem, Khan, Javed Ali, Alharbi, Lubna Abdulaziz, Raza, Asaf, Khan, Wahab, and Ahmad, Ijaz. "An efficient approach for crops pests recognition and classification based on novel DeepPestNet deep learning model." *IEEE Access* 10 (2022): 73019–73032.
12. Mallick, M. D. Tausif, Biswas, Shrijeet, Das, Amit Kumar, Saha, Himadri Nath, Chakrabarti, Amlan, and Deb, Nilanjan. "Deep learning based automated disease detection and pest classification in Indian mung bean." *Multimedia Tools and Applications* 82, no. 8 (2023): 12017–12041.
13. Jothiaruna, N., Joseph Abraham Sundar, K., and Ifjaz Ahmed, M. "A disease spot segmentation method using comprehensive color feature with multi-resolution channel and region growing." *Multimedia Tools and Applications* 80, no. 3 (2021): 3327–3335. https://doi.org/10.1007/s11042-020-09882-7
14. Joseph Abraham Sundar, K., Vaithiyanathan, V., Manickavasagam, M., and Sarkar, A. K. "Enhanced singular value decomposition based fusion for super resolution image reconstruction." *Defence Science Journal* 65, no. 6 (2015): 459–465.
15. Jothiaruna, N., Sundar, K. J. A., and Karthikeyan, B.. "A segmentation method for disease spot images incorporating chrominance in comprehensive color feature and region growing." *Computers and Electronics in Agriculture* 165 (2019): 104934.
16. Patrick, Mensah Kwabena, Akoto-Adjepong, Vivian, Adu, Kwabena, Mighty, Abra Ayidzoe, Elvis, Asare Bediako, Nyarko-Boateng, Owusu, Boateng, Samuel, Esther, Fobi Donkor, Faiza, Umar Bawah, Nicodemus, Songose Awarayi, Nimbe, Peter, Isaac, Kofi nti, Abdulai, Muntala, Remember, Roger Adjei, Opoku, Michael. "Dataset for crop pest and disease detection." *Mendeley Data*, V1 (2023). https://doi.org/10.17632/bwh3zbpkpv.1
17. Yeh, Jui-Feng, Wang, Sung-Yi, and Chen, Yi-Ping. "Crop disease detection by image processing using modified AlexNet." In *2021 IEEE 3rd Eurasia Conference on Biomedical Engineering, Healthcare and Sustainability (ECBIOS)*, Taipei, Taiwan, pp. 159–160. IEEE, 2021.
18. Paymode, Ananda S., and Malode, Vandana B. "Transfer learning for multi-crop leaf disease image classification using convolutional neural network VGG." *Artificial Intelligence in Agriculture* 6 (2022): 23–33.
19. Kumar, Sachin, Pal, Saurabh, Singh, Vijendra Pratap, and Jaiswal, Priya. "Performance evaluation of ResNet model for classification of tomato plant disease." *Epidemiologic Methods* 12, no. 1 (2023): 20210044.
20. Bhojanapalli, Srinadh, Chakrabarti, Ayan, Glasner, Daniel, Li, Daliang, Unterthiner, Thomas, and Veit, Andreas. "Understanding robustness of transformers for image classification." In *Proceedings of the IEEE/CVF International Conference on Computer Vision*, pp. 10231–10241. 2021.
21. Bazi, Yakoub, Bashmal, Laila, Rahhal, Mohamad M. Al, Dayil, Al Reham, and Ajlan, Al Naif. "Vision transformers for remote sensing image classification." *Remote Sensing* 13, no. 3 (2021): 516.

13 Optimizing Deep Learning Models for Real-Time Plant Disease Detection

Enhancing Agricultural Productivity

Midhun P Mathew, Abhinand S, Alan Sebastian, Amal Sasikumar, and Adheena Thomas

13.1 INTRODUCTION

Several authors tested the following models: ResNet50, VGG19, DenseNet121, and MobileNet v2, and reported high-performance results. Such models as a rule use image datasets such as PlantVillage [1] and PlantDoc [2], which contain healthy and unhealthy plant photos [3]. The identification and categorization of plant diseases are crucial activities of modern farming, since these diseases are known to adversely affect crop growth and agricultural productivity as a whole. In view of this, it can be expected that further development of measures to combat hunger or rising food demand in regions experiencing population growth would ensure economic development [4]. We have already pointed out earlier that the world population is projected to increase to about 10 billion people in the year 2050 [5], which means that wiser agricultural approaches and better plant cultivation will have to evolve toward addressing the feeding of billions of people worldwide. The recognition or identification of plant diseases has been performed manually by farmers or agricultural specialists through the naked eye examinations of leaves, stems, and even fruits of the infected plants. It is important to note that this manual method of diagnosing plant diseases is characterized by being time wasting, cumbersome, and error prone particularly in instances of dealing with subtle or trivial cases.

The emergence of machine learning (ML) and deep learning (DL), especially Convolutional Neural Networks (CNNs), has changed the method of identifying and classifying plant diseases. CNNs, which are a distinct category of DL applicable to images, have been proven to yield better results in image classification processes. With the use of CNNs, there is an opportunity for researchers and agricultural technologists to automate the task of determining the different types of diseases that affect

DOI: 10.1201/9781003594093-13

plants through images of leaves with a great level of precision, speed, and reliability [6]. CNNs have the ability to obtain and learn the hierarchical features from the raw images, thus giving them the capability of differentiating between healthy and unhealthy plants. Consequently, there is a growing demand for CNN-based systems as useful instruments that are applied in the early detection of plant diseases for appropriate action to be taken in good time. Many types of Dos have been used in the recent past to develop effective plant disease identification systems that require an average of three minutes to detect all types of diseases affecting different crops. These systems have been applied effectively in detecting infections in chief crops such as tomato, mango, and cucumber, among others. Bacterial spots, blight, powdery mildew, and other diseases can do a great deal of damage if left untreated on time.

Automated disease detection systems not only cut down the manual inspection time and labor process but also reduce the chances of human mistakes, hence giving the farmers accurate information on the preventive measures regarding plant diseases [7, 8]. Timely spotting makes it possible to perform increased control measures aimed at specific pathogens and minimizes the chances of resorting to the global restoration option, which could be a chemical fume solution that may be harmful to the environment as well as the people. However, despite the success that CNNs have shown in detecting plant diseases, some challenges remain. One of the foremost issues is to ensure that the CNN algorithms have cross-domain adaptability on aspects such as lighting, angles, and background, which exist in agricultural practices. Another daunting task is the issue of training CNNs on big datasets, especially when the datasets include a variety of diseases, though in small or imbalanced proportions. Image data sets containing pictures of plants with disease at various angles, scales, and rotations, and with or without flipping the images, can be used to enhance the application of CNN models. In addition, transfer learning can be an effective technique aimed at improving these problems and making the classifiers more accurate in various environments.

The authors have, in this chapter, attempted to apply CNNs in detecting diseases in plants by using images of leaves [9, 10]. The outlined methods and models seek to improve on the existing plant disease detection methods in terms of their accuracy and scalability. With advanced image processing and DL techniques, our models are able to accurately identify a large number of diseases affecting various crops. This automated process not only enables farmers to make efficient decisions about disease control but also allows more efficient use of chemicals, which helps to reduce the adverse effects on the environment. The use of these systems allows farmers to adopt more effective strategies based on data analysis, which in turn will help them to achieve better performance in crop production and improve the sustainability of agriculture.

The harvesting of crops is always a task that has been extremely tiresome and labor intensive. However, the voice reading technology powered through artificial intelligence (AI), along with sensors, is allowing farmers to cut back on their hard work. These days, speaking out loud can be a type of contemporary farming. Enduring foresight through DL and working in harsh environments seems to be a benefit that AI is going to offer. Moreover, AI-powered devices assist farmers in scanning leaves and capturing images of them; these images can be uploaded onto mass storage systems for disease classification. These technologies all enhance smart farming, especially across controlled conditions such as greenhouses, where all-encompassing evaluation can avert the potential for plants getting infected. For example, a sophisticated detection system comprising AI devices integrated into a greenhouse allows for the interface to self-detect and identify the causes of plant diseases. This self-detection allows for essential measures such as changing the environmental settings or changing the local treatments to be administered in time. Moreover, apart from Internet of Things (IoT) technologies, TinyML has created new possibilities for the implementation of CNN models in low-powered devices such as microcontrollers. This involves the use of IoT devices that require low power while enabling the execution of complex algorithms in real time. It has also been possible to deploy these systems in the field using, for example, Raspberry Pi or microcontrollers as the power and network connectivity in these areas might be weak. Therefore, an edge-based solution of deploying AI in agriculture enables a farmer to obtain instant feedback and allows him/her to make timely decisions to improve the efficiency of the farm.

In this chapter, we review the use of CNN techniques for the detection of plant diseases with specific emphasis on tomato and chrysanthemum crops [11–14]. Also discussed are the issues that involve deploying DL models in poor environments and possible solutions that utilize image processing, optimization of the model, and IoT integration. Additionally, this work focuses on the implementation of agricultural machine-learning models in low-powered devices and demonstrates how CNNs can improve agriculture.

The chapter is divided into the following sections. Section 13.1 deals with the introduction, followed by Section 13.2, which explores the concept of convolution in DL, followed by image enhancement techniques used in DL and the results generated.

13.2 CONCEPT OF CONVOLUTION IN DEEP LEARNING

One of the most significant breakthroughs in DL, especially for computer vision tasks such as object recognition, image segmentation, and image classification, is CNNs. Ever since they were proposed, CNNs have been the most used architecture for processing image data, allowing machines to understand complexities embedded in various images. In the area of detecting plant diseases, CNNs are absolutely vital since they are able to automatically learn representations from images, while in traditional ML models, feature extraction was done manually, requiring a lot of time and being very error prone. By eliminating the need for manual feature extraction, CNNs provided a significant leap in improvement in the effective scaling of systems used for detecting plant diseases, which is a great technological advantage in modern farming.

The design of a CNN is interesting in the sense that it learns hierarchically organized levels of features right from the primary image data, which is shown in Figure 13.1. That is, instead of having to have features handed to them, CNNs have the capacity to learn low-order features like shapes and objects right through to even high-level features such as edges and texture. CNN architectures can typically be divided into three major layers as follows:

1. **Convolutional Layers**: The convolutional layers are the basic building blocks of a CNN. The convolutional filters (or kernels) are used on the input images that seek to highlight local structures, such as edges, textures,

colors, and complex forms, by scanning the image. For plant disease detection, it plays an important role since it can detect the signs of biological diseases like spots, lesions, and discoloration changes on leaves without manual "help." Each filter in a convolutional layer detects only a single pattern, which makes it possible for the model to learn multiple patterns when trained on different leaf images. The deeper the network gets, the more complex combinations it can build, resulting in greater accuracy when classifying images of leaves.

2. **Pooling Layers**: Each of the convolutional layers creates feature maps, which have spatial dimensions that pooling layers work on. This helps in reducing the number of parameters of the networks and thus enhances the computational efficiency of the model. One of the most used pooling techniques is max pooling, which takes out from a feature map the most useful information and gets rid of less useful data. In plant disease detection, the pooling layers reduce the focus of the model to the regions of the diseased leaves that provide the most relevant features, thus also helping in lowering the computational cost. Hence, the ability of the CNN to work with large datasets improves, which is shown in Figure 13.2.
3. **Fully Connected Layers**: This definition refers to the layers that are most commonly placed near the end of a CNN and that have a role as a decision-maker of the network. Both these layers specialize in tasks learned from convolutional and pooling processes that help structure the data into a number of

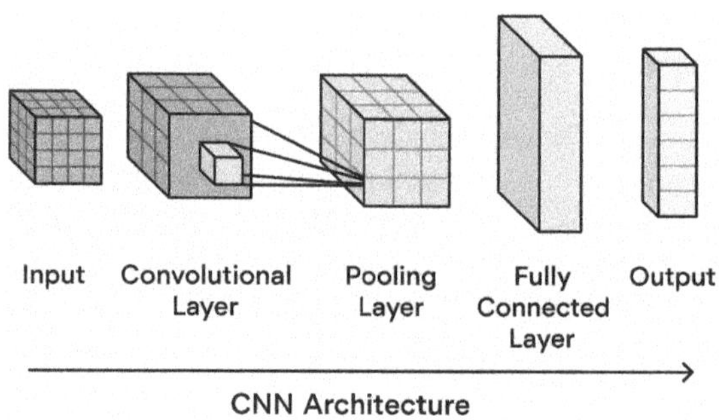

FIGURE 13.1 CNN architecture [4].

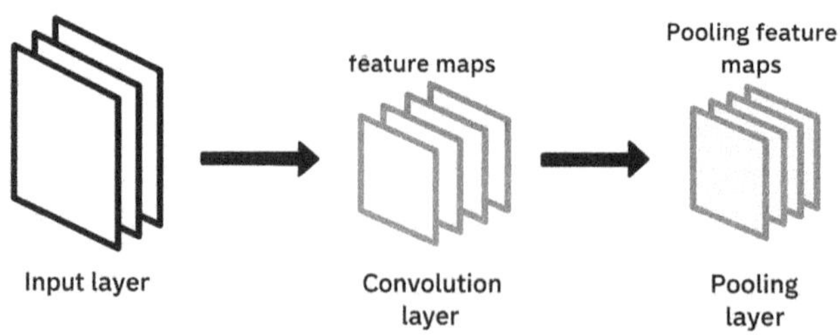

FIGURE 13.2 Pooling layers [7].

subclasses. For example, in a plant disease detection task, the fully connected layers may be responsible for classifying the leaf as healthy or diseased, and if diseased, identifying the specific type of disease. Since the fully connected layers take all the neurons from the previous layer, it allows the model to base its prediction on all the features that we learned through the model, which then ensures that there is a final output, usually a score that predicts the likelihood of each class, which is shown in Figure 13.3.

13.2.1 Performance in Plant Disease Detection

CNNs have proven to be effective in the area of plant disease recognition, specifically for the classification of certain conditions from the images of leaves. For instance, in the case of tomato leaves, CNNs were used to detect such conditions as early blight, bacterial spot, and yellow leaf curl virus. This task has seen the development of multiple CNN architectures that differ in their performance characteristics relating to accuracy, speed of computation, and the complexity of the model [12].

13.2.2 Popular CNN Architectures

Some different approaches to CNN that have been employed in the detection of plant diseases are as follows:

1. VGG-16: VGG-16 is composed of 16 layers, and due to the ease of its architecture and the depth of the layers, it has a substantial application for image classification. The ease of its architecture further seeks its application for a DL system as it allows the system to comprehend many more features, thus seeing even the slightest changes in the plant's diseases.

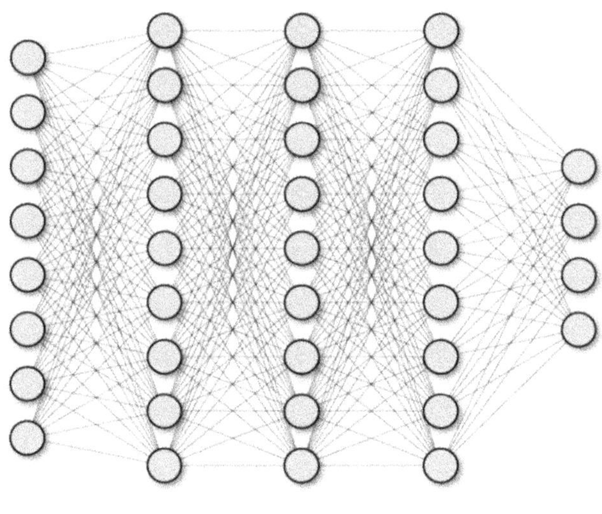

FIGURE 13.3 Fully connected layers [10].

2. ResNet: ResNet (Residual Networks) solves this problem with the introduction of skip connections, thereby remembering information from the residuals of other layers. This is critical in helping solve the vanishing gradient problem that is very rampant in very deep networks. Hence, due to their provision, CNN ResNet networks perform very well for large-scale plant disease detection, where the accuracy needed is high [3].
3. Inception: The Inception Neural Network utilizes multi-scale convolutions, which allow the network to comprehend features at different scales at the same time. It is very important for the case of plant diseases, as symptoms tend to vary in size across different leaves [13].

These architectures have been extensively tested in agricultural tasks, and each implementation has its advantages and disadvantages in terms of accuracy, speed, and processing requirements. For instance, it is common knowledge that deeper architectures, such as ResNet, have been observed to achieve greater accuracy compared to lightweight architectures such as MobileNet, which are more suited for low-resource devices.

13.2.3 Benefits of CNNs in Agriculture

One of the most notable benefits of CNNs when used in plant disease detection is the vast amount of data they are capable of handling and the relevance of features that are learned, which makes them ideal for agricultural use cases. Plant disease datasets contain large amounts of images of the same leaf but captured in different environments and with different disease severity and stages. These variations are what CNNs are able to learn and generalize well when faced with different data or conditions. The challenge of having to source raw images to manually build sample datasets is turned on its head. This is because there is no need to learn raw image features to extract features that need to be fed into the system, and this greatly simplifies the process of building a plant disease detection system. Performance of models that are based on CNNs gets better with the advent of newer and more diverse datasets, particularly in recognizing initial signs of diseases. It is an important skill in agriculture, because many times spraying of agrochemicals is avoided based on early warning of the plant diseases, and timely applications are made. In addition to that, the nature of the problems that need to be solved in the agricultural domain forms a good case for the adoption of CNNs because the data flow from the farms, greenhouses, and agricultural research centers is ever-increasing.

Ensemble Learning Approaches: In some recent works, it has been reported that the use of ensemble learning can enhance the performance of CNNs in plant disease detection tasks. Ensemble learning is the process of using a number of models and combining their predictions to enhance the accuracy of the prediction. For example, in the case of the Ensemble stack deep neural network (ESDNN) that has been proposed, the outputs of several CNN models (ResNet, VGG, and Inception, among others) are used to make a prediction. In this approach, pretrained CNN models from other feedforward neural networks are used, and thus patterns that are learned are easily generalizable to the new task.

13.2.4 Applications in IoT and Smart Agriculture

IoT is the seamless connection of devices over the internet and the range in which satellite or deep space elements communicate with the Earth. In smart agriculture, taking it to the next level, leveraging IoT monitoring is important, which has given rise to a new type of CNNs focusing on agricultural applications, structured on connecting wires on the plants and monitoring growth physics images. For instance, low-power devices such as a Raspberry Pi can deploy lightweight CNN models that can detect symptoms of plant diseases. This helps to monitor the growth of the plants and helps in taking the actions required to maintain stability at a prime condition [1]. The aforementioned technique improves the effectiveness of processes such as plantation and irrigation, improving overall yields while using the same space and resources. This model can be further enhanced with sensor networks, along with a satellite image, providing a more up-to-date point of view on a broader scale. The rational choice model is often criticized for assuming too many variables as constant and for ignoring the informal social burdens during decision-making. Figures 13.4 and 13.5 show the concept of IoT in agriculture.

13.2.4.1 Challenges and Future Directions

As much as CNNs have been proven to be efficient in detecting diseases such as blight in plants, a number of obstacles prevent those models from being used in an actual farm setting. First, the cost of computing power to execute CNNs on extensive datasets or on microcontrollers is expensive. While models such as ResNet can deliver results, their implementation on these edge devices needs heavy optimization, such as quantization and pruning. And, because many factors exist, such as the orientation of lighting, leaves covering other leaves, or changing ecology throughout the year, it exposes the model to chances of making false detections. To keep these weaknesses at bay, researchers are looking to bolster CNNs with data augmentation, domain adaptation, and image enhancement techniques. Moreover, Tiny Machine Learning (TinyML) will allow deploying CNN models on devices that are extremely

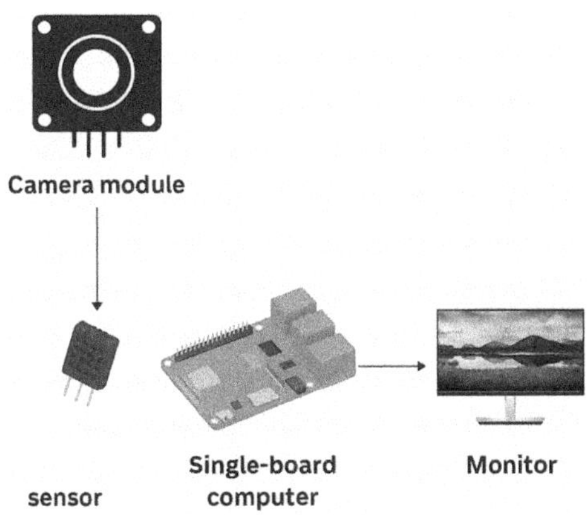

FIGURE 13.4 Block diagram.

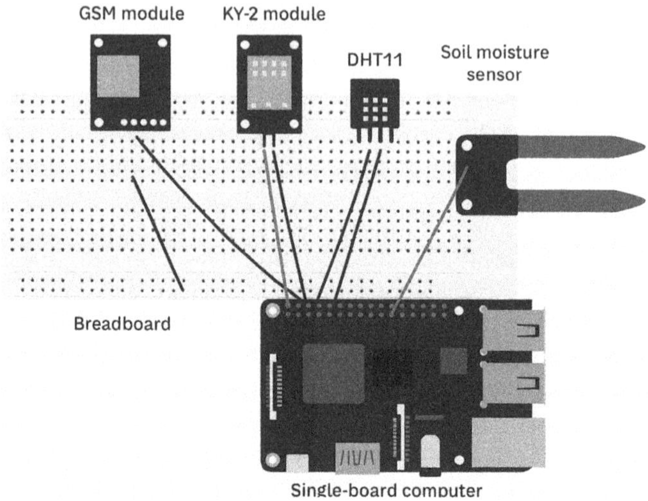

FIGURE 13.5 System schematics [6].

resource-limited while retaining their accuracy. There is a possibility that in the future, research can optimize the CNN architecture, making real-time applications possible for plant disease detection systems that would be efficient and convenient.

13.3 IMAGE ENHANCEMENT TECHNIQUES

Techniques of image enhancement are crucial in pre-processing the images before the analysis is done by CNNs [15–17]. These methods are intended to modify the visual quality and emphasize important regions of interest in the input images, so that the CNN can concentrate on fine yet important details that are critical for tasks such as plant disease recognition. Since diseases, for the most part, are represented by mild variations in the texture, color, or shape of the leaves, there is an absolute need to improve the clarity of enhanced images to automate the detection with a higher level of accuracy. In farming, for example, these techniques allow working in suboptimal image acquisition settings, such as insufficient light or the presence of noise, which results in a clear and uniform formation of the image input for the CNN. Figure 13.6 shows the leaf diseases of tomato, early blight.

In practical agricultural environments, image noise is presently a problem caused by changes in lighting circumstances and camera imperfections, and even such environmental factors as dust or moisture have their influence. This is because the noise will tend to cover up important visual information in the image, thus making it hard for the CNN to detect basic features that correspond to plant diseases. The technique of Gaussian filtering should be able to smooth the image owing to its operations performed on a given pixel and its adjacent ones while still conserving important features of the image by cutting down on high-frequency noise. It works best for agricultural images where natural leaf vein textural images gets mixed with noise.

FIGURE 13.6 Diseased plant leaf.

The fine spots and lesions characteristic of disease are made evident through the application of Gaussian filtering. Mathematical formula for the Gaussian filter is given as follows:

$$G(x,y) = \frac{1}{2\pi\sigma^2} \exp\left(-\frac{x^2+y^2}{2\sigma^2}\right) \tag{13.1}$$

Median filters are better than the most common Gaussian filters, which employ the method of averaging in median filtering. This is especially useful in removing salt and pepper window noise when working with images, which is a typical disturbance in digital pictures. The Mathematical representation of median filter is given as follows:

$$I_{\text{filtered}}(i,j) = \text{median}\left(\{I(x,y) : (x,y) \in \Omega(i,j)\}\right) \tag{13.2}$$

This type of filtering will leave the edges of the diseased regions on the leaves intact, enabling the CNN to recognize crucial features. In plant disease detection, the reduction of image noise improves the model's ability to detect small, localized symptoms such as discoloration, spots, and even fungal growths on the leaves that otherwise would have been missed. Image segmentation refers to the process of partitioning a digital image into multiple segments. This is useful, for example, in isolating diseased leaf regions from the background, which consists of the remaining leaf portions. This also allows CNNs to concentrate on the pertinent parts of the image, effectively increasing both precision and recall.

Thresholding: It is among the most common segmentation techniques and involves the division of an image that has been captured into diseased and healthy areas depending on the intensity with which each of the pixels has been set. Two simple classes are formed; the first includes all pixels "above" a certain level (e.g., necrotic spots/diseased area) while the second includes all those below (healthy area). This can be helpful in enhancing clear variations in leaf coloration, such as necrotic spots appearing on a leaf.

Masking: Masking works by creating a binary mask image that only contains pixels that reside within the area of interest; for example, the infected area of the leaf and the rest are set to a value of zero. This helps in concentrating more on the crucial parts and, therefore, less emphasis is put on areas of the background that are irrelevant, thus improving the overall accuracy of the categorization of images.

U-Net for Pixel-level Segmentation: U-Net is a modification of the CNN used for semantic segmentation. For example, a U-Net network can tell whether a pixel of an image belongs to a sick or healthy image. In the case of plant pest infection, this method can be combined with additional coverage and can emphasize stronger, even the smallest sick areas, and thus allow detection of a budding infection, which otherwise would have gone unnoticed. Segmentation ensures that the CNN is trained and tested on the most relevant portions of an image with regard to whether the area under the network is diseased or healthy, hence enabling the network to improve in distinguishing diseased regions from healthy regions. Data augmentation is one of several methods used to artificially increase the size and variability of a data set by randomly altering the images. This is useful for agricultural technologies where large datasets cannot be obtained because it is seasonal or due to agricultural land that is not very diverse.

Geometric Transformations: Rotating, flipping, zooming, and shifting move the image to a different direction and size. A CNN for diseased crops could also have images of planted diseased beans at different angles, which, when rotated, provide a better view. Therefore, the neural network can predict unseen images.

Color Jittering: Altering the brightness, contrast, or color of the image assists in replicating different portions in the surrounding environment, making the CNN robust to changes due to light, shadows, or exposure of the camera.

Random Cropping: This technique focuses on cropped image patches and targets smaller parts of the input image. For instance, targeting a diseased area on a leaf enables the CNN to capture microscopic details of the disease, which refines its classification abilities. Increasing data, which increases the variety of images, reduces overfitting as the CNN has to deal with multiple images during training, which in turn increases its performance in regard to real-life problems.

Generalization in the unseen data is ensured through the use of pre-processing techniques such as removing noise and unnecessary data, enhancing visibility of certain parts, and data division. This is important, especially in complex areas such as IoT-based monitoring of farms, where images can be influenced by noise from the sensor, the environment, or pollution. Making use of these steps allows the CNN to always focus on the crucial details of the images, irrespective of disturbances from the environment and other changes, which improves the function as a whole. This procedure is important because it allows the data set to have uniformity throughout, irrespective of exposure/lighting conditions that may skew the model. Such skewness may affect the prediction, and therefore normalization ensures that this does not happen. This is especially useful in agricultural datasets, where the images may be taken in different conditions, such as sunlight and background shading. Correction of such inconsistencies enables the CNN to train effectively.

Global Normalization: This approach modifies all pixel values in the image so that they are all in the same value range, between 0 and 1, for example. Normalizing an image set by exposure and lighting conditions guarantees that the CNN focuses on the true object's features (leaf texture or color) instead of other features created by variations that are not relevant to the actual leaf being analyzed.

Color Space Conversion: Another method of normalization is changing images from RGB to other color models like HSV or YUV. For example, in the detection of plant disease, color space conversion can be effective because certain disease symptoms may appear in specific channels, therefore facilitating diagnosis. For example, when converted from the RGB to the HSV color space, models become sensitive to hue and saturation, which could, for example, allow for easier detection of diseased leaves that are turning yellow or brown. The process of image normalization promotes the homogeneity of the dataset, making it possible for the CNN to recognize essential features such as changes in texture or pigmentation associated with plant diseases instead of memorization of the lighting conditions. In many agricultural images, on the leaves, the difference between diseased and healthy parts may not be clear-cut at first. The use of contrast enhancement techniques helps these regions to become clearer, allowing for easier identification of symptoms of the disease by the CNN.

Histogram Equalization: This approach normalizes the intensity distribution of the image so that the original mass of pixels most present in the image is spread out across the entire range of connecting intensities. It improves the contrast in some parts of the plant's leaves or any other portions of the image, which are faint or discolored and cannot be seen well at first [18–21].

13.4 RESULT AND DISCUSSION

The incorporation of CNNs in the detection of plant diseases is revolutionizing agricultural practices by providing a more rapid and precise alternative to the cumbersome classical methods. With food security emerging as one of the serious issues of concern worldwide, the capability of detecting plant diseases on occurrence is necessary to reduce the amount of loss that the crops would incur. The manual methods involve a lot of work as well as problems associated with inaccuracy. With the help of the CNN, the timely diagnosis and treatment of disease reduces the reliance on chemical applications, thereby making it possible for low-input farming methods. These networks make it possible to detect signatures of disease, such as discolored spots or patches on crop leaves, after subjects are trained on extensive data sets, making it possible to be used across various types of crops and locations without bulk feature proper extraction. Noise filtration and image standardization are examples of pre-processing that serve to improve the results by enhancing the quality of the images, while data enhancement helps to promote the strength of the model to withstand differences in environments. In the past, there have been some developments in assessing and evaluating if a certain plant species has been infected with a disease using CNN networks. However, there are still several challenges that they have not resolved. One such difference is being able to use similar devices that capture images of leaves in the same light and climate settings.

A CNN model can perform poorly with images that were captured in poor lighting or with too much noise in the environment (e.g., a lot of dust or even rain). Further, most of the models are trained on clean, consistently labelled images in a controlled environment. Such conditions might not always apply in the real-world agricultural settings and reduces the accuracy of agricultural deployable models trained with the images collected from a perfect environment. Lack of detailed and varied datasets for particular crops or diseases is another major problem. Other datasets, such as PlantVillage, have become popular, but they do not represent every possible type of infection of the crop. Thus, this incapacitates the capability of the model to accurately predict the type of infection variety of that particular crop type.

In addition, computing constant or limitations is a challenge to deploying such models on devices such as Raspberry Pi with a lower compute capacity. While models, for example MobileNet models, have been made, ensuring accurate recognition in real-time scenarios still remains a hurdle. To address these challenges, future work should focus on several directions: (1) transfer learning and leveraging pretrained models (e.g., ImageNet) and fine-tuning them for plant disease detection can reduce the need for large, labeled datasets and improve generalization. (2) Hybrid Models: the usage of other techniques like SVMs or decision trees along with CNNs may make the models more robust and less sensitive to noise. (3) Mobile and Edge Computing: altering the designs of CNNs for low-powered hardware would enable farmers to have disease-detecting systems even in remote areas with poor network coverage. (4) Model Pruning: model pruning and quantization could make it possible for complex models to be used on devices with limited resources, broadening applications in the real world. One more important aspect is the development of more comprehensive and diverse datasets, which contribute to a wider variety of diseases and environmental conditions to be scanned in future work. Table 13.1 shows the summary of the dataset used and its performance.

TABLE 13.1
Summary of the Existing Methods Based on IoT in Agriculture

Reference	Architecture Type	Dataset Used	Advantage	Disadvantages	Research Gap
[3] Dhannya J. & V. Sangeetha	IoT-based Deep Learning Model	Not Specified	Real-time disease detection	Limited dataset details	Need for scalability & real-world validation
[4] Neda Fatima et al.	IoT-based Deep Learning	Not Specified	Smart greenhouse integration	Energy consumption of IoT devices	Optimization for real-time inference
[5] George Routis et al.	Machine Learning on SBCs	Not Specified	Lightweight ML for IoT	Processing power constraints	Improved model efficiency on SBCs
[6] Akram Abdullah et al.	Deep Learning	Not Specified	Tomato disease classification	Limited to a single crop	Generalization across multiple crops
[7] Tanilal Bhavsingh Thakur & Amit Kumar Mittal	ML in Cloud IoT	Not Specified	Cloud-based classification	Dependence on internet connectivity	Edge computing for resilience
[8] Aniqa Dilawari et al.	Deep Metric Learning	Video Dataset	Near-duplicate video detection	High computation cost	Lightweight feature extraction
[9] L. Manjusha & V. Suryanarayana	Deep Learning	Image Dataset	Duplicate image removal	Dataset-specific approach	Generalization for diverse datasets
[10] Bhaskar Mehta et al.	Machine Learning	Visual Spam Dataset	Spam detection	False positives in detection	Robust detection algorithms
[11] Yi Zhang et al.	CNN	Image Pairs Dataset	Learning duplicate images	Computationally expensive	Efficient model for large-scale datasets
[12] A. Usha Ruby et al.	Modified ResNet50	Wheat Leaf Dataset	Improved classification accuracy	High model complexity	Model simplification without accuracy loss
[13] Vinay Gautam et al.	Ensembled DNN	Mango Leaf Dataset	High classification accuracy	Computational overhead	Lightweight ensemble models
[14] Sumitra Samal & Vijayant Verma	Self-developed CNN	Tomato Leaf Dataset	Custom model for detection	Model training cost	Comparison with existing architectures

(*Continued*)

TABLE 13.1 (CONTINUED)

Reference	Architecture Type	Dataset Used	Advantage	Disadvantages	Research Gap
[18] Puja Singla et al.	Deep CNN	Not Specified	Convolutional model efficiency	Lack of dataset details	More dataset diversity
[19] Nicholas James et al.	Distributed Deep Learning	Raspberry Pi Cluster	Parallel processing	Limited processing power	Optimization for embedded devices
[20] Toan Khac Nguyen et al.	DNN on Raspberry Pi	Chrysanthemum Dataset	On-device disease recognition	Latency in processing	Low-latency model implementation
[21] Mitali V. Shewale & Rohin D. Daruwala	High Performance DL	Not Specified	Early detection of diseases	Model complexity	Balancing accuracy and efficiency
[1] Ankita Muhuri & Suryakanta Sethy	IoT for Agriculture	Not Specified	Smart agriculture	Limited AI integration	Deep learning-based IoT improvements
[2] U. Meenakshi et al.	Raspberry Pi-based System	Soil Parameters Dataset	Embedded analysis	Processing constraints	Low-power computation
[15] M. P. Mathew et al.	CNN-based DL	Bell Pepper Dataset	Accurate classification	Limited dataset	More real-world testing
[16] M. P. Mathew et al.	CNN	Bell Pepper Dataset	Bacterial spot identification	Needs dataset expansion	More diverse disease detection
[17] M. P. Mathew et al.	DL with Optimization	Bell Pepper Dataset	Comparison of optimizers	Optimizer dependency	Hybrid optimization techniques

13.5 CONCLUSION

Efforts remain in progress to ensure that CNNs are usable in practice. Operational challenges such as changes in light and temperature can affect the performance measures of the model. Hybrid techniques such as transfer learning could perhaps give better adaptability, while the emergence of low-power technologies such as TinyML and edge computing makes it practical to start deploying CNNs on mobile devices for on-field utilization. Furthermore, surgical approaches have started to evolve, for example, ESDNN, which increases the number of interstage connections between a constellation of CNNs, resulting in better classification outcomes. Optimally, it has been proven to be helpful in detecting disease in crops such as mangoes and tomatoes. Also, convolutional networks combined with the web can enhance real-time applications for smart farms using disease detection algorithms that can carry out automatic control depending on the disease rate. With the development of algorithms for the architecture of CNNs and the optimization of the used hardware, these will contribute toward enhancing food security, increasing crop yield, and promoting the sustainability of farming systems.

REFERENCES

1. Hughes, D. P., & Salathé, M. (2016). An open access repository of images on plant health to enable the development of mobile disease diagnostics. *arXiv preprint arXiv:1511.08060.*
2. Singh, D., Jain, N., Jain, P., Kayal, P., Kumawat, S., & Batra, N. (2020). PlantDoc: A dataset for visual plant disease detection. *Proceedings of the 7th ACM IKDD CoDS and 25th COMAD*, 5–7 January 2020, Hyderabad, India. ACM. https://doi.org/10.1145/3371158.3371196
3. Dhannya J. and Sangeetha V. "An IoT-Based Model for the Classification of Plant Diseases using Deep Learning." Research Scholar, Karpagam Academy of Higher Education, Coimbatore, and Professor, Dept. of Computer Science, Karpagam Academy of Higher Education, Coimbatore.
4. Neda Fatima, Siddiqui Salman Ahmad, and Ahmad Anwar. IoT-based Smart Greenhouse with Disease Prediction using Deep Learning. *International Journal of Advanced Computer Science and Applications* 12, no. 7 (2021).
5. George Routis, Michailidis Marios, and Roussaki Ioanna. "Plant Disease Identification Using Machine Learning Algorithms on Single-Board Computers in IoT Environments." Department of Computer Science, National Technical University of Athens; School of Electrical Engineering, National Technical University of Athens, 2024.
6. Abdullah, A., Amran, G. A., Tahmid, S. M., Alabrah, A., AL-Bakhrani, A. A., & Ali, A. (2024). A Deep-Learning-Based Model for the Detection of Diseased Tomato Leaves. *Agronomy*, 14(7), 1593. https://doi.org/10.3390/agronomy14071593
7. Thakur, T. B. and Mittal, A. K., "Real Time IoT Application for Classification of Crop Diseases using Machine Learning in Cloud Environment," *International Journal of Innovative Science and Modern Engineering (IJISME)*, 6(4), 1–5, Jan. 2020, doi: 10.35940/ijisme.D1186.016420
8. Dilawari Aniqa, Iqbal Sajid, Syed Farial, and Ilyas Qazi Mudassar. "Deep Metric Learning for Near-Duplicate Video Retrieval Leveraging Efficient Semantic Feature Extraction." Department of Computer Science and Information Technology, University of Home Economics, Lahore, Pakistan; King Faisal University, Saudi Arabia; University of Regina, Canada, 2024.

9. Manjusha L. and Suryanarayana V. "Detect/Remove Duplicate Images from a Dataset for Deep Learning." M.Tech Student, Dept. of CSE, Ramachandra College Of Engineering, A.P, Eluru, India; Professor & HOD, Dept. of CSE, Ramachandra College of Engineering, A.P Eluru, India.
10. Mehta Bhaskar, Nangia Saurabh, Gupta Manish, and Nejdl Wolfgang. "Detecting Image Spam using Visual Features and Near Duplicate Detection." Google Inc., Zurich, Switzerland; IIT Guwahati, Assam, India; L3S Forschungszentrum, Hannover, Germany.
11. Zhang Yi, Zhang Yanning, Sun Jinqiu, Li Haisen, and Yu Zhu. "Learning Near Duplicate Image Pairs using Convolutional Neural Networks." School of Computer Science and Engineering, Northwestern Polytechnical University, Xi'an, China.
12. Usha Ruby A., George Chellin Chandran J., Chaithanya B. N., Swasthika Jain T. J., and Patil Renuka. "Wheat Leaf Disease Classification Using Modified ResNet50 Convolutional Neural Network Model." *Multimed Tools Appl* 83, 62875–62893 (2024). https://doi.org/10.1007/s11042-023-18049-z
13. Gautam V., Ranjan R.K., Dahiya P., and Kumar A.. "ESDNN: A Novel Ensembled Stack Deep Neural Network for Mango Leaf Disease Classification and Detection." Multimed Tools Appl 83, 10989–11015 (2024). https://doi.org/10.1007/s11042-023-16012-6
14. Samal S. and Verma V.. "Tomato Plant Leaf Diseases Detection and Classification Using an Advanced Self-Developed CNN Model." Received: 6 May 2023 / Revised: 1 February 2024 / Accepted: 14 March 2024 © The Author(s), under exclusive license to Springer Science+Business Media, LLC, part of Springer Nature 2024.
15. Mathew M. P., Elayidom S., and Jagathyraj V., "Disease Classification in Bell Pepper Plants Based on Deep Learning Network Architecture," *2023 2nd International Conference for Innovation in Technology (INOCON)*, Bangalore, India, 2023, pp. 1–6. https://doi.org/10.1109/INOCON57975.2023.10101269
16. Mathew M. P., Elayidom S., and Jagathiraj V. P. "Bacterial Spot Classification in Bell Pepper Plants Based on Convolutional Neural Network Architecture". In: Pundir, A. K. S., Yadav, A., Das, S. (eds) *Recent Trends in Communication and Intelligent Systems. ICRTCIS 2023. Algorithms for Intelligent Systems*. Springer, Singapore, 2023. https://doi.org/10.1007/978-981-99-5792-7_7
17. Mathew M. P., Elayidom S., V. P. Raj, and Mahesh T. Yamuna. "A Comparison of Optimization Techniques DeepLearning Models Based on Bell Pepper Leaves Diseases Classification." *2023 Annual International Conference on Emerging Research Areas: International Conference on Intelligent Systems (AICERA/ICIS)*, Kanjirapally, India, 2023, pp. 1–7. https://doi.org/10.1109/AICERA/ICIS59538.2023.10420205
18. Singla Puja, Kalavakonda Vijaya, and Senthil Ramalingam. "Detection of Plant Leaf Diseases using Deep Convolutional Neural Network Models." *Multimedia Tools and Applications*, *83*(24), 64533–64549.
19. James Nicholas, Ong Lee-Yeng, and Leow Meng-Chew. "Exploring Distributed Deep Learning Inference Using Raspberry Pi Spark Cluster." *Future Internet* 14 (8) (2022), 220.
20. Nguyen Toan Khac, Dang L. Minh, Do Truong-Dong, and Lim Jin Hee. "The First Study of White Rust Disease Recognition by Using Deep Neural Networks and Raspberry Pi Module Application in Chrysanthemum." *Inventions*, *8*(3) (2023), 76.
21. Shewale Mitali V. and Daruwala Rohin D. "High Performance Deep Learning Architecture for Early Detection and Classification of Plant Leaf Disease." *Journal of Agriculture and Food Research* 14 (2023): 100675.

14 Machine Learning-Driven Millet Processing for Quality Enhancement

Abubakar Kabir Adam, Tithli Sadhu, and Idris Kaida Zubairu

14.1 INTRODUCTION

The global population has experienced an increasing rate of malnutrition and food insecurity, most of which surpasses the economic well-being of individuals, specifically in developing and underdeveloped countries. Thus, several factors led to the current food insecurity situation, which includes climate change compounded with economic instability [1]. It is therefore imperative to transform the food system into a sustainable way to achieve global food security. Millets belong to the Poaceae family and are considered climate-resistant plants largely grown in India, China, Niger, Nigeria, and other tropical regions [2]. India ranks as the highest millet-producing country with 42%, Niger (11%), and China (8%), while other countries share the remaining 39% in 2022–2023 [3]. The global interest in millet production increased because of its potential to ensure sustainability pertaining to agricultural development and reduce the impact of climate change [4]. Millets offer the potential to enhance balanced diets and elevate the nutritional well-being of global populations facing malnourishment [5]. Millets are a sustainable source of nutrients that enhance nutritional food security and support the economic stability of the populations [6]. They are regarded as a powerhouse of nutrients with high dietary fiber, phytochemicals, proteins, and mineral content, thereby being called a nutricereal. Millet is considered a smart cereal grain with its important potential as gluten-free and low glycemic index. It is a future food crop to alleviate food insecurity and synergistically cope with the micronutrient deficiencies affecting vulnerable populations. Major varieties include foxtail millet, finger millet, proso millet, pearl millet, and kodo millet, which have shown a potential to meet global nutritional demands. The Food and Agriculture Organization (FAO) declared 2023 the "International Year of Millets," highlighting millet's potential to address global food demand.

Millets have remarkable investment capacity while subject to food product developments. Although millets can substitute other cereal crops, especially gluten-containing cereals, few millet-based foods are available in grocery and urban communities. Many foods developed from millet are processed through traditional

DOI: 10.1201/9781003594093-14

methods in rural areas, depending on the types of dishes to be prepared. Subaitha et al. (2024) stated that despite the health potentials of millet, its inclusion in the diet of many populations is limited, citing the preparation and storage methods as the major challenges [6]. In addition, the processing methods significantly impact the quality characteristics of the millets. The lack of contemporary advanced techniques to showcase the full potential of millet has been identified as a barrier to the development of millet food varieties.

Techniques, including germination, fermentation, milling, baking, and boiling, alter the quality characteristics of the food across the value chain. This depends upon the functional and chemical properties of the food products, which are interconnected with the transformation in the physicochemical properties during processing. Therefore, to meet global nutritional needs, studies have explored advanced and minimal processing techniques to enhance nutrient availability in food. Attempts were also made to measure the post-nutritional composition of processed millet and other cereals. Yet, despite this research focus, the objectives for prediction models are compelling for nutritional composition and quality characteristics of cereal grains, including millets. This study aims to understand the use of machine learning (ML) models to predict millet quality characteristics during processing.

14.2 MILLET VARIETIES AND NUTRITIONAL COMPOSITION

14.2.1 Finger Millet

Finger millet, as in Figure 14.1a, is an important variety found in developing countries such as India, Nigeria, Zambia, and Nepal, accompanied by certain Asian nations, including China [7]. Finger millet is a nutritious food used widely in porridge, flour, and bakery products like bread and cakes [8]. It is also widely accepted for producing liquor, breakfast, chappati, and ragi. In addition, it possesses an important source of nutrients for the body's functions, including protein, dietary fiber, vitamins, and minerals, as depicted in Table 14.1.

14.2.2 Pearl Millet

Pearl millet, as in Figure 14.1b, ranks as the most important millet species widely available in India and subtropical countries of Africa. It is a unique hardy grain that can be cultivated in temperate climatic conditions, which has recently been widely utilized in commercial-scale food industries. Pearl millet has more nutritional value than other varieties, as depicted in Table 14.1, thus making it useful in many traditional beverages, breakfast foods, and cookies produced from this millet. The grain has huge food potential because of its lack of gluten and nutritional content, such as dietary fiber, than some staple foods, including rice [11]. The production of this millet expands to non-traditional practices as iron and zinc-biofortified varieties have been developed to improve global nutritional security [12].

14.2.3 Proso Millet

Proso millet, as in Figure 14.1c, is another variety that appears in different colors, including cream, yellow, brown, white, orange, black, to red, and is smaller compared to other millets such as pearl [13]. The global production of proso millet is ranked third after pearl and foxtail millets and is used in flour blends for bread-making, flakes, steamed, and boiled. Proso millet is known for its exceptional health benefits and rich nutritional profile, comparable to other staple crops [14]. It is a rich source of essential amino acids, vitamins, minerals, phenolic compounds, and starch, as depicted in Table 14.1. Proso millet contains antioxidants, beta-glucan, and lecithin, which support the body's health [15].

14.2.4 Kodo Millet

Kodo millet originated from India, known popularly as *arakalu*, cultivated extensively across diverse agricultural climatic regions. Kodo millet is regarded as a variety of millet resistant to drought and a resilient crop that can grow over a wide land coverage [16]. The color of this millet, in Figure 14.1d, differs from dark grey to light red, which is covered by a hardy husk [17]. It forms part of the main staple foods and is a vital source of nutrients with high protein, dietary fiber, and reduced fat content, and possesses a speedy digestibility property, as depicted in Table 14.1.

14.2.5 Foxtail Millet

Foxtail millet, in Figure 14.1e, is another significant type and stands as the world's second-largest millet crop in production. It is an ancient cereal grain that is considered to have originated in Asian countries such as India and China [9]. Foxtail millet saves the developing world with six million tons of food for many populations in subtropical areas and southern Europe. Like other millet varieties, Foxtail is also nutrient-dense with no gluten content in the grain, as depicted in Table 14.1. Moreover, besides its nutritional benefits, its glycemic index is reduced compared to other staple cereal crops, and it is often considered for wider food applications [18]. Foxtail millet is used in various foods, such as porridge, pancakes, muffins, cookies, and flatbread [19].

14.2.6 Barnyard Millet

Barnyard millet, in Figure 14.1f, is a versatile grain used for food and forage. Recently, barnyard millets have shown an upsurge in production worldwide, with India as the biggest producing country, and becoming one of the most important minor millets in less-developed nations [20]. This millet is an important source of dietary fiber and a functional protein source. It comprises low and slowly digestible carbohydrates, which increases its unique food application to the modern world, especially for people with sedentary lifestyles. Barnyard is also advised for individuals with cardiac conditions and diabetes because it generates significant levels of resistant starch.

FIGURE 14.1 Millet varieties (a) finger millet; (b) pearl millet; (c) proso millet; (d) kodo millet; (e) foxtail millet; (f) barnyard millet.

TABLE 14.1
Nutritional Composition of Millet Varieties

Millet Variety	Carbohydrates (%)	Protein (%)	Total Lipid (%)	Total Ash (%)	Dietary Fiber (%)	Iron (mg/g of N)	Zinc (mg/g of N)	Thiamine-B1 (mg)	Riboflavin-B2 (mg)	Niacin-B3 (mg)
Finger millet	66.82	7.16	1.92	2.04	11.18	4.62	1.51	0.37	0.17	1.34
Pearl millet	61.78	10.96	5.43	1.37	11.49	6.42	2.95	0.25	0.20	0.86
Proso millet	70.04	12.50	1.10	3.6	8.5	3.3	1.81	0.41	0.28	4.50
Kodo millet	66.19	8.92	2.55	1.72	06.39	2.34	-	0.29	0.20	1.49
Foxtail millet	60.09	12.30	4.30	3.77	7.06	4.22	2.68	0.59	0.11	3.20
Barnyard millet	65.55	6.20	2.20	3.8	7.6	9.3	2.6	0.33	0.10	4.20

Source: Sharma et al. (2021) and Devi et al. (2014) [9, 10]

14.3 MILLET PROCESSING

Millet processing involves transforming its primary constituents, including the germ, the starch-containing endosperm, and the protective pericarp. Various methods are used to convert millets into processed or semi-processed products, including traditional and advanced techniques. Despite technological advancements, traditional methods are still employed in many developing countries to process millet, especially where it is considered a staple food. However, these traditional processing methods are energy-consuming and ineffective in achieving target millet product development [21]. Purposely, the methods are used to produce local dishes. Some common traditional methods used in millet processing include decortication, malting, fermentation, roasting, flaking, and grinding (see Figure 14.2). Logically, it was explored that these methods have contributed to limiting the significant importance of millet food diversity. This is attributed to the lack of technical know-how among the local processors for manipulating various processing parameters to enhance the efficient utilization of millet. Processing millet poses a significant challenge because its tough outer layer demands abrasive force to soften it for subsequent handling. The advanced techniques may include non-thermal processes. Millet processing, such as microwave drying, hydrothermal, ohmic, and infrared heating, is gaining recognition from many food industries. The non-thermal treatments of millet processing involve cold plasma, high-pressure, ultrasonic, pulsed electric, and enzyme treatment, which are widely employed to efficiently eliminate the potential barriers to millet utilization.

14.3.1 Traditional Methods

Millets are primarily subjected to traditional methods at the local level, which are inefficient and resource-consuming. Dehulling, drying, soaking, malting, germination, fermentation, milling, and grinding are common traditional methods used during millet processing. These methods, used mostly in rural communities, are not feasible for large commercial scale and remain a basic requirement for secondary processing of the millet grain.

14.3.1.1 Soaking

Soaking is a pretreatment method used to loosen the hull of millet grains. The method involves simultaneous water absorption and swelling influenced by the soaking temperature and time. As the temperature increases, the water absorptivity also increases, which affects the diffusion rate of the water by which the grain is immersed. However, once the grain reaches saturation, the absorptivity remains constant. The purpose of this method in millet treatment is to loosen the hull, thereby increasing the hardness of the endosperm during drying, which enhances the effectiveness of dehulling operation. Insufficient understanding of the necessary soaking time and temperature can negatively impact the physicochemical properties of millet, highlighting the drawbacks of relying on traditional techniques.

14.3.1.2 Drying

Drying involves evaporating moisture from millet grains and minimizing water activity to prevent deterioration and improve dehulling efficiency. In rural areas, millet is mainly spread out in open, exposed spaces to ensure direct contact with air and sunlight. This is reported to have numerous disadvantages, including inefficient drying of the whole millet grains, time-consuming, insect infestation, and other factors that

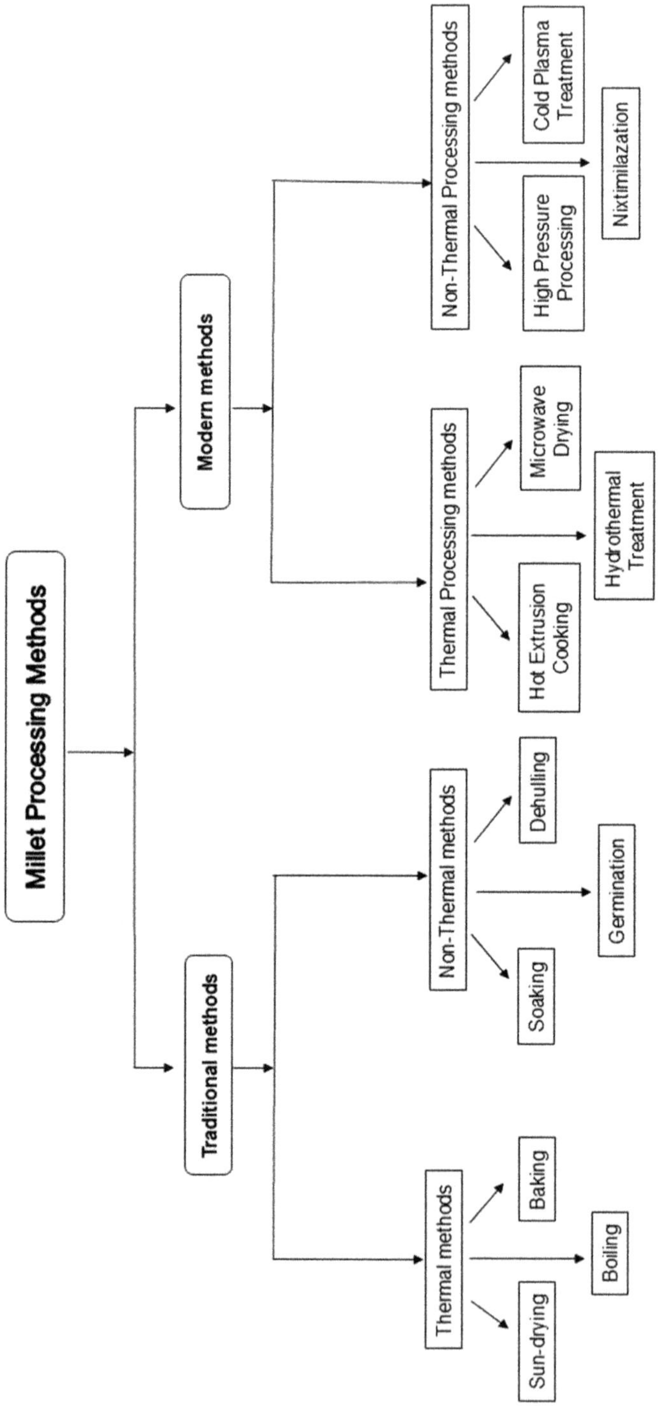

FIGURE 14.2 A schematic diagram of different types of millet processing methods.

affect the final quality of the millet products. However, fluidized bed drying is a suitable drying technique, which emerges due to its high rate of drying and efficient results during contact between the millet grains and the medium. In addition, food industries have used other drying methods such as hybrid drying, supercritical carbon dioxide, and microwave drying for grain, including millets [2]. Millets are harvested with a moisture content of 15–20% and then dried to bring the moisture level down to 10–12% or lower to ensure proper storage [2]. This increases the endosperm's hardness, which makes it easier to dehull from the millet grain. Furthermore, drying affects the physical qualities of the millet grain, such as size and shape, enhancing further processes. Moreover, drying is an important measure to reduce water activity that impacts and prevents microbial activity, thereby maintaining the quality properties of the cereal grains and extending the shelf life.

14.3.1.3 Dehulling

Dehulling involves hitting the millet grains onto the hard surface using a force that leads to cracks in the outer covering to remove the endosperm from the hull. Traditionally, millet dehulling involves using a mortar and wooden or stone surfaces to remove the hulls. Various factors can affect the efficiency of the dehulling process, including the contact surfaces, amount of force employed, moisture content, size, and shape. In major millets, the increase in moisture content will help remove the pericarp and affect the endosperm's outer layer [2]. Higher moisture level increases the hull's brittleness and softness, making it easier to remove through abrasion, compression, or shearing. As earlier stated, to obtain an efficient dehulling process in millet processing, pretreatment is employed, such as soaking and drying, to facilitate a smooth and rapid removal of the husk.

14.3.2 Modern Techniques

The modern techniques employed in millet processing encompass thermal and non-thermal methods. These methods can be classified based on the primary and secondary processing techniques. The common primary thermal treatments, including infrared drying, ohmic heating, and microwave drying, are used in large-scale industries for millet processing. Moreover, flaking, superfine grinding, extrusion cooking (cold and hot temperature processing), and baking are secondary thermal processing techniques used in developing millet products. Recent advances highlighted novel methods used in millet processing, including high-pressure processing, pulsed electric field, cold plasma, and ohmic heating to transform millet food applications.

14.3.2.1 Microwave Drying

Microwave drying is a pretreatment that utilizes an electro-thermal method, converting electrical energy into thermal energy. The generated heat serves as the medium of drying. Electromagnetic waves cover a frequency range of 300 MHz to 300 GHz, with 915 MHz and 2.45 GHz being the most used for efficient drying. Rao et al. (2021) reported that drying foxtail millet using a microwave oven increases the rate of dehulling efficiency and subsequently varies depending on exposure time and power level [22].

14.3.2.2 Hydrothermal Treatment

Hydrothermal treatment involves subjecting the grains to hot water or high-pressure steam for a determined time. This process helps to loosen the hull in a cereal grain, thereby increasing the efficiency of the milling operation. The method is comprised of steaming, soaking, and drying. It was observed that the water absorption rises and remains unchanged after saturation. The hydrothermal method is an important technique to reduce the breakage of the cereal grains. As reported by Rathore and Singh (2017), the impact of the hydrothermal process on some millet-like pearl is positive in enhancing the decortication operation [23].

14.3.2.3 Extrusion

Extrusion involves chemical and physical changes resulting from certain processing conditions. The milled grain can first be mixed in the barrel to obtain a uniform powder and then pasted by subjecting it to shear forces, pressure, and high temperature, accompanied by protein unfolding and gelatinizing starch [18].

When the paste is allowed to flow from the outlet, the pressure decreases, causing water to evaporate with immediate structural expansion and a drop in temperature.

Extrusion cooking is used to produce a variety of millet products and also plays an important role in the inactivation of enzymes and microorganisms in millets. It can also help to destroy antinutritional factors and natural toxic substances in millet products while improving protein digestibility [24].

14.3.2.4 Superfine Grinding

Superfine grinding is an advanced industrial technology that has developed rapidly in the past two decades. Superfine grinding is widely used in food industries, besides other applications such as in metallurgy and medicine [25]. The technique, in comparison to the traditional methods, improves the fineness of the flour and increases the nutritional bioavailability. Research has been conducted on the effect of superfine grinding on the properties of millet starch and concluded that the method can affect the structure of starch granules and reduce their size [26]. However, the temperature during gelatinization decreases but increases the solubility of the starch granules.

14.3.2.5 Cold Plasma Treatment

Plasma describes a state of matter consisting of partially ionized gas. Its temperature is defined by the interaction between electrons and ions. The effectiveness of cold plasma treatment on the millet grain depends on the parameters, including grain constituents, excited states of the ions, and atom species. Sun et al. (2022) studied the morphological changes of a proso millet starch structure treated under combined treatments (cold plasma and ultra-high-pressure techniques) [27]. The isolated starch does not show a varied change due to the combination treatments but instead improves starch digestibility and enhances pasting properties. Furthermore, Jaddu et al. (2022) highlighted that the functional properties of millet flour are enhanced when subjected to cold plasma treatments [28].

14.3.2.6 High-Pressure Processing

High-pressure processing (HPP) is a cutting-edge non-thermal technique that applies consistent pressure to food materials to enhance their quality. This technique plays a crucial role in the food industry for processing cereal grains, irrespective of their shape or size [29]. HPP shows a minimal effect on the nutritional composition, sensitive compounds, pigments, and other vitamins compared to heat treatments. The processing duration of high pressure for food materials ranges from a few seconds to 20 minutes, utilizing pressures between 100 and 800 MPa, and can occur at low and high temperatures [2]. A separate study by Zhang et al. (2022) shows that high-pressure processing treatment alters the distribution of millet protein, resulting in growth in particle size, which in turn influences the hydrophobic properties of millet compounds [30].

14.3.2.7 Nixtamalization

Nixtamalization involves dipping the cereal grain into alkaline solutions (lime, lye, and wood ashes). The soaked grain is followed by rinsing with water, which is then ground to remove the hulls [2]. The efficacy of the treatment rests on the types of alkaline solutions that facilitate the solubility of the grain's outer covering, resulting in the removal of the husk. The importance of the process was traced in the enhancement of certain pretreatments, including dehulling, which resulted in improving the quality characteristics of the grains [31]. The resulting incorporation of the calcium hydroxide causes a drop in the antinutrients of cereal grains, mostly including tannins, trypsin, and phytate. It is due to the activation of phytase enzymes, which speed up phytic acid breakdown.

14.3.3 Challenges in Processing

The processing of millet encounters challenges, which directly impact the quality of the final products. Considering conventional ways of millet processing, such as dehulling, drying, and fermentation, were limited in terms of their intensive utilization due to the expense of labor and other resource consumption, yielding inefficiencies in large-scale operations [2]. The outcome shows a reduction in the nutritional content and compromises the presumed quality of the processed millet [21]. In addition to the previous reason, millet grain is also surrounded by a harder outer covering, which poses a barrier that may need an abrasive stress to facilitate the efficiency for further processing. While these cost a lot for effective optimization of the resources, millet, which is conventionally processed especially in rural areas, has also been hindered for high-capacity utilization due to low level of technical know-how for modern processing equipment [2]. Overall, addressing these challenges is crucial for improving processing efficiency and maximizing the nutritional benefits of millet products.

14.4 MODELLING FOR NUTRITIONAL AND QUALITY EVALUATION IN MILLET PROCESSING

The nutritional composition of millet before and after processing requires different approaches. A thorough understanding of the influences of pretreatment and post-processing on the nutritional composition of the food product as a holistic approach is needed. Recently, various approaches, including retention factors, have been employed to calculate the post-processing nutritional composition of food products. This is based on estimates derived from the analytical composition data of a food sample. This technique is frequently employed in the food industry for nutritional labeling on packaging and is utilized in dietary surveys carried out by the United States Department of Agriculture (USDA). These approaches have shortfalls in their applications. Similarly, kinetic models have a bottleneck in addressing the food processing parameters, because the methods used are resource-consuming and entail experimental difficulties. Therefore, these challenges can be addressed using non-destructive and scalable approaches to predict the post-processing food processing parameters for achieving high quality. An important approach is using ML to interpret analytical data. In food industries, ML predictive models are developed and trained to learn the multi-parametric transformation of statistical data and changes that occur during the processing of raw foods using different techniques.

14.4.1 Predicting Processing Parameters with ML: Case Study I on Advancements in Cold Plasma Treatment and ML Applications for Millet Seed Priming

ML is a branch of artificial intelligence that explains the patterns of analyzed data. ML can be used to analyze the data from current signals, emitted spectrums, visual images, and many more, making it widely used to predict conformational changes in food products [32]. A study was conducted by Shelar et al. (2022) on the application of ML in cold plasma treatment of millet germination and priming of seeds (see Figure 14.3) [33]. The study highlighted the application of plasma treatment to improve the millet seed quality, while predicting the changes that occurred during processing. They found out that since seed is vulnerable to physical and chemical modifications during germination, priming of the millet grain improves the desired quality characteristics and efficiency of the processing. The large analytical data from cold plasma processing of the millet was analyzed using artificial neural network (ANN) for optimizing the process. They concluded that an ANN-ML model improved millet grain characterization based on quality parameters, directly affecting the final product's nutritional composition. Recently, new models have emerged that transform input data and give real-time information related to the impact of processing methods on biological materials.

Artificial Neural Networks Model general equation:

$$y = N\sum_{i=1}^{N} w_i X_i + b \tag{14.1}$$

Where:

- f = activation function (like ReLU, sigmoid)
- y = neuron output
- X_i = features of input
- w_i = weights
- b = bias

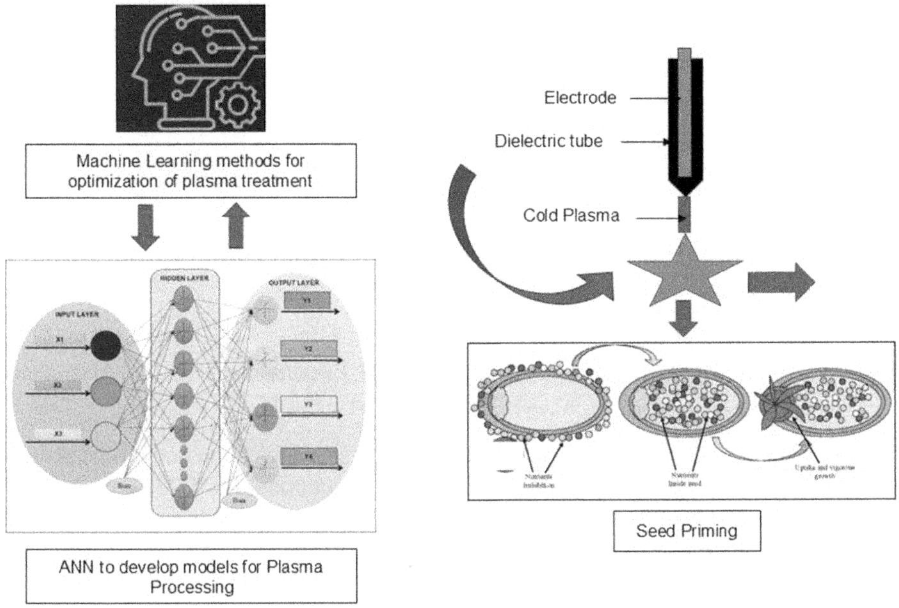

FIGURE 14.3 Machine learning approach for millet seed priming using cold plasma processing.

14.4.2 Case study II: HSI and ML as a Non-Destructive Method for Proso Millet Grain Detection and Classification

As stated earlier, millet has different varieties, and quality evaluation is critical for developing safe and nutritious foods. Generally, the quality of any food product depends on the characteristics of the raw material. The classification of cereal grains, including millet, is very important for food processors and researchers. However, the most up-to-date methods of seed classification include a visual assessment of the morphological features, protein electrophoresis, and phenotype identification [34]. The methods stated were considered limited following the time constraints, labor, complexity, and non-specificity with a subjective approach [35]. A suitable and non-destructive approach is highly required to classify the millet, especially with some varieties having different nutritional compositions from others and differing in texture and appearance. Considering the aforementioned, with some other reasons, Ekramirad et al. (2024) developed a machine-learning model for sorting and grading of millet cultivars with specificity to proso millet variety [35]. The outcome of the study has established a potential knowledge of using Hyperspectral Imaging (HSI) in classifying different proso millet cultivars (see Figure 14.4). During the study, Gradient tree boosting and RF classifiers were used for the seed categorization, and the best classification performance was recorded at accuracy rates of 99.4% and 99.1%, respectively. The study concluded that the ML model is a reliable tool for classifying proso millet cultivars and could have a substantial impact on the millet processing industry by facilitating the selection of high-quality varieties. Linear regression model general equation:

$$y = \beta_0 + \beta_1 X_1 + \beta_2 X_2 + \cdots + \beta_n X_n + \varepsilon \tag{14.2}$$

Where:

- y = predicted outcome (like nutritional value)
- X_1 = independent variables
- β_1 = weights
- ε = residuals

14.4.3 Case Study III: Predicting Nutrient Profile in Food – Using SCS and PINS Models

ML models have been developed to address attributes related to the nutritional profiles of a food. In a study carried out by Naravane and Tagkopoulos (2022) on predicting nutrient profiles using ML, Solid Content Scaling (SCS) and Process-Invariant Nutrient Scaling (PINS) models were used for the assessment [36]. The idea was to select the best-performing model for cross-validation based on six regressor types: Random Forest, Decision Trees, Shrinkage and Selection Operator, Least Absolute, Multilayer Perceptron, and Elastic Net. The models were developed to predict the nutrient profile of processed foods, including starch and its components, protein, fat, dietary fiber, phenolics, soluble sugars, and phytic acid. The models were trained based on the USDA data set for different categories of foods, including cereal grains. SCS is employed to scale the content of various compositions if the food's water content remains constant. The composition of the raw and processed food is measured on a 100 g weight scale and helps to set a gauge for preserving the nutritional content of the food. However, the SCS model shows a bias in the extent of the solid concentration depending upon the processing method. In contrast, the PINS model helps to reduce the biases experienced in the SCS model. The model is developed to identify the nutritional composition and processing influence on the product, which is set on the true weight of the processed food corresponding to raw food at 100 g. The separation of the composition using the PINS model was according to the wet and dry basis of the food product. All the models are arranged to share similar inputs (nutritional composition) for the raw food. The limitation of their study is the exclusion of other processing parameters datasets such as temperature, time, and moisture content. The predicted output (nutrient content) is obtained by multiplying the retention factor and the overall nutrient content of the sample. Root mean square error, the coefficient of determination, Pearson's correlation coefficient, and Spearman's rank correlation coefficient were used to assess the performance of the six regressor models. The study concluded that SCS and PINS models are important tools for predicting carbohydrates, proteins, and other essential nutritional components of foods, including cereal grains.

Solid Content Scaling general equation:

$$Cs = \frac{C_r}{1 - W_r} \tag{14.3}$$

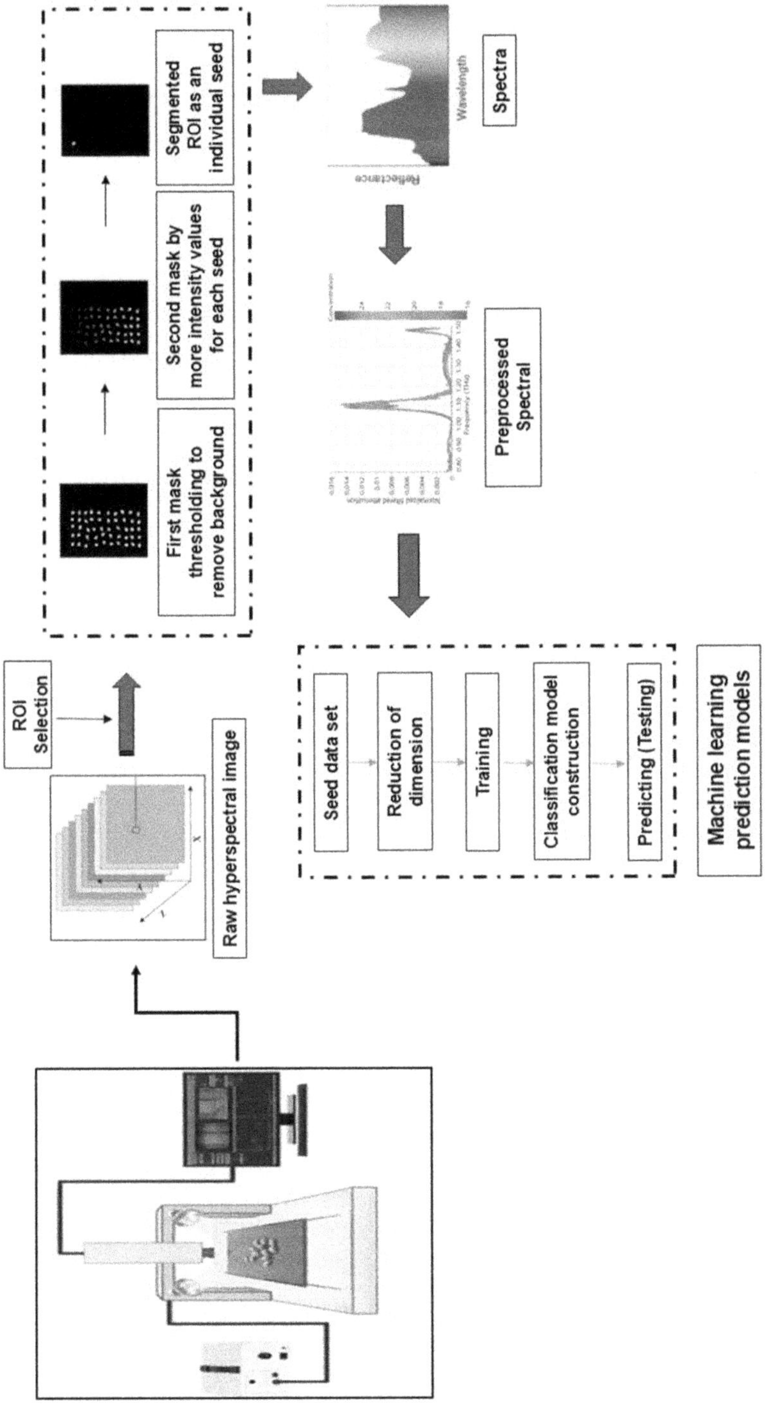

FIGURE 14.4 Hyperspectral image data analysis for proso millet classification using machine learning.

where:

- C_s = solid content
- C_r = concentration of solids
- W_r = water content

Process-Invariant Nutrient Scaling general equation:

$$N_p = \frac{N_r x W_r}{W_p} \tag{14.4}$$

where:

- N_p = nutrient content
- N_r = nutrient content
- W_r = weight of the raw product
- W_p = weight of the processed product

General criteria for the Decision Tree include Gini Impurity and Entropy:

$$\text{Gini} = 1 - \sum_{i=1}^{C} P_i^2 \tag{14.5}$$

$$\text{Entropy} = -\sum_{i=1}^{C} P_i \log_2 (P_i) \tag{14.6}$$

Random Forest general equation:

$$y = \frac{1}{N} \sum_{i=1}^{N} f_i(X) \tag{14.7}$$

where:

- N = number of trees
- $f_i(X)$ = prediction from the i – th decision tree

14.4.4 Case Study IV: ML Prediction of Millet Food Processing Using Foodprox Model

The criteria for classification of processed food remain a challenge for many food industries, with available methods having minimal coverage to differentiate the influence of the degree of processing on the food product quality. This affects food choices, limited knowledge of consumer behavior toward product acceptability, and public health concerns about processed foods. In the year 2021, Menichetti and co-inventors developed FoodProx, an ML algorithm that predicts the impact of food processing on product quality [37]. The study used a NOVA scale for the classification of food processing as follows: raw or minimally processed (NOVA 1), like cereal grains including millets, and other food sources; semi-processed or processed additives (NOVA 2), such as salts; processed foods (NOVA 3), such as cookies, biscuits, and dairy product; and ultra-processed foods (NOVA 4). The FoodProX was developed as an ML classifier that used nutritional parameters as inputs and was trained to predict the degree of different processing techniques on food products. They concluded that FoodProX is a reproducible and scalable means of predicting the influence of processing measures on the nutritional constituents of any food product. FoodProx (FPOs) are also recognized to predict the relationship between processing conditions and nutritional content variables such as starch, protein, minerals, and dietary constituents of food products. It helps in quantifying the overall diet quality of individuals; it is ultimately a measure that can be used to correlate the degree of processing characterizing individual diets and multiple diseases.

14.5 CONCLUSION

Integrating ML into millet processing offers a powerful means to enhance its quality and nutritional value. The millets, being nutrient-dense and resilient to climate variations, hold significant promise in tackling the challenges of global food security and nutrition. This chapter examined traditional and modern processing techniques, emphasizing their impact on millet's quality characteristics. The application of ML models, such as SCS and PINS, offers innovative solutions for accurately predicting and optimizing the nutritional composition of millet during processing. These models are valuable for assessing biochemical parameters such as starch, protein, amylose, dietary fiber, phenolics, oil, phytic acid, and soluble sugars in grain germplasm. Moreover, these advanced models enable food processors to make informed decisions, ensuring that nutritional integrity is maintained while minimizing resource consumption. Additionally, the exploration of non-destructive techniques, such as hyperspectral imaging, further highlights the potential of ML to revolutionize quality assessment practices in the food industry. Despite the promising advancements, challenges remain in the widespread adoption of these technologies, particularly in resource-limited settings. Continued research and investment in ML applications and millet processing methods are essential to overcome these barriers. Future studies should focus on the scalability of ML-driven approaches, their integration into existing food systems, and the exploration of their effects on diverse millet varieties. Ultimately, harnessing the capabilities of ML enhances the processing efficiency of millets and aligns with broader sustainability goals. Advancing millet as a nutritious food source through innovative processing will be vital in the global fight against food insecurity and malnutrition. By embracing these modern technologies, the food industry can unlock the full potential of millets, contributing to improved health outcomes and greater food security for populations worldwide.

REFERENCES

1. Food and Agriculture Organization of the United Nations. *The state of food security and nutrition in the world 2023*. Rome: FAO; 2023.

2. Singh SM, Joshi J, Rao PS. Technological advancements in millet dehulling and polishing process: An insight into pretreatment methods, machineries, and impact on nutritional quality. *Grain & Oil Sci Technol.* 2024;7(3):186–195. https://doi.org/10.1016/j.gaost.2024.05.007
3. FAS-USDA. Production trends, millet [Internet]. 2024 [cited 2024 Jan 23]. Available from: https://www.fas.usda.gov/data/production/commodity/0459100
4. Kheya SA, Talukder SK, Datta P, Yeasmin S, Rashid MH, Hasan AK, et al. Millets: The future crops for the tropics - Status, challenges and future prospects. *Heliyon.* 2023;9(11):e22123. https://doi.org/10.1016/j.heliyon.2023.e22123
5. Kaur G, Ahmadzadeh-Hashemi S, Amir S, Khan ZS, Gulsunoglu-Konuskan Z, Karimidastjerd A, et al. From millet to marvels: An improved future food through innovative processing. *Future Foods.* 2024;9:100367. https://doi.org/10.1016/j.fufo.2024.100367
6. Subaitha ZA, Priyadarshini S, Yoha K, Moses J. Impact of post-harvest processing techniques on the glycemic index of millets. *Food Chem Adv.* 2024;4:100636. https://doi.org/10.1016/j.focha.2024.100636
7. Zubairu IK, Mansir A, Ahmad T. Evaluation of breakfast cereals produced from finger millet, wheat, soybean, and peanut blends. *Int J Food Sci.* 2019;2(1):17–24. https://doi.org/10.47604/ijf.984
8. Abioye VF, Babarinde GO, Ogunlakin GO, Adejuyitan JA, Olatunde SJ, Abioye AO. Varietal and processing influence on nutritional and phytochemical properties of finger millet: A review. *Heliyon.* 2022;8:e12310. https://doi.org/10.1016/j.heliyon.2022.e12310
9. Sharma R, Sharma S, Dar BN, Singh B. Millets as potential nutri-cereals: A review of nutrient composition, phytochemical profile and techno-functionality. *Int J Food Sci Technol.* 2021;1–16. https://doi.org/10.1111/ijfs.15044
10. Devi PB, Vijayabharathi R, Sathyabama S, Malleshi NG, Priyadarisini VB. Health benefits of finger millet (*Eleusine coracana* L.) polyphenols and dietary fiber: A review. *J Food Sci Technol.* 2014;51:1021–1040. https://doi.org/10.1007/s13197-011-0584-9
11. Kumar R, Dhiman M, Sharma L, Dadhich A, Kaushik P, Sharma MM. Nanofertilizers: The targeted nutrient supplier and enhance nutrients uptake by pearl millets (*Pennisetum glaucum*). *Biocatal Agric Biotechnol.* 2022;45:102524. https://doi.org/10.1016/j.bcab.2022.102524
12. Taylor J. Millet pearl: Overview. In *Elsevier eBooks.* 2016. pp. 190–198. https://doi.org/10.1016/b978-0-12-394437-5.00011-5
13. Narciso JO, Nyström L. The genetic diversity and nutritional quality of proso millet (*Panicum miliaceum*) and its Philippine ecotype, the ancient grain "kabog millet": A review. *J Agric Food Res.* 2023;11:100499. https://doi.org/10.1016/j.jafr.2023.100499
14. Das S, Khound R, Santra M, Santra D. Beyond bird feed: Proso millet for human health and environment. *Agriculture.* 2019;9(3):64. https://doi.org/10.3390/agriculture9030064
15. APEDA. Indian millets [Internet]. 2024. Available from: https://apeda.gov.in/IndianMillets
16. Ravikesavan R, Jeeva G, Jency JP, Muthamilarasan M, Francis N. Kodo millet (*Paspalum scorbiculatum* L.). In *Elsevier eBooks.* 2023. pp. 279–304. https://doi.org/10.1016/b978-0-323-90537-4.00019-3
17. Bunkar DS, Goyal SK, Meena KK, Kamalvansh V. Nutritional, functional role of kodo millet and its processing: A review. *Int J Curr Microbiol Appl Sci.* 2021;10(1):1972–1985. https://doi.org/10.20546/ijcmas.2021.1001.229
18. Yang T, Ma S, Liu J, Sun B, Wang X. Influences of four processing methods on main nutritional components of foxtail millet: A review. *Grain Oil Sci Technol.* 2022;156–165. https://doi.org/10.1016/j.gaost.2022.06.005
19. Rao MV, Sunil C, Venkatachalapathy N. Effect of microwave and hot air radiofrequency treatments on physicochemical and functional properties of foxtail millet flour and its protein isolate. *J Cereal Sci.* 2023;114:103774. https://doi.org/10.1016/j.jcs.2023.103774

20. Renganathan VG, Vanniarajan C, Karthikeyan A, Ramalingam J. Barnyard millet for food and nutritional security: Current status and future research direction. *Front Genet.* 2020;11. https://doi.org/10.3389/fgene.2020.00500
21. Birania S, Rohilla P, Kumar R, Kumar N. Post-harvest processing of millets: A review on value-added products. *Int J Chem Stud.* 2020;8(1):1824–1829. https://doi.org/10.22271/chemi.2020.v8.i1aa.8528
22. Rao MV, Akhil KG, Sunil CK. Effect of microwave treatment on physical and functional properties of foxtail millet flour. *Int J Chem Stud.* 2021;9:2762–2767. https://doi.org/10.22271/chemi.2021.v9.i1am.11641
23. Rathore S, Singh K. A critical optimization study on hydrothermal treatment for decortication of pearl millet to improve its consumption efficiency. *J Food Meas Charact.* 2017;11:1501–1515. https://doi.org/10.1007/s11694-017-9529-y
24. Li X. *Effect of extrusion on the characteristics of soybean dregs compound powder and its application [dissertation].* Harbin: Harbin University of Commerce; 2020.
25. Zhao XY, Ao Q, Yang LW, et al. Application of superfine pulverization technology in biomaterial industry. *J Taiwan Inst Chem Eng.* 2009;40(3):337–343. https://doi.org/10.1016/j.jtice.2008.10.001
26. Muttakin S. Tailoring physicochemical and sensorial properties of defatted soybean flour using jet-milling technology. *Food Chem.* 2015;6. https://doi.org/10.1016/j.foodchem.2015.04.104
27. Sun X, Saleh AS, Lu Y, et al. Effects of ultra-high pressure combined with cold plasma on structural, physicochemical, and digestive properties of proso millet starch. *Int J Biol Macromol.* 2022;212:146–154. https://doi.org/10.1016/j.ijbiomac.2022.05.128
28. Jaddu S, Pradhan RC, Dwivedi M. Effect of multipin atmospheric cold plasma discharge on functional properties of little millet (*Panicum miliare*) flour. *Innov Food Sci Emerg Technol.* 2022;77. https://doi.org/10.1016/j.ifset.2022.102957
29. Muntean M, Marian O, Barbieru V, Cătunescu GM, Ranta O, Drocas I, et al. High pressure processing in food industry – characteristics and applications. *Agric Agric Sci Procedia.* 2016;10:377–383. https://doi.org/10.1016/j.aaspro.2016.09.077
30. Zhang Y, Zhang X, Zhang Z, et al. Effect of high hydrostatic pressure treatment on the structure and physicochemical properties of millet gliadin. *LWT-Food Sci Technol.* 2021;154. https://doi.org/10.1016/j.lwt.2021.112755
31. Pandey S, Waghmare R. Nixtamalization and its applicability for the processing of food grains particularly maize. *Indian Food Ind Mag.* 2019;1:55–59.
32. Mesbah A, Graves DB. Machine learning for modeling, diagnostics, and control of non-equilibrium plasmas. *J Phys D Appl Phys.* 2019;52:30LT02. https://doi.org/10.1088/1361-6463/ab1f3f
33. Shelar A, Singh AV, Dietrich P, Maharjan RS, Thissen A, Didwal PN, et al. Emerging cold plasma treatment and machine learning prospects for seed priming: A step towards sustainable food production. *RSC Adv.* 2022;12(17):10467–10488. https://doi.org/10.1039/d2ra00809b
34. Wang X, Li Z, Zheng D, Wang W. Nondestructive identification of millet varieties using hyperspectral imaging technology. *J Appl Spectrosc.* 2020;87:54–61.
35. Ekramirad N, Doyle L, Loeb J, Santra D, Adedeji AA. Hyperspectral imaging and machine learning as a nondestructive method for proso millet seed detection and classification. *Foods.* 2023;13:1330. https://doi.org/10.3390/foods13091330
36. Naravane T, Tagkopoulos I. Predicting nutrient profiles in food after processing. *bioRxiv.* 2022. https://www.biorxiv.org/content/10.1101/2022.09.28.509827v1
37. Menichetti G, Ravandi B, Mozaffarian D, Barabási A. Machine learning prediction of food processing. *medRxiv.* 2021. https://doi.org/10.1101/2021.05.22.21257615

15 IoT and AI Synergy in Precision Agriculture

S. Vijayanand and S. Jagir Hussain

15.1 INTRODUCTION: THE DIGITAL REVOLUTION IN AGRICULTURE

As Internet of Things technology and artificial intelligence (AI) burst into the agriculture sector, the conventional mode of agriculture is going through tremendous change. These digital technologies have revolutionized agriculture, giving it a new look with precision agriculture. This new way of farming looks like it could help solve the problems of global food security, better resource management, and environmental sustainability [1]. It is not incremental improvement but paradigmatic change in how IoT and AI converge in agriculture. Farmers can now make data-driven decisions, optimize resource usage with an increase in productivity and reduce environmental impacts. Markets and Markets mentions that according to a new study, the global smart agriculture market size is expected to grow from USD 12.9 billion in 2021 to USD 20.8 billion by 2026, at a Compound Annual Growth Rate (CAGR) of 10.1%. This rapid growth is due to the rapid growth and adoption of smart farming across the world. This smart farming style includes IoT sensors, drones, autonomous vehicles, and AI analytics within a product to change the face of agricultural value chains of every level – farmer's unit or a giant agro-industrial enterprise. All these technologies are now available for real-time monitoring of crop and soil health, weather conditions, and the application of water, fertilizer, and pesticides.

Besides crop production, the innovative agricultural revolution in the context of digital development integrates every other activity. Livestock management not only concerns animal welfare but also extends to supply chain maximization and direct engagement with consumers. There are smartphone-based animal identification, wearable and blockchain-based traceability systems, and AI-based demand forecasting, among other technological developments that shape the entire ecology of farming.

15.2 THE INTERNET OF THINGS ECOSYSTEM IN CONTEMPORARY FARMING

15.2.1 Sensor Networks: The Core of Farm Intelligence

Advanced sensor technologies deployed throughout fields will provide real-time data on soil, crop health, and environmental conditions, which will form the basis of smart

DOI: 10.1201/9781003594093-15

farming systems. Sensors measure a wide range of parameters, including but not limited to the following:

- Soil moisture and temperature
- Nutrient levels
- PH levels
- Electrical conductivity
- Leaf wetness
- Atmospheric conditions: temperature, humidity, and air pressure

A study by Talaviya et al. (2020) reported that IoT-based soil monitoring systems can improve water use efficiency by as much as 50% and decrease fertilizer usage by around 20% [2]. The scope of sensor networks goes far beyond just resource optimization. It helps farmers make real-time decisions since it offers continuous granular data about field conditions, thus making agriculture a data-driven science.

Precision Irrigation: Soil moisture sensors, incorporated with weather forecasts and crop water requirements, will actually be the triggers for the automated irrigation systems. This would prevent overwatering, which causes root diseases and nutrient leaching while conserving water.

Targeted Fertilization: Nutrient sensors provide a detailed basis for making soil fertility maps. These can then be used by farmers for targeted fertilizer application in areas required, hence reducing the involved costs and further reducing the negative impacts that excess runoff causes on the environment.

Early Disease Detection: The leaf wetness sensors that measure temperature and humidity can figure out when fungi-based diseases are most likely to happen. This lets people take action quickly, which reduces the need for pesticides that aren't required as often because they are used correctly.

Yield Prediction: By correlating past records of yield with the sensor output, AI models will accurately predict yield, which is helpful in supply chain management and financial planning.

Microclimate Management: Sensors allow you to design the optimal microclimate for each species of plant for maximum growth and quality in the greenhouse.

Advances in sensor technology continue to expand the possibilities under this category. For example, hyperspectral sensors can now sense minute changes in plant physiology that might represent stress or disease through visible symptoms. Similarly, advanced soil sensors give real-time data related to microbial activity, offering insights into soil health that have not been available until now. Moreover, the integration of such sensor networks with other IoT devices and AI is creating these powerful synergies. For example:

- Drone-mounted sensors can quickly scan vast areas, complementing ground-based sensor networks.
- IoT-enabled farm equipment can modify operations in real time, such as the depth of seeding and the amount of fertilizer applied, based on real-time sensor data.

- AI can analyze the sensor data, along with satellite imagery and weather forecasts, to provide holistic farm management recommendations.

15.2.2 Agricultural Drones: Eyes in the Sky

Equipped with a multispectral camera, an unmanned aerial vehicle conducts comprehensive aerial surveillance, thus enabling detailed crop monitoring and precision application of resources. Drones would be used for high-resolution field maps, crop health assessment via Normalized Difference Vegetation Index (NDVI), targeted pesticide/fertilizer spraying, and monitoring the irrigation system. According to research by Krishna et al. (2021), crop stress can be detected up to two weeks before its visual inspection using the help of drone-based multispectral imaging, which thus allows for interventions before damage is caused [3]. The above characteristic of early detection in crops makes this kind of technology a huge leap forward in enhancing crop management and boosting productivity in a farm in general. Agriculture not only uses drones for crop monitoring but also for other tasks. Drones now perform even more mundane tasks like cattle management, fence inspections, and even pollination in some highly experimental deployments. Drone technology, with its advancements in flight time, payload capacity, and AI-driven image analysis, will expand its role in precision agriculture, enhancing precision farm operations across all scales and geographies.

15.2.3 Smart Farming Equipment: From Tractors to Harvesters

The use of IoT-enabled farm machinery optimizes operations through real-time data exchange and autonomous functioning, thus making them more efficient and less prone to human error. Examples of such include the Global Positioning System (GPS)-guided, self-driving tractors or robotic harvesters that can track yields and smart sprayers that modulate application rates according to the needs of the crop. In one instance, research conducted at John Deere showcased how their autonomous tractors managed to cut fuel consumption by as much as 8% while boosting productivity up to 14%. Such innovations in farm machinery not only contribute to the improvement of operational efficiency but also ensure sustainable farming methods through resource wastage reduction and lesser impact on the environment. Furthermore, the integration of IoT into farm equipment is enabling large-scale precision agriculture, enabling farmers to manage larger areas more accurately and with less labor. These technologies pave the way for further innovations, including swarm robotics, which facilitates multirobot collaborative field operations, and AI-driven predictive maintenance systems, which anticipate failures and proactively prevent equipment failure.

15.3 AI: THE BRAIN BEHIND PRECISION AGRICULTURE

15.3.1 Crop Management with the Assistance of Machine Learning Models

Applying complex algorithms for predicting crop yields, making disease diagnoses, and optimizing planting and harvesting schedules has revolutionized crop management with the help of machine learning. Recently, there has been truly commendable work in this field. With deep learning models, it is now possible to predict crop yields with an accuracy rate above 90%, which ensures better resource allocation decisions and market planning by farmers. Convolutional Neural Networks (CNNs) have shown significant potential in early crop disease detection [4, 5], which can immediately prevent crop loss. Reinforcement algorithms are also being applied to optimize crop rotation strategies to promote better soil health as well as long-term productivity [6]. AI-driven approaches are paradigmatic shifts from incremental improvements within agricultural decision-making systems. For example, machine learning models can help combine different types of data, like past crop yields, weather reports, soil sensor inputs, and satellite images, to give a full picture of things that can't be seen otherwise. Some advanced systems even offer field-specific recommendations, such as suggesting optimal planting dates, fertilizer application rates, and even the best times for crop sales in a given market. As these models learn and become more intelligent year by year with each successive growing season, their accuracy and utility are likely to increase exponentially. The involvement of AI in crop management is also opening up a new era of "digital agronomy," whereby agronomists and AI systems are cooperatively developing highly customized crop management strategies. This marriage between human know-how and machine intelligence promises to unlock solutions to the deeper challenges of agriculture – from adaptation to climate change to yield maximization under stress.

15.3.2 Computer Vision in Agricultural Monitoring

AI and machine vision bring about radical changes in agriculture through image recognition; they can rapidly identify the attack of pests, diseases, and crop stress. Recent applications include automatic weed detection that reduces the use of herbicides by 90%, real-time fruit ripeness for optimal harvest [7], and livestock monitoring for health and behavior analysis [8]. Such technologies also go beyond basic recognition. AI models now predict crop diseases before their visible symptoms show up by analyzing leaf reflection patterns. In viticulture, AI-powered drones assess grape quality and predict wine characteristics pre-harvest. The integration of AI with robotics is also giving rise to autonomous harvest systems for delicate operations such as soft fruit picking. As these technologies advance further, the role of AI in precision agriculture will only increase, where one can envision completely autonomous farms with unprecedented efficiency in resource consumption.

15.3.3 Predictive Analytics for Farm Operations

AI-based models utilize historical as well as real-time data to predict the weather patterns, market trends, and need for resources so that farmers can make data-driven decisions. Other areas applied include seasonal weather forecasting for long-term planning, crop price prediction for strategic planting decisions, and pest and disease outbreak prediction for proactive management. An AI-based predictive model can bring crop yield predictions up to 30% above that of traditional methods, according to a study by Liakos et al. (2018) [9]. This will have major implications for the improvement of accuracy in forecasting and has significant future applications in AI, transforming the way of agriculture planning and managing risks in crop selection, resource allocation, and market timing.

15.4 KEY APPLICATIONS OF IoT AND AI IN PRECISION FARMING

15.4.1 Soil Health Management through AI

The combination of sensor data with AI analysis gives a full picture of the soil's makeup, nutritional levels, and microbial activity, which lets the soil be managed in a way that is accurate and lasts. New developments include AI-driven precision fertilizer recommendations that can reduce fertilizer overuse by up to 40%, machine learning models for predicting soil organic carbon content, which constitutes an important parameter in climate change mitigation, and IoT-based monitoring systems for conservation agriculture against soil erosion [10]. It is thought that advanced hyperspectral imaging techniques could help find early signs of soil degradation and nutrient deficiency so that steps can be taken right away [11]. Blockchain technology is also integrated into systems of soil management to ensure transparency and traceability for sustainable farming practices, from soil treatment to crop production [12].

15.4.2 Intelligent Irrigation: Optimal Use of Water

These smart irrigation systems use AI to look at things like soil moisture, weather, and crop water needs. They then use this information to help the systems automatically distribute water so that it is used as efficiently and as little as possible. Advances in the field include AI-controlled drip irrigation systems that can decrease water usage as much as 60% yet retain or even improve crop yields. Crop water stress predictive models allow for scheduling of irrigations based on forecasts rather than measurements. Satellite imagery overlaid with ground-based sensor data on a large scale makes effective use of water resources [13]. Machine learning algorithms are now deployed to optimize irrigation patterns, given historical yield data and real-time plant health indicators that improve water usage efficiency [14]. Other developments with IoT-enabled smart valves and pumps designed to accurately allow the control of irrigation to each zone prevent wastage of water while applying the correct quantity of water needed in a crop of interest [15].

15.4.3 Animal Welfare: Predictive Health Monitoring

Wearable IoT devices, coupled with AI analytics, monitor the behavior and vital signs of the animals so that early health issues and stress factors in livestock can be detected. It includes AI-enabled lameness detection systems for dairy cows, which can detect lameness early and potentially improve treatment outcomes by up to 80%. There is facial recognition in individual animal identification and tracking, whereas there have been models based on poultry farm outbreaks that have reduced the mortality rate by up to 50% [16]. Wearable tech is adding more and more advanced biosensors to keep an eye on animals' metabolisms in real time. This way, nutritional deficiencies and health problems can be detected before they show any symptoms.

Moreover, AI-driven acoustic assessment systems are being developed for the evaluation and decoding of calls, which could provide an indirect means of evaluating animal welfare and emotional states in livestock [17, 18].

15.4.4 Hyperlocal Weather Forecasting and Climate Adaptation

Using on-farm IoT weather stations, AI-enhanced weather prediction models produce highly localized and accurate forecasts and aid climate-adaptive farming practices. Machine learning models for microclimate prediction have improved predictions by up to 40% in some farm areas. Other important areas include AI-driven climate risk assessment tools for long-term planning in agriculture and combining IoT weather data with crop growth models to make real-time yield predictions [19]. The strength of IoT and AI is when they are considered together in agriculture. Farm Management Information Systems (FMIS) today encompasses information systems that integrate data streams from multiple sources to provide holistic support for farmers' decisions [20]. New developments include blockchain-based systems for secure data sharing and traceability in the agricultural supply chain or cloud-based platforms that consolidate data streams from multiple farm sensors, satellites, and other sources into holistic farm management. Further, AI-powered digital assistants have emerged that provide personalized recommendations to farmers based on integrated data analysis [21]. Researchers are thus beginning to develop more advanced algorithms in advanced machine learning that predict and mitigate the impact of extreme weather events on crop yield, hence helping farmers adapt to climate change. There are also edge computing technologies integrated into IoT devices that enable real-time data processing and decision-making even in areas with limited connectivity to the internet [22–24].

15.5 CHALLENGES IN IMPLEMENTING IoT AND AI IN AGRICULTURE

Data security and privacy issues, especially for small-scale farmers, are problems that make it hard for these technologies to be widely used. Other problems include the digital divide, the need to build up the skills of farmers in developing countries, and the fact that different IoT devices and platforms need to be able to talk to each other [25]. This, however, brings with it challenges such as increased reliance on the providers of these technologies, which creates concerns for data ownership and farmer freedom. This issue also has the potential to further isolate industrial small-scale farms from others, given the incredibly wide-ranging inequalities within the agricultural sector [26]. Electronic waste that leaks from destroyed or broken IoT devices represents another environmental issue. Additionally, because modern systems are harder to integrate with farm skills, radical changes in the way farms are managed and decisions are made may face ferocious opposition from traditionalist farmers. In order to make sure that the benefits of these smart farming technologies are more widely available and last longer for all sizes of farms, policymakers, agricultural extension services, and farmers themselves need to work together [27].

15.6 FUTURE TRENDS: THE ROAD AHEAD FOR SMART FARMING

Other emerging technologies that will continue to shape the agricultural landscape include edge computing, which would process farm data in real time to reduce latency and improve response times; 5G networks capable of supporting more complex and responsive IoT ecosystems in rural settings; applications of quantum computing for complex simulations and optimization problems in agriculture; and bioinformatics and AI, which would accelerate crop breeding and genetic improvement [28]. Advanced robotics and autonomous systems will revolutionize farming operations through precision harvesting robots and swarm robotics for managing large farms [29, 30]. Vertical farming and controlled environment agriculture (CEA) technologies are advancing at a rapid pace, driven by AI-driven climate control and IoT sensors that optimize resource use in urban and peri-urban areas. Using Clustered Regularly Interspaced Short Palindromic Repeats (CRISPR) gene editing and AI-powered genomic analysis to change genes will speed up crop development and make it easier to make climate-resilient and high-yielding varieties [31, 32]. Blockchain technology is up for consideration for better traceability and transparency within the agriculture supply chains from the farm to the table. Another area of advanced research in nanotechnology has focused on developing smart nanomaterials for precision nutrient delivery and pest control with a potential reduction in the environmental impact of agricultural inputs [33].

15.7 CONCLUSION

The convergence of IoT and AI will transform the farming paradigm. Data-driven, optimized resource consumption, and decision-making will unlock opportunities for sustainable, productive, and resilient agricultural practices with both technologies. Future generations will depend a lot on smart farming technologies to ensure food security in the face of climate change and growing global populations. The integration of these advanced technologies offers the opportunity for significant reduction in agriculture's environmental footprint while increasing yields and improving farmer livelihoods. Precision in application, optimum use of resources, and early detection of crop and animal health issues can result in efficient and environmentally friendly farming practices. Furthermore, the data from these systems can contribute broadly to agricultural research and policymaking by providing insights into better ways of managing land sustainably. However, this potential would depend on building up the challenges in ensuring equitable access, data governance, and capacity building. Efforts are required to ensure that smallholder farmers and those in developing regions can also benefit from these advances by bridging this crucial digital divide in agriculture. Governments, technology providers, research institutions, and farming communities will have to collaborate to build inclusive smart farming ecosystems. In fact, this is to be supported with more ethical considerations about AI and IoT in agriculture, most importantly about data ownership, privacy issues, and the autonomy of farmers. Technological innovation should not erase traditional knowledge and ways of farming, which would result in a moribund and unsustainable system of farming. The key to the future of agriculture lies in respect and deep digital technical potential that considers complex relationships between humans, technology, and the environment. These steps are going to help build a future where agriculture feeds the world and gives positive contributions toward environmental stewardship and rural development.

REFERENCES

1. Wolfert, Sjaak, Ge Lan, Verdouw Cor, and Bogaardt Marc-Jeroen. "Big data in smart farming–a review." *Agricultural Systems* 153 (2017): 69–80.
2. Talaviya, Tanha, Shah Dhara, Patel Nivedita, Yagnik Hiteshri, and Shah Manan. "Implementation of artificial intelligence in agriculture for optimisation of irrigation and application of pesticides and herbicides." *Artificial Intelligence in Agriculture* 4 (2020): 58–73.
3. Krishna, Kowligi R. *Agricultural Drones: A Peaceful Pursuit*. Apple Academic Press, 2018.
4. Chlingaryan, Anna, Sukkarieh Salah, and Whelan Brett. "Machine learning approaches for crop yield prediction and nitrogen status estimation in precision agriculture: A review." *Computers and Electronics in Agriculture* 151 (2018): 61–69.
5. Ferentinos, Konstantinos P. "Deep learning models for plant disease detection and diagnosis." *Computers and Electronics in Agriculture* 145 (2018): 311–318.
6. Gautron, Romain, Maillard Odalric-Ambrym, Preux Philippe, Corbeels Marc, and Sabbadin Régis. "Reinforcement learning for crop management support: Review, prospects and challenges." *Computers and Electronics in Agriculture* 200 (2022): 107182.
7. Koirala, Anand, Walsh Kerry B., Wang Zhenglin, and McCarthy Cheryl. "Deep learning–Method overview and review of use for fruit detection and yield estimation." *Computers and Electronics in Agriculture* 162 (2019): 219–234.
8. Norton, Tomas, Chen C., Larsen Mona Lilian Vestbjerg, and Berckmans Daniel. "Precision livestock farming: Building 'digital representations' to bring the animals closer to the farmer." *Animal* 13, no. 12 (2019): 3009–3017.
9. Liakos, Konstantinos G., Busato Patrizia, Moshou Dimitrios, Pearson Simon, and Bochtis Dionysis. "Machine learning in agriculture: A review." *Sensors* 18, no. 8 (2018): 2674.
10. Balafoutis, Athanasios, Beck Bert, Fountas Spyros, Vangeyte Jurgen, Van der Wal Tamme, Soto Iria, Gómez-Barbero Manuel, Barnes Andrew, and Eory Vera. "Precision agriculture technologies positively contributing to GHG emissions mitigation, farm productivity and economics." *Sustainability* 9, no. 8 (2017): 1339.
11. Padarian, José, Minasny Budiman, and McBratney Alex B.. "Machine learning and soil sciences: A review aided by machine learning tools." *Soil* 6, no. 1 (2020): 35–52.
12. Pimentel, David, and Burgess Michael. "Soil erosion threatens food production." *Agriculture* 3, no. 3 (2013): 443–463.
13. Adeyemi, Olutobi, Grove Ivan, Peets Sven, and Norton Tomas. "Advanced monitoring and management systems for improving sustainability in precision irrigation." *Sustainability* 9, no. 3 (2017): 353.
14. Goldstein, Anat, Fink Lior, Meitin Amit, Bohadana Shiran, Lutenberg Oscar, and Ravid Gilad. "Applying machine learning on sensor data for irrigation recommendations: revealing the agronomist's tacit knowledge." *Precision Agriculture* 19 (2018): 421–444.
15. Segarra, Joel, Buchaillot Maria Luisa, Araus Jose Luis, and Kefauver Shawn C. "Remote sensing for precision agriculture: Sentinel-2 improved features and applications." *Agronomy* 10, no. 5 (2020): 641.

16. Nuffel Van, Annelies Ingrid Zwertvaegher, Pluym Liesbet, Van Weyenberg Stephanie, Thorup Vivi M., Pastell Matti, Sonck Bart, and Saeys Wouter. "Lameness detection in dairy cows: Part 1. How to distinguish between non-lame and lame cows based on differences in locomotion or behavior." *Animals* 5, no. 3 (2015): 838–860.
17. Khanna, Navneet, Pusavec Franci, Agrawal Chetan, and Krolczyk Grzegorz M. "Measurement and evaluation of hole attributes for drilling CFRP composites using an indigenously developed cryogenic machining facility." *Measurement* 154 (2020): 107504.
18. Berckmans, Daniel. "General introduction to precision livestock farming." *Animal Frontiers* 7, no. 1 (2017): 6–11.
19. Salam, Abdul, and Shah Syed. "Internet of Things in Smart Agriculture: Enabling Technologies." In Adam Drobot, Elfed Lewis (eds.) *2019 IEEE 5th World Forum on Internet of Things (WF-IoT)*, pp. 692–695. IEEE, 2019.
20. Feng, Puyu, Wang Bin, Liu De Li, Waters Cathy, and Qiang Yu. "Incorporating machine learni-ng with biophysical model can improve the evaluation of climate extremes impacts on wheat yield in south-eastern Australia." *Agricultural and Forest Meteorology* 275 (2019): 100–113.
21. Chlingaryan, Anna, Sukkarieh Salah, and Whelan Brett. "Machine learning approaches for crop yield prediction and nitrogen status estimation in precision agriculture: A review." *Computers and Electronics in Agriculture* 151 (2018): 61–69.
22. Kamilaris, Andreas, Fonts Agusti, and Prenafeta-Boldó Francesc X. "The rise of blockchain technology in agriculture and food supply chains." *Trends in Food Science & Technology* 91 (2019): 640–652.
23. Wolfert, Sjaak, Ge Lan, Verdouw Cor, and Bogaardt Marc-Jeroen. "Big data in smart farming–a review." *Agricultural Systems* 153 (2017): 69–80.
24. Rose, David C., Sutherland William J., Parker Caroline, Lobley Matt, Winter Michael, Morris Carol, Twining Susan, Ffoulkes Charles, Amano Tatsuya, and Dicks Lynn V. "Decision support tools for agriculture: Towards effective design and delivery." *Agricultural Systems* 149 (2016): 165–174.
25. Wiseman, Leanne, Sanderson Jay, Zhang Airong, and Jakku Emma. "Farmers and their data: An examination of farmers' reluctance to share their data through the lens of the laws impacting smart farming." *NJAS-Wageningen Journal of Life Sciences* 90 (2019): 100301.
26. Tzounis, Antonis, Katsoulas Nikolaos, Bartzanas Thomas, and Kittas Constantinos. "Internet of things in agriculture, recent advances and future challenges." *Biosystems Engineering* 164 (2017): 31–48.
27. Salemink, Koen, Strijker Dirk, and Bosworth Gary. "Rural development in the digital age: A systematic literature review on unequal ICT availability, adoption, and use in rural areas." *Journal of Rural Studies* 54 (2017): 360–371.
28. Brewster, Christopher, Roussaki Ioanna, Kalatzis Nikos, Doolin Kevin, and Ellis Keith. "IoT in agriculture: Designing a Europe-wide large-scale pilot." *IEEE Communications Magazine* 55, no. 9 (2017): 26–33.
29. Shi, Weisong, Cao Jie, Zhang Quan, Li Youhuizi, and Lanyu Xu. "Edge computing: Vision and challenges." *IEEE Internet of Things Journal* 3, no. 5 (2016): 637–646.
30. Saiz-Rubio, Verónica, and Rovira-Más Francisco. "From smart farming towards agriculture 5.0: A review on crop data management." *Agronomy* 10, no. 2 (2020): 207.
31. Skoki, Arian, Lerga Jonatan, and Štajduhar Ivan. "ML-Based Approach for NFL Defensive Pass Interference Prediction Using GPS Tracking Data." In *2021 44th International Convention on Information, Communication and Electronic Technology (MIPRO)*, Opatija, Croatia, pp. 1038–1043. IEEE, 2021.

32. Harfouche, Antoine L., Jacobson Daniel A., Kainer David, Romero Jonathon C., Harfouche Antoine H., Mugnozza Giuseppe Scarascia, Moshelion Menachem, Tuskan Gerald A., Keurentjes Joost J. B., and Altman Arie. "Accelerating climate resilient plant breeding by applying next-generation artificial intelligence." *Trends in Biotechnology* 37, no. 11 (2019): 1217–1235.
33. Walter, Achim, Finger Robert, Huber Robert, and Buchmann Nina. "Smart farming is key to developing sustainable agriculture." *Proceedings of the National Academy of Sciences* 114, no. 24 (2017): 6148–6150.

16 Turkey's Road Map in the Smart Agriculture 5.0 Process

Seval Selimoglu and Resül Yazici

16.1 INTRODUCTION

Climate change causes significant problems in agriculture and its related areas on a local and global scale. Among these issues, issues like the endangerment of ecosystems and the decline in livelihoods and food security have grown in significance for both the current and upcoming generations. Because of these connections, sustainable agriculture is crucial to meeting the food needs of both the current and future generations in a timely and uncompromising manner. Climate change affects agriculture, especially through changing precipitation conditions and more frequent extreme weather events. Due to supply chain disruptions and decreased crop yields, these changes result in higher production costs and inflation. The following is a summary of the fundamentals of sustainable agriculture, which attempts to lessen the effects of these issues: minimizing the negative effects of waste on the environment in order to maximize the utilization of inputs like water, fertilizer, and energy; boosting resistance to illnesses, pests, and climate change in order to ensure biodiversity; and preserving the soil and its structure and practices like crop rotation, covered cultivation, and decreased tillage. In order for Smart Agriculture 5.0 projects to be successful, they must first receive support in terms of training and design. This will increase their adaptation to the Industry 5.0 process and their capacity to use and manage new technologies.

Smart Agriculture 5.0 includes increased use of technology in agriculture; in other words, data-based agriculture through technologies such as smart farms, Internet of Things (IoT) devices, drone use, and remote monitoring systems. With the climate crisis, data-based management in agriculture is vital to increase productivity and ensure sustainability. The strategies to be followed in this process require the use of sensors to monitor soil moisture levels, pH levels, and other key factors. In summary, Industry 5.0 and Smart Agriculture 5.0 aim to adopt environmentally friendly agricultural methods and sustainable agricultural practices and are of vital importance. How this can be done with smart agricultural applications in the Industry 5.0 process in Turkey is the subject of this theoretical study. In the study, the conditions for making agriculture more efficient and sustainable are given by integrating technological developments into the sector despite the difficulties of adapting them to the

DOI: 10.1201/9781003594093-16

agricultural sector. In the first section, the concept and importance of agricultural economy are given, and in the next section, the applications and general problems in agricultural economy in Turkey are given. Then, the general framework and effects of Smart Agriculture 5.0 and Industry 5.0 applications, their importance in solving problems in agriculture, and the relationships between them are mentioned. In the final section, after looking at the applications and strategies of Industry 5.0 and Smart Agriculture 5.0 in Turkey, solution suggestions are given.

16.1.1 The Concept of Agricultural Economics and Its Importance

The concept of economics, in its simplest terms, includes how human needs can be met with limited resources – how choices can be made between alternative uses of resources. Preferences are available not only from the perspective of consumers but also from the perspective of producers, firms, or the general economy. In this case, the main problem that the consumer, producer, firm, agriculture, and industry sectors and the economy in general seek a solution for emerges as the effective use of existing limited resources. The branch of agricultural economics, which has a very close relationship with the technical branches of agriculture and the economy, emerges as a mixture of many branches of statistics, mathematics, business, economics, and social sciences, as well as basic information on agricultural techniques. The branch of agricultural economics, which undertakes the function of effectively conducting and directing agricultural activities, including the production and processing of plant and animal products, in the light of economics, at macro and micro levels, is, in other words, a discipline that applies the principles and analytical models of economics to agricultural problems [1].

This perspective of agricultural economics aims to increase the safety and reliability of food by developing measures against many problems, such as increasing population, rapid urbanization, industrialization, climate change, war, and epidemic diseases in the world. In this context, mechanization in agricultural areas has occurred gradually. This situation has made the timing of mechanization different in each society. Today, agriculture is still done only based on human power in some settlements. The increase in mechanization in agriculture offers the opportunity to do more agriculture with less labor and increased agricultural production. With the increase in agricultural production, the raw materials produced for factories have also started to increase. In fact, mechanization in agriculture has become an element that changed all societies, not only economically but also socially [2]. In short, increasing production volumes by implementing modern agricultural techniques on a global and local scale has become increasingly important each passing day. Failure to keep up with this trend poses a significant threat to both farmers who produce in a traditional manner and their countries.

16.1.2 Overview of Agricultural Economic Practices and Problems in Turkey

Turkey is a country with high agricultural production potential due to its geographical location and various climatic conditions. However, agricultural productivity remains below its potential due to migration from villages to cities, land fragmentation, deficiencies in irrigation infrastructure, and various problems in accessing modern technology and information. The causes and consequences of these problems will be summarized, and solution proposals will be developed in the following headings:

- Rural to Urban Migration: As a result of the failure to improve social and economic conditions in rural areas, migration from rural areas to urban areas has led to a decrease in population in rural areas and thus has negative effects on the agricultural workforce.
- Land Fragmentation: The division of agricultural lands through inheritance from generation to generation creates serious pressure on production. Fragmented fields make it difficult to use modern agricultural equipment effectively, which increases production costs because economies of scale cannot be utilized.
- Deficiencies in Irrigation Infrastructure: Water is of critical importance in agricultural production, but the inadequacy of irrigation infrastructure and the widespread use of surface irrigation methods lead to water loss and ineffective use of resources.
- Inadequate Access to Modern Technology and Information: Farmers have inadequate access to modern agricultural technologies and information on their use. This deficiency prevents the use of economies of scale; as a result, it limits productivity and increases production costs.

Due to these problems in Turkey and the inadequacy or inefficiency of agricultural support policies, the traditional production method poses a significant threat to farmers and consumers. According to Sayın and Sayın [3], with the idea that the economic and social problems originating from the agricultural structure will be eliminated through the community helping each other, producer organization models have come to the agenda for various purposes in our country. Farmers have not been in a strong position in establishing producer organizations, but when we look at the increase in both the number of organizations and the number of members since the early years of the Republic, it can be said that organizations in our country have developed. However, this development or diversity has not been able to solve the problems. Due to new consumer demands, improving farmers' access to information and enhancing their ability to adapt to changing conditions is crucial. Their information needs and adaptive capacity must be enhanced. Failure to do these will lead to failure.

In general, failure in the agricultural economy can be grouped under two main headings. The first is the individual deficiencies of farmers in decision-making and application techniques on what, when, how much, and how to produce. The second is the failure to deliver the products produced to the relevant markets on time. In other words, the failure to create temporal and spatial property benefits in the products and the failure to obtain high added value. Therefore, the deficiencies in the production and marketing process of the sector will negatively affect both micro actors and the macroeconomic structure more with climate change. Therefore, a perspective that can plan and implement all stages is required in the solution. In short, the inadequacy in any weak link of the food chain, no matter how strong all the other links forming the chain are, will be completely reflected in the agricultural economy. In this context, the management of Smart Agriculture 5.0, Industry 5.0, and sustainability processes undertakes an important task in solving the problems in agriculture.

16.2 CONCEPTUAL FRAMEWORK AND IMPACTS OF SMART AGRICULTURE 5.0 AND INDUSTRY 5.0 APPLICATIONS

Industry 4.0 has digitalized production processes with automation, big data, artificial intelligence (AI), the IoT, and robotic systems. These applications have provided high efficiency and low cost in production. However, Industry 4.0 has been insufficient due to reasons such as global warming and the increase in environmental pressures caused by human-centered production. Industry 5.0, which has come to the fore especially with the developments in automation and AI applications, prioritizes environmental sustainability and encourages the reduction of carbon footprint, circular economy, energy efficiency, localization, and gives more importance to customized production processes. With the adaptation of Industry 5.0 applications to Smart Agriculture 5.0, the use of humans and technology together increases the efficiency of agricultural production, and environmental and social sustainability is also ensured. In the future, with the more effective use of advanced applications of AI, robotics, and biotechnology in agriculture, a more flexible, resilient, and environmentally friendly ecosystem will be formed in agricultural production.

Recent technologies, especially in areas such as digitalization and AI, are profoundly transforming our ways of working and trading. While this transformation creates many opportunities, it also brings with it some risks. The rapid spread of digital technologies will increase productivity, while causing the workforce to lose jobs in some areas or evolve into jobs that require new skills [4]. Industry 5.0 goes beyond the digitalization and automation-focused character of the Fourth Industrial Revolution and includes a human-centered approach, designing industrial processes to minimize natural resource consumption and reduce carbon footprint. In other words, the flexibility provided by automation and AI allows production to be optimized according to individual consumer needs; Environmental, Social, and Corporate (ESG) can be established in accordance with the principles of governance.

Smart Agriculture 5.0 includes both environmental and social sustainability goals by integrating traditional agricultural methods with advanced technology and AI to increase the efficiency of agricultural activities. In this context, it is aimed to effectively implement the following technological innovations to both reduce costs and increase efficiency by reducing environmental impacts:

- Use of big data and AI applications aimed at monitoring and optimizing agricultural processes with various data.
- Developing proactive rather than reactive approaches through the use of IoT devices in areas such as irrigation, soil analysis, and plant health.
- Drone and sensor technologies that provide data for remote monitoring of fields and precision agriculture applications.

These developing technologies enable the diversification and development of alternative agricultural opportunities today. Applications integrated with advanced technology and AI have led to the importance of

> Hydroponic (soilless), Aeroponic (plant roots are suspended in the air), and Aquaponics (soilless plants and fish together) agricultural cultivation. These systems, which provide different solutions to different problems encountered and enable agriculture even in problematic areas, have offered new solutions for producers. Another advantage of these technologies is that vertical farming production, which increases the efficiency of the area unit, can be applied with these systems. In addition, these new modern systems can be designed to support continuous production throughout the year. The biggest advantage of these methods, which have started to be used in our country as well as all over the world, is that they allow farmers to have more control over production, save labor and time, and turn their businesses into more profitable ones [5].

16.3 THE IMPORTANCE OF THE INTERACTION BETWEEN INDUSTRY 5.0 AND SMART AGRICULTURE 5.0 IN SOLVING PROBLEMS IN AGRICULTURE

Various factors that lead to food crises, such as changes in food production and increases in product prices, create serious problems for many segments of society. It is very important for decision-makers to correctly interpret these emerging problems and create policies that will alleviate the effects of the problems. Although the policies that countries will implement are important for protecting existing resources and food supply, organizations with various goals, such as reducing hunger, increasing food production, and ensuring food security, play an important role in the food crisis [6]. In this context, it is possible to make better use of existing land, create scalable systems, and even produce mobile systems with alternative agricultural methods and different systems of these methods. It is possible to produce with alternative agricultural methods even in areas where the land structure is not suitable or the water quality in the area is low. Water saving can be achieved with alternative agricultural methods, as well as earliness, quality, and yield increases. Since climate control is provided, production can be made in every season, independent of external effects, and the loss of time between harvests is prevented. These methods provide more product diversity within the same area [7].

In cases where environments suitable for alternative agricultural methods cannot be provided, excessive agricultural practices, excessive grazing, deforestation, incorrect irrigation practices, etc., turn fertile lands into arid lands. Soil degradation, which is directly related to food and water security, carbon emissions, biodiversity loss, and political and social risks, has a significant impact on our economies and societies. For businesses, soil degradation can mean reduced profitability due to reduced productivity and increased costs. This also increases risks related to market access. Due to all these, we see that the concept of sustainable agriculture is increasingly on the agenda of actors involved in the agricultural process [8]. In short, sustainable agricultural practices require various measures and actions to strengthen the foundations of the economy, increase the economic sustainability of farmers and stakeholders, optimize productivity, and adapt to market conditions. After this general overview of sustainability, when the historical development of agricultural activities is summarized in the next paragraph, it will become more apparent why the practices of the process are important.

The most fundamental characteristic of the time in which the first revolution, defined as Agriculture 1.0, was experienced was the dominance of a production method dominated by low efficiency and labor. Toward the end of the 1950s, synthetic pesticides, fertilizers, and machines providing greater effects brought production costs to lower levels, and thus the Agriculture 2.0 time period, known as the "Green Revolution," emerged. Farming productivity has increased with lower costs and new practices. Global Positioning System (GPS) signals have become available to everyone; the agriculture 3.0 process began in the 1990s and is now known more as the concept of "Precision Agriculture." GPS technology, manual steering applications, harvesting machines, Variable Rate Application (VRA) systems, and fertilization process monitoring and tracking have become the most widely used technological systems during this time [9].

The agricultural sector started to undergo a parallel process in the 2010s that was comparable to the industry's transformation with Industry 4.0. The use of smart technologies, such as sensors, detectors, microprocessors, autonomous decision systems, and cloud-based information and communication technologies, in the agricultural industry is commonly referred to as "Agriculture 4.0," "Smart Agriculture," and "Digital Agriculture." The storing and analysis of big data allows for the tracking, guidance, and future projections of the entire process from field to table, thanks to internet-based portals and a variety of algorithms. The collaboration of various players in the food and agricultural value chain, and hence the significance of the ecosystem, is another aspect of Agriculture 4.0 [10]. Agriculture 4.0, which uses contemporary production technologies, data exchange, in short, modern automation systems, reflects the effects of the Industry 4.0 process, which affects almost every moment of life. Industry 5.0, which has come to the agenda in most developing countries before these changes are fully felt, is expressed as productive AI applications; it will affect the environment, social, and institutional relations in various ways. "This new age claims that artificial intelligence will penetrate life together with humans and position humans at the center of the universe in order to increase the capacity of human-robot cooperation and human resources" [11].

In agricultural science, research on the Fourth and Fifth Agricultural Revolutions has advanced to a high level. In their agricultural production, all industrialized countries use the techniques and tools of Agriculture 4.0, a more advanced stage of agricultural growth. Farming 5.0, a new phase of agricultural technologies, is currently being developed. However, there are a number of contentious questions in Agriculture 5.0, and their answers are merely hypothetical. Industry 5.0's development of bioeconomics brings to light the fascinating facets of sustainable agriculture, including resource optimization, product customization, the development of novel and distinctive products, and the application of unique decision-making systems via robotic systems [12]. In short, "this 5th industrial revolution is human-centric, sustainability-oriented, and

TABLE 16.1
Importance of Industry 5.0 and Smart Agriculture 5.0 Applications in Solving Problems in Agriculture

Fundamental Problems in Agriculture	Solutions to Be Provided by Industry 5.0	Solutions to Be Provided by Smart Agriculture 5.0
Climate Change and Environmental Sustainability	Renewable energy, optimum resource use, reducing carbon footprint	Production methods that reduce the effects of climate change
Low Productivity in Agriculture	Production of autonomous agricultural machines, artificial intelligence-based forecasting systems	Data-based decision-making, precision agriculture techniques
Decrease in Soil Fertility	Development of artificial intelligence-supported soil analysis	Biotechnology-supported soil management, organic fertilizer use
Decrease in Water Resources	Optimizing water use with data analysis of machines and sensors	Smart/drip irrigation systems and rainwater harvesting irrigation
Excessive Use of Pesticides and Chemicals	Use of targeted spraying and biological control methods	Precise spraying with drones, biotechnological pest management
High Cost of Agricultural	More efficient use of resources with artificial intelligence-supported cost analysis	Optimum fertilizer or drugs and water use
Reduction in Labor Force, Intelligence-Supported Decision Systems	Due to migration from rural areas, reducing workload, thanks to human-machine cooperation	Autonomous tractors, artificial intelligence-supported decision systems

resilience-focused. Advancements in technology and digital development enable increased human-machine interaction. For example, AI and robots interact more with consumers and workers. Industry 5.0 perspective emphasizes using technology for Human Welfare and, thus, the rising importance of eco-friendly operations" [13]. From another perspective, the importance of the interaction between Industry 5.0 and Smart Agriculture 5.0 in solving problems in agriculture is summarized in Table 16.1.

16.4 OVERVIEW OF INDUSTRY 5.0 – SMART AGRICULTURE 5.0 APPLICATIONS AND STRATEGIES IN TURKEY

The Ministry of Agriculture and Forestry of the Republic of Türkiye is experimenting with some of the agricultural technologies by putting them into practice in the light of scientific studies, evaluating the results, and introducing them to farmers. Examples of these are OTAK (Development of Domestic Automatic Steering and Control System) Project, Farm Management System Development Project, Unmanned Aerial Vehicle and Image Processing Based Precision Agriculture Applications Project, and Development of Smart Measurement Platform Prototype for Small Animal Breeding Project. As seen in the examples, the Ministry has been putting forward innovative studies in the light of scientific developments in recent years [14].

The Ministry supports the digital transformation in the sector by adopting smart agricultural practices, as can be seen in the 2024–2028 Strategic Plan and other studies on its website. Thus, the Ministry aims to increase agricultural productivity, provide support to farmers, and promote sustainable agriculture by using various technologies. The issues included in the Strategic Plan of Agriculture and Forestry (2024: 37) [15] can be summarized as follows: in order to raise the standard of living for our farmers, we must establish an efficient agricultural sector that is socially, environmentally, and economically sustainable; offers the nation's citizens a sufficient and balanced diet; takes supply and demand into account; is extremely productive; and is socioeconomically developing.

- Agricultural Technologies and Information Systems: The use of modern machinery compatible with climate change will be encouraged, the use of shared machinery will be expanded, and AI applications for production planning will be developed.
- Agricultural Research and Development: Research and development activities will be carried out to ensure accessibility to safe food and quality agricultural products, to ensure the sustainable use of agricultural and ecological resources, to protect local genetic resources and develop new varieties, and to develop alternative control methods.

Within the scope of these objectives, the Strategic Plan for Agriculture and Forestry (2024: 123) includes the following strategies under the title of disseminating innovative practices in agriculture to create a resilient and development-ready agricultural sector:

- Agricultural-industry integration and advanced technology applications in agriculture will be expanded.
- Smart agricultural practices will be expanded in the agricultural sector with the use of technological developments such as product verification, yield estimation, AI, geographic information systems, and algorithms.
- Applications such as soilless agriculture and vertical agriculture will increase.
- The needs of our producers will be met with environmentally friendly, efficient technologies developed with domestic and national resources.

In a study titled "New Trends in Agriculture and Forestry" [16], the restructuring of the agriculture-food sector toward the application of high technologies, the management of quality systems in plant and animal production, smart agriculture, prioritizing the development of standards in farm management and increasing efficiency in agricultural production; geographic information systems, remote sensing, image processing, big data analysis in agriculture, the establishment of decision support systems, food traceability system, encouraging supportive food reliability and follow-up activities for cooperation in these areas are among the issues that can be addressed in the projects to be implemented by our Ministry and in the mitigation of the effects of the Kahramanmaraş earthquake on February 6, 2023, on the agriculture and forestry sector and in the reconstruction process.

In addition to these micro-scale approaches of the Ministry, it is necessary to emphasize once again that Industry 5.0 and Smart Agriculture 5.0 applications offer great opportunities to make the agricultural sector in Turkey more efficient, sustainable, and innovative. However, in order for this transformation to be fully successful, the digital infrastructure must be strengthened, and farmers must be adapted to technology. For these, independent regulatory and supervisory structures must be established and operated. In short, it is inevitable for Turkey to adopt a holistic strategy for the Smart Agriculture 5.0 ecosystem to increase its global competitiveness.

16.5 CONCLUSION AND RECOMMENDATIONS

Problems with a nation's macroeconomic structure will inevitably have an impact on its microelements. On the other hand, it is certain that the macro-indicators will be the result of issues that arise in the agriculture sector for a variety of reasons. In particular, the failure to benefit from economies of scale in the sector, the dependence of agricultural activities on seasonal conditions, the inability of farmers to follow market dynamics, and the fact that fundamental problems have not been solved both threaten food security and prevent farmers from earning the income to meet their minimum living conditions. This leads to chronic problems such as farmers abandoning agriculture and villages and food price increases, causing inflation. Farmers in Türkiye are typically fragile due to the way the agriculture sector is organized and the failed methods they have encountered. Therefore, the technological solutions to be offered to them should be continuous, durable, and easy to use, and the technical services provided should be fast. This sector should be provided with ease of access to data in the internet infrastructure that constantly struggles with nature. The technical infrastructure of the farmer should not remain idle; continuity should be ensured in smart agricultural practices. For this, the efforts of the Ministry, which supports farmers legally and voluntarily, and other stakeholders supporting agriculture should be strengthened with producer-public cooperation and university-industry cooperation [17]. Gürsoy and Çolak [18] also draw attention to gender inequality in their studies and offer the following suggestions: it is highly probable that smart agricultural practices will develop and become widespread in the coming years. We would like to emphasize that while these practices and policies are being produced, it is vital to proceed with sensitivity to the gender line that deeply cuts through the social structure to achieve the hoped-for change. There is a chance to improve the current state of disadvantage if smart agricultural practices are developed in a way that will improve women's status in the home and in the production process in terms of identity, decision-making power, access to services, and property. In addition to these points, we recommend that the applications that are created, particularly for our nation, focus on young farmers rather than middle-aged and older farmers, and on mobile devices rather than PCs.

By putting these viewpoints and the study's recommendations into practice, farmers can become more knowledgeable about agricultural production planning, marketing, and profitability, as well as sustainable agriculture, which will benefit the nation's economy in the following ways:

- Farmers can switch to sustainable agricultural methods and continue these methods, thanks to their access to modern agricultural technologies and information on their use.
- Farmers can increase their productivity and profits by saving labor and time because they have more control over production.
- Protection of natural resources and increasing soil fertility.
- Ensuring that renewable energy sources are used more widely.
- Expanding the output of organic farming and making markets more accessible.

To put it briefly, as the agricultural sector becomes more digitally advanced through Smart Agriculture 5.0 and Industry 5.0 applications, farmers will be able to use sensor networks and IoT technologies in agricultural areas to increase the efficiency of their operations and give them access to the information and decision support they need. They will also be able to access alternative and online markets more easily.

REFERENCES

[1] Rehber, E. (1998). *Vertical integration in agriculture and contract farming. Department of Agricultural Economics*, Uludag University (Vol. 46), Turkey, Working Paper.
[2] Aydınbaş, G. (2023). A study on smart agriculture (agriculture 4.0) from an economic perspective. *BILTÜRK Journal of Economics and Related Studies*, 5(2), 63–86.
[3] Sayın, B., & Sayın, C. (2004). Türkiye'de tarımsal üretici örgütlenmesi. Avrupa Birliğine uyum hazırlıkları ve tarımsal üretici birlikleri kanunu. Türkiye VI. Tarım ekonomisi kongresi, 16–18.
[4] Aktaş Çimen, Z. (2024). Dijital Ticaret ve E-İhracat. Dijital Kalkınma içinde, Editors, *Hidayet Gizem Ünlü Ören and Nihat Altuntepe*, 55–68, Ankara: Gazi Kitabevi.
[5] Bingöl, B. (2019). Alternatif Tarım Yöntemleri; Aeroponik, Akuaponik, Hidroponik. *Harman Time Dergisi*, 7(82), 34–42.
[6] Dağ, M. M., & Akbay, C. (2022). Sürdürülebilir tarımsal uygulamalar ile küresel gıda krizine karşı alternatif çözümler. *Tarım Ekonomisi Araştırmaları Dergisi*, 8(2), 101–113.
[7] Aktaş Çimen, Z. (2024). Dijital Ticaret ve E-İhracat. Dijital Kalkınma içinde, Editör, *Hidayet Gizem Ünlü Ören ve Nihat Altuntepe*, 55–68, Ankara: Gazi Kitabevi.
[8] SKD. (2018). Sürdürülebilir Tarım: Dünya'da ve Türkiye'de Güncel Durum, https://www.skdturkiye.org/blog/surdurulebilirtarim-dunyada-ve-turkiyede-guncel-durum
[9] Ercan, Ş., Öztep, R., Güler, D., & Saner, G. (2019). Tarım 4.0 ve Türkiye'de uygulanabilirliğinin değerlendirilmesi. *Tarım Ekonomisi Dergisi*, 25(2), 259–265.
[10] Saygılı, F., Kaya, A. A., Çalışkan, E. T., & Kozal, Ö. E. (2018). Türk tarımının global entegrasyonu ve tarım 4.0. *Izmir Ticaret Borsası, Yayın*, 98, 98.
[11] Yücebalkan, B. (2020). Endüstri 4.0'dan endüstri 5.0'a geçiş sürecine genel bakiş. *Pearson Journal*, 5(9), 241–250.
[12] Haloui, D., Oufaska, K., Oudani, M., & El Yassini, K. (2024). Bridging industry 5.0 and agriculture 5.0: historical perspectives, opportunities, and future perspectives. *Sustainability*, 16(9), 3507.
[13] Zengin, Y. (2023). Going green-industry 5.0, supply chain and demanding customer. *Uluslararası Ekonomi ve Siyaset Bilimleri Akademik Araştırmalar Dergisi*, 7(17), 21–30.

[14] Yılmazer, E. B., & Tunalıoğlu, R. (2024). Çiftçiler ve Tarımı Destekleyen Kurum Teknik Elemanlarının Bakış Açısıyla Akıllı Tarım Uygulamalarının Değerlendirilmesi. *Tarım Ekonomisi Araştırmaları Dergisi*, 10(1), 70–89.
[15] Agriculture and Forestry. (2024). T.C. Tarım ve Orman Bakanlığı 2024-2028 Stratejik Plan. https://www.tarimorman.gov.tr/SGB/Belgeler/stratejikplan.pdf
[16] Agriculture and Forestry. (2023). Tarim ve Ormancılık Alanında Yeni Trendler. https://www.tarimorman.gov.tr/ABDGM/Belgeler/%C4%B0ZLEME%20VE%20DE%C4%9EERLEND%C4%B0RME/Tar%C4%B1m+ve+Ormanc%C4%B1l%C4%B1k+Alan%C4%B1nda+Yeni+Trendler.pdf
[17] Yılmazer, E. B., & Tunalıoğlu, R. (2024). Çiftçiler ve Tarımı Destekleyen Kurum Teknik Elemanlarının Bakış Açısıyla Akıllı Tarım Uygulamalarının Değerlendirilmesi. *Tarım Ekonomisi Araştırmaları Dergisi*, 10(1), 70–89.
[18] Gürsoy, Ö. B., & Çolak, E. (2023). Akıllı tarım literatürünün toplumsal cinsiyet perspektifinden Türkiye bağlamında değerlendirilmesi. *Ahi Evran Üniversitesi Sosyal Bilimler Enstitüsü Dergisi*, 9(1), 185–203.

17 Internet of Things (IoT) in Sustainable Agriculture

Nikhil Kumar Goyal, Monika Dandotiya, Monika Kumari, Shikha Sharma, and A. Anushya

17.1 INTRODUCTION

The current and future growth challenges facing agriculture worldwide in light of demographic pressures from rising population growth, coupled with the issue of changing climates and higher demand for more sustainable operations, now mandate farmers to produce far more food without further detrimental impact on their environment, conserving ever scarcer natural resources. Under this situation, one technology appears highly capable of revolutionizing the future of modern farming – the Internet of Things, or IoT. IoT refers to the network of physical devices, vehicles, home appliances, and other items embedded with sensors, software, and other technologies that collect and exchange data over the internet. Applied to agriculture, IoT allows farmers to monitor and manage their crops, livestock, and resources in real time through data-driven decision-making (Figure 17.1).

It represents a farming management approach utilizing the use of IoT devices in observing, measuring, and responding to field conditions. This change is grounded in precision agriculture, wherein optimization of inputs such as water, fertilizers, and pesticides for higher crop yields and minimizing waste become goals. Smart sensors allow the tracking of soil moisture, weather patterns, crop health, and livestock behavior. All this would help them approach farmers with actionable information that could aid farmers in making decisions toward more effective farming, productivity, and environmental footprint [1]. Smart irrigation represents one of the most typical uses of IoT in the agriculture sector. This process uses moisture sensors and the information available from weather forecasting for the automation and optimization processes for the proper distribution of water, and to avoid cropping and watering at inappropriate moments. Livestock monitoring is another significant application involving the real-time monitoring of health, behavior, and location of animals through a wearable sensor. This enables timely intervention that prevents the spread of illness and generally enhances herd management [2].

Despite all these benefits of IoT to agriculture, its implementation has not been without challenges. Connectivity is the main problem, especially in rural and remote areas. The IoT infrastructure costs too much, and the integration process into traditional farm

DOI: 10.1201/9781003594093-17

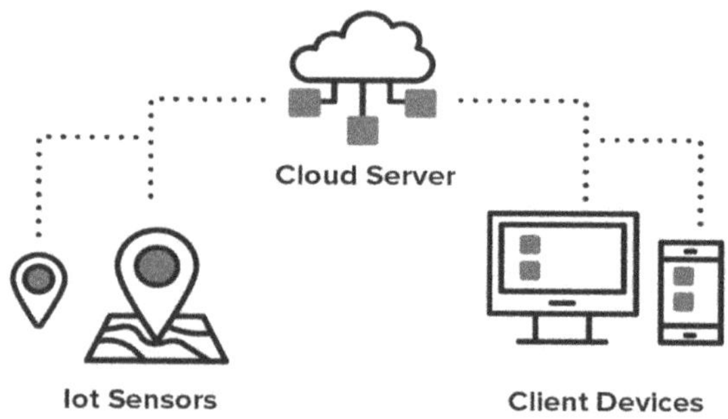

FIGURE 17.1 Connection of servers.

practices is complicated, among other huge barriers that many face in farms, especially in developing regions. As more sensitive data are collected and transmitted over the IoT networks, data privacy and cybersecurity concerns have become increasingly relevant. Despite these shortcomings, IoT has been a grand promise in sustainable agriculture, where the growing need is to address the food needs of surging populations. It would help alleviate food insecurity globally while minimizing environmental impacts from agriculture through effective resource utilization and increased productivity. This chapter provides an overview of promising applications in agriculture based on IoT applications that have enormous impact potential in changing farm practices and ensuring a bright future for agriculture [3].

17.1.1 Definition and Overview of IoT in Agriculture

The IoT in agriculture combines automated instruments such as sensors with drones, along with automated systems, to obtain and exchange real-time operational data for farming purposes. Through software-based technology and internet connectivity, farmers gain access to track environmental conditions and resource management as well as perform automated remote tasks. The accumulated information consists of soil moisture levels and temperatures, together with crop health assessments, livestock relocation behavior, and equipment functionality to provide enhanced farming practice insights.

17.1.1.1 Applications

During recent times, agricultural technology development has led to a transformation in traditional farming by enhancing productivity, resource-performance improvements, and environmental sustainability practices. Water conservation technology combined with crop health protection is known as smart irrigation technology. The automation process for water distribution is managed through advanced sensors, which detect exact crop needs assessment along with soil condition analysis by utilizing real-time soil moisture observation. The water management method reaches peak plant growth to reduce water loss while advancing agricultural yields and maintaining soil health.

The use of drones helps to improve crop conditions by utilizing a system that merges drone sensors and high-resolution imaging to acquire complete crop-related data. A particular system allows farmers to remotely check crops for nutritional deficiencies and pest attacks, along with growth issues. When farmers execute variable-rate fertilizer application with target pest control operations, they get exact solution areas to optimize harvest yields with environmental safeguards. The evolution of wearable technology features as transformative equipment designed to monitor livestock wellness, together with their health condition. The tracking devices monitor significant health indicators together with animal behaviors and movement data to both discover diseases in early stages and enhance breeding techniques and animal feeding procedures. These technologies provide animal health data instantly to help producers achieve better productivity with enhanced standards of care as well as welfare benefits.

The connection of GPS technology to automated equipment produces better agricultural operations when used in planting and harvesting operations. Self-governing systems operate with high precision, as a result of which agricultural operations need less manpower while using resources more efficiently. When farmers use precise GPS navigation systems, they reduce fuel consumption while preserving the soil structure, ensuring whole field coverage and better land use, together with environmental protection practices. Together, these new technologies amount to a paradigm shift in agriculture. The current agricultural industry lets producers respond to growing food market needs by applying sustainable resource management practices. Future food production systems will experience a total transformation because of these advancing technological developments.

17.1.1.2 Goals and Benefits

A data-driven model of an efficient farming mechanism stands as the essential idea for implementing IoT technology in agricultural practices. Through this system, farmers will obtain high-quality decisions that reduce waste or optimize fertilizer and water usage for maximal crop yields. Technology creates this shift from conventional farmland to contemporary farming systems that resolve both food production needs and natural protection concerns.

17.1.2 Importance of IoT in Modern Agriculture

The world population has grown significantly; therefore, there is a future need for an additional production rate of food in subsequent decades. Hence, agronomic systems have to yield more output with fewer inputs, given the increasing food demands coupled with the issue of scarcity of raw materials and climate change. Recent advancements in IoT technology apply impetus to modern agriculture to bring about effective, affordable, as well as sustainable agricultural systems. Several wireless devices, in conjunction with sensors, alongside data analysis tools, and integrated with the help of IoT technologies, help to define the productive and eco-friendly way of agricultural methodologies. The most significant value creation is observed in the sphere of precision agriculture, where farmers can get the maximum result using minimal resources. So, the surveyed soil conditions can inform IoT sensors on how and when water, fertilizers, and pesticides must be applied to the particular sections. Intelligent usage of IoT helps in using fewer agricultural inputs, which in turn will help in minimizing sustainable negative impacts on the environment. Integration of smart machinery with machines and automated instruments lowers the chances of manual interference, thus leading to scaling up of operation and elimination of human-related uncertainties. Since the delivery of the interventions in agriculture can be made precise due to the methods in precision farming, the farming result is optimized with less expense and environmental impacts acknowledged.

The usability of the IoT for farmers can be seen in the fact that it may provide them with useful data for various decision-making processes. Some connected devices obtain measurements of several variables as soon as they are tested, and this data may be used in decision-making in agriculture. Farmers are in a position to have all the information about the status of the soil as well as the condition of the crops, pest attacks, and the state of weather, and they are able to develop the most appropriate strategies on the right sequence of crops to plant and how to fight pests and diseases as well as the right time to harvest. This method creates various points that integrate farming strategies, which benefit from modern tendencies in the climate. As with other industry production devices, IoT systems in animals aim at monitoring the general health and behavior of livestock. The tracking of animals on a continuous basis in order to be able to spot diseases ensures that no use of antibiotics is needed, which makes the animal care sustainable and healthful to the farm livestock. Business operations become more effective with the use of IoT systems, whereby risk management can be seen as a main purpose of such systems. The surveillance capacities of these technologies cover weather information and soil and crop conditions, within which they are able to come up with alerts of threats as risky events, pest and disease breakouts, and severe weather incidents. It becomes even crucial when the climate becomes erratic because it helps farmers to introduce preventive measures before situations that may cause extensive harm to farming become uncontrollable. This chapter also affirms that IoT-based risk management plays a role in improving the stability of the food production system and solidarity in cases of risk factors, given the vulnerability of the sector to negative environmental factors.

The IoT technology needed for environmental stewardship is applied in developing sustainable agriculture, which is an integral part of its functions. Sustainable agriculture requires proper management in terms of water usage, greenhouse gases, and land management. IllegalStateException is obtained through IoT technologies by means of a specific resource measurement with monitoring features to make the farmers 'operations sustainable. Automated irrigation systems and other input use machines/technologies to reduce the emission of gases from agricultural processes as part of water conservation practices. Information acquired by IoT devices within the home assists farmers in efficient land management and prevents soil degradation, ensuring long-term use [4]. That is why the application of IoT technologies results in an innovative approach that satisfies the need for food quantity as well as solves the issue of environmental protection. IoT is one of the facilitators of change and advancement in the field of farming and implements the occupation precisely through data-oriented operations and risk minimization, along with environmentally friendly solutions. These are so because their evolution is hinged on application by a multitude of people, as this guarantees future food needs' satisfaction coupled with the planet's preservation.

17.1.3 Key Contributions

The application of IoT in the agriculture system, a key aspect of the IoT, creates an essential avenue of addressing concerns on resource depletion, alongside environmental degradation and increased food consumption demand. By getting connected through IoT technologies, firms are able to improve organizational productivity and reduce the negative effects on the environment as they maintain the right ethic of farm operations. In this section, the use of IoT technology that is most important for sustainable agriculture is discussed, and future uses considering machine learning (ML) and artificial intelligence (AI) technology are also considered. One of the ways through which IoT technology improves the farming business is through conserving water supply. Under the IoT system, smart irrigation uses soil moisture combined with weather stations and computerized controllers for the management of water. It guarantees that the particular systems deliver the correct amount of water as and when needed to avoid the negative effects of over-irrigation, resulting in erosion of soils and loss of nutrients, all of which are cases of waste. The method of distribution assists in promoting wellness for agricultural crops, as it also helps control water resources that are decreasing in the agricultural producing regions [5].

They include stepping up energy efficiency as a sub-element that helps the IoT technologies support agricultural processes. This is because, through the combination of automated systems that are controlled by IoT devices, it is possible to control energy-intensive greenhouse lighting as well as control the air circulation in the greenhouses, which is referred to as ventilation, together with the use of machines. The organization gets data on a real-time basis, with the data gathered from the environment added to the production conditions, so the resources are provided as and when needed. Such operations provide extraordinary power reduction integrated with decreased operating cost, with minimal invasive influence on the agricultural environment. Momentum to carry out sustainable resource efficiency that positively improves economic sustainability while at the same time achieving environmental sustainability. The use of IoT technology to lower chemical inputs shows its benefits toward sustainable farming methods. Many modern sensors, together with advanced imaging systems, monitor crops accurately to identify health conditions and pest activities and detect nutrient problems. Based on data measurements, this strategy enables accurate pesticide and fertilizer distribution, which decreases such chemicals while maintaining optimal pest and nutrient measures. The implementation of precise agricultural methods serves to reduce environmental pollutants, along with lowering operation expenses, which results in better ecological health. Animal husbandry tracks livestock through IoT devices at an increasing rate. Modern wearable sensors provide real-time observation of biological measurements as well as body movement activities, which enable prompt medical interventions after detecting illness symptoms. Through this technology, animals receive better care while infectious diseases remain confined, and antibiotics need to be used less frequently. IoT-based livestock monitoring finds support from consumers and governments because it helps establish sustainable and ethical animal farming practices [6].

In the future, the development of the possibility of future growth of IoT in the agricultural sector is expected to rise even higher with the integration of ML and AI. These technologies enhance the analytical capabilities of IoT systems for predicting crop yield, managing resources efficiently, as well as studying the impacts on the environment. Such a large amount of data collected by the IoT devices can be analyzed by AI algorithms to identify patterns and forecast climate change or

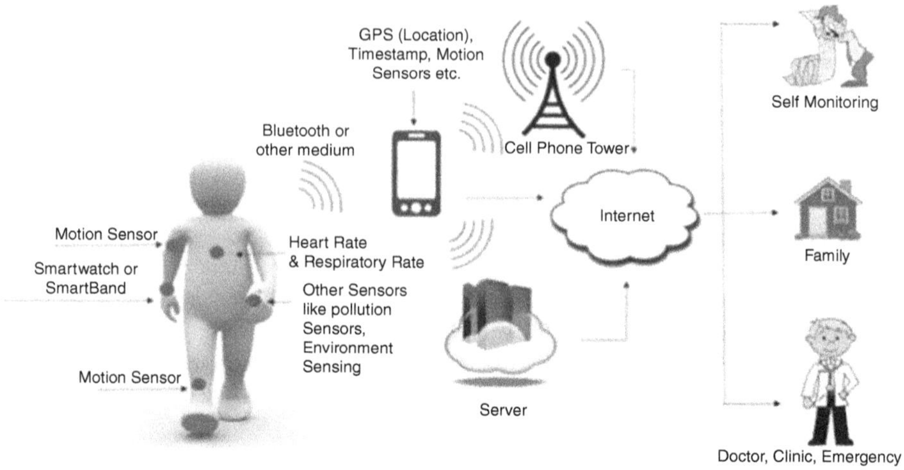

FIGURE 17.2 Layers of IoT.

offer the right planting, irrigation, and harvesting times. This capability helps farmers to make decisions in advance and to avoid as many unfavorable occurrences as possible; it can also help to enhance sustainable agriculture even further. To sum up, the usage of IoT technologies in the agriculture sphere is a revolutionary approach to solving the problem of food security. Ways that have mainly been made possible by the IoT include water conservation, energy conservation, chemical use reduction, improvement of animal production, and big data support for modern farming. Further developments of IoT along with AI and ML advancements will lead the way to even more innovations in the future resource utilization and environmental, as well as preserve and cultivate the agricultural sustainability of the global population's food stability.

IoT is instrumental in developing a robust and environmentally aware agricultural system because it equips farmers with the ability to improve the efficiency of resources (Figure 17.2).

17.2 PRECISION AGRICULTURE WITH IoT

Agriculture utilizes the potential of information technology to optimize crop and soil health. This integration of IOT will enable farmers to make more data-centric decisions in order to implement more efficient, low-cost methods and reduce negative effects on the environment. Soil and weather conditions about crops, along with machinery and performance, may be available instantly by this IoT [7].

17.2.1 Key Components of IoT in Precision Agriculture

Innovations in contemporary agriculture have greatly changed traditional farming methods, allowing farmers to improve productivity, maximize the use of resources, and ensure sustainable agriculture. Major technological interventions involve the incorporation of sensors, unmanned aerial vehicles (UAVs), smart machines, weather stations, and IoT data analytics platforms.

1. Sensors: Sensors optimally located within farms constantly record important soil and environmental factors such as soil water content, temperature, humidity, pH, and nutrient levels. This constant flow of information helps farmers make proper decisions, taking timely actions to improve crop quality and yield. By providing accurate irrigation and nutrition control, sensor networks help with optimal resource use and better crop production [8].
2. Drones and Unmanned Aerial Vehicles (UAVs): Drones and UAVs with high-resolution imaging capabilities offer complete aerial monitoring of vast agricultural fields. These systems take real-time measurements of crop health, growth rates, pest infestations, and soil conditions, enabling early detection of abnormalities. The scalability and efficiency of drone technology allow farmers to cover large areas rapidly, resulting in prompt corrective measures and optimized field management.
3. Intelligent Tools: The merging of IoT and Global Positioning System (GPS) technologies within autonomous agricultural vehicles has transformed the processes of seeding, fertilization, and reaping. Such smart tools perform with high precision, reducing waste and ensuring seeds, fertilizers, and fuel are used in an optimal manner. Autonomous harvesters and tractors improve working efficiency, lessen dependency on manpower, and ensure eco-friendly cultivation by conserving resources [9].
4. Weather Stations: Field weather stations are critical in recording meteorological data at localized levels, such as temperature, precipitation, wind speed, and humidity, in real time. With the ability to supply precise and instant weather forecasts, these stations allow farmers to adjust their farm activities – like irrigation scheduling, pesticide application, and harvesting – to conditions prevailing at the time. Provision of weather information in real time reduces the risks posed by extreme weather phenomena and enhances farm management decision-making.
5. IoT Data Analytics Platforms: The sheer volume of data created through IoT-connected devices requires strong analytics platforms that can process and analyze intricate data sets. IoT data analytics platforms process trends in soil health, crop development, weather, and equipment performance, providing actionable information for optimizing agricultural practices [10]. The platforms not only improve decision-making but also support predictive analytics, enabling farmers to foresee future problems and take preventive action to enhance overall agricultural output.

17.2.2 Applications of Precision Agriculture with IoT

The interlinking of advanced technologies for precision agriculture has transformed agricultural process management to foster greater efficiency of resources, crop productivity, and more sustainable farm practices. Advanced technological uses range from intelligent irrigation systems to variable rate technology (VRT), robotic pest and disease management, outcome monitoring and predictions, and auto crop scouting. Smart irrigation systems are important in water conservation while providing maximum irrigation. The systems function on real-time soil moisture information gathered using sensor networks. Smart irrigation delivers water only when and where it is required, avoiding over-irrigation and under-irrigation, protecting crop health while minimizing water use. Such accurate management of water resources is especially worth the investment in areas experiencing water shortage, leading to environmental preservation as well as enhanced agricultural yields. VRT allows for site-specific application of fertilizers based on the unique nutrient needs of varying field zones.

Through analyses of soil samples, crop needs, and the environmental factors that affect the growth of crops, VRT is able to put the right amount of fertilizer in the right location and at the right time. It minimizes wastage of nutrients, does not pollute the environment, and enhances the health of the soils for enhanced production of crops. This is because VRT not only improves productivity but also has a positive effect on sustainable agriculture by reducing the adverse effects of the use of fertilizers on the environment. Pest and disease control systems employ the application of remote sensing equipment and information analysis that would determine the initial signs of pest invasion and crop disease. This control systems can be effectively applied at the crucial stage of the development of the pest or the disease outbreak. Consequently, there is minimal use of chemical pesticides. Hence, less pollution, cost of production is reduced, and foods are made safer for the consumer. The real time and result monitoring and forecasting technologies utilize near-real-time data and a sophisticated process of analyzing it to help in determining exact yields. On the health of the plant, the rate of growth, and climatic conditions, these technologies offer valuable information on likely yields to the farmers. It is particularly useful when it comes to planning the harvest cycles, using the resources, supplying to the market, and so on, which, in turn, reduces wastage of food and brings about competency in general farm management.

Advanced scouting through the use of drones, or what is widely known as UAVs, has been adopted in advanced cropping scouting. Being equipped with highly calibrated cameras and lenses, it is possible to fly over extensive agricultural lands and get accurate images of the crop state. Basically, it helps in the early detection of such negative factors as nutrient deficiencies, pest invasion, and water stress. Once early notice is made, solutions are sought, and the crop is made healthy with optimum input standards. For this reason, automated scouting is significantly faster than the traditional manual method of crop monitoring by orders of magnitude. Altogether, these new applications of precision agriculture technologies provided a more comprehensive approach to the modern issues of farming. They contribute further to making efficiency, sustainability, and productivity that will bring out a sturdy and environmentally friendly farm segment.

17.2.3 Benefits of Precision Agriculture with IoT

Precision agriculture technologies have remained one of the most transformative approaches toward conventional farming practices through determining productivity, sustainability, and resource utilization. The first benefit is increased resource utilization and effectiveness of the company. It also means that through high technologies, it is possible to apply the necessary inputs such as water, fertilizers, and pesticides in accurate measurements. In providing inputs at the exact point and time of need, farmers are in a position to minimize wastage, excessive amounts of agrochemicals on the crops and consequently minimize environmental degradation. Such a specific approach is useful in environmental conservation through proper utilization of water, enhancement of soil fertility, as well as prevention of the pollution of natural systems. Other benefits that precision agriculture helps in achieving include precision farming, which increases crop production.

In the following sections, live data and various monitoring techniques that can be used to detect pests, lack of nutrients, and poor weather in a farm will be discussed. Such help is offered in a way that does not compromise timing to provide proper attention to the plants and avoid losses. When it comes to stock resources, accurate application means "right intensity crop inputs" get to the right crops in the right quantities, thereby improving crop quality and drastically boosting production. Precision farming is also useful in controlling operational cost, which is a major factor [11]. Resources must be properly utilized and inefficient and excessive use of agrochemicals reduced so that there is less polluting of the atmosphere in the course of farming. Precision farming is sustainable in the long run because it helps improve soil health and use of water and impacts the environment less. By adopting these sustainable approaches, agriculture becomes more climate resilient, and the world has been striving to achieve its food security and conservation of the environment.

17.2.4 Challenges in Implementing IoT for Precision Agriculture

Despite the numerous benefits associated with the use of IoT technologies in the agricultural field, some issues act as hurdles to the expansion of use, especially among the small- and medium-scale farmers. Some of them include high initial costs of investment, problems connected to the access, matters to do with handling data and privacy, and one vital one that requires the user to have technical expertise. These challenges cannot be ignored to prevent the proper adoption as well as distribution of IoT solutions into diverse farming climates. The high initial costs associated with the adoption of IoT technologies may also be a challenge in farming. This includes massive technology expenses in sensors, drones, automatic instrumentation, data analysis algorithms, and communication systems for IoT systems. "Cahs operating" large-scale farms may easily be able to absorb these expenses because in the long run the overall output may increase and thus the expenses offset by the increased production. Nevertheless, for small- and medium-sized farming businesses, the initial cost seems unmanageable, which restricts their ability to acquire and use such superior technology. Therefore, such IoT innovations should be made affordable by devising appropriate and financially feasible products or financial assistance, including subsidization by the government or pooling of resource sponsors. This further worsens the question of IoT technologies' practical application on farms, especially those located in rural areas, where connectivity is likely to be poor.

IoT networks mainly rely on fast and reliable internet connections to collect, forward, and analyze data as they are being generated. However, almost all those corners of the world that are associated with the agricultural profession lack the necessary digital equipment that creates interruptions in connections. Poor internet connections hamper the efficient transfer of data, reduce the functionality of automations, and restrict farmers' decision-making. Broadening the existing rural internet access and experimenting with other ways of getting connected, such as satellite internet delivery, are the two crucial steps that have to be taken to overcome this deficiency. Security and privacy concerns are caused by the large amount of data and data traffic in IoT devices. The continuous monitoring of data on the quality of the soil, the status of crops, the weather conditions, as well as the working of the machinery, requires functionality in the management of huge amounts of data [12].

17.2.5 Future Trends in Precision Agriculture with IoT

The future of precision agriculture involves integrating advanced technologies like ML, AI, and robotics; enhancing predictive capabilities; and decision-making. AI will analyze data to predict diseases, while robotics will automate complex tasks. Additionally, satellite imagery and remote sensing will improve data accuracy and coverage. As IoT devices become more affordable, their adoption will likely standardize precision agriculture globally [13, 14] (Table 17.1).

TABLE 17.1
Advantages and Disadvantages of Sensors

Sensors	Advantages	Disadvantage
Red/green/blue (RGB)	Portable, common, relatively inexpensive, provide high-quality TGB images for ortho-imagery and digital surface model. Easy to integrate with different platforms	Wider bandwidth and lower spectral resolution lack geometric and radiometric calibration
Multispectral	Multiple wavelengths allow geometric reconstruction and radiometric calibration	Relatively costly, image processing relatively complex compared to RGB, radiometric, and atmospheric corrections may be required
Microwave	High spatial resolution, useful under low visibility weather conditions (e.g., cloud cover)	Sensitive to surface roughness
Thermal	Sense a wide range of temperatures, relatively cheaper	Thermally sensitive, vignetting effects, low spatial resolution radiometric corrections required, sensitive to changes in surface roughness
LiDAR	Generates dense cloud points, high-quality data, direct geometric measurements	Higher cost, limited compatibility with UAVs, requires ground filtering corrections
Hyperspectral	High spectral resolution area across various wavelengths of the visible and near infrared spectrum	Higher cost than multispectral, low signal-to-noise ratio, limited compatibility with standard UAVs and flight control software packages, radiometric and atmospheric corrections may be required

17.3 IoT IN SMART IRRIGATION

Irrigation, a crucial agricultural practice, traditionally relied on manual or semi-automated methods that applied water uniformly, often leading to inefficiencies. IoT-enabled smart irrigation optimizes water distribution by using real-time data, ensuring crops receive the appropriate amount of water at the right time [15].

17.3.1 Traditional Irrigation

In this method, water is directly applied to the fields and permitted to flow over the soil surface through gravity. Although its initial installation and running costs are low, flood irrigation is very inefficient. Much of the water is wasted by evaporation, deep percolation outside the root zone, and surface runoff. These losses lower water-use efficiency and may cause erosion of the soil, leaching of nutrients, and waterlogging, all potentially damaging to crop health and to soil quality. Though still widely used in areas with limited resources because of its simplicity, modern systems are being marketed to maximize water savings and yields. Sprinkler irrigation systems mimic natural rainfall by delivering water from a network of pipes and spray nozzles located above the crop canopy. The technique provides relatively uniform water distribution across the field and is adaptable to a range of crops and soils.

Yet, conventional sprinkler systems usually distribute water uniformly without taking into account spatial differences in soil moisture levels and crop water needs. Thus, some parts might receive too much water, resulting in wastage of water and possible plant stress, while other parts remain insufficiently irrigated. Sprinkler irrigation is also susceptible to water losses due to wind drift and evaporation, especially under windy or hot weather conditions. Even with these constraints, sprinklers provide a more regulated irrigation method than flood irrigation and are commonly used in small- and large-scale agriculture. Manual watering, commonly employed on small farms and home gardens, uses buckets, watering cans, or hoses to supply water to the plants directly. Manual watering is imprecise, and the water application is usually inconsistent and can either under-water or over-water crops. The inefficiencies of human-powered irrigation, coupled with the increasing need for labor-saving technologies, have contributed to its decline in favor of more sophisticated irrigation systems. Finally, although older methods of irrigation like flood irrigation, sprinkler irrigation, and watering by hand have been an integral part of world agriculture for generations, they tend to be characterized by important inefficiencies in the consumption of water, labor requirements, and adaptability for particular crops (Table 17.2).

17.3.2 Smart Irrigation System Parts

The coupling of sophisticated technologies with contemporary irrigation systems has greatly improved the efficiency, accuracy, and sustainability of water management in agriculture. Through the use of real-time information from sensors, automatic controls, and cloud-based systems, farmers can maximize irrigation timetables to suit the precise requirements of crops while preserving valuable water supplies. The principal elements of such contemporary systems are soil moisture sensors, weather

TABLE 17.2
Comparison: Traditional Irrigation vs. IoT-Enabled Smart Irrigation

Aspect	Traditional Irrigation	IoT-Enabled Smart Irrigation
Water Application	Uniform and manual; fixed schedules	Targeted, data-driven; automated and adaptive
Water Efficiency	Low; significant water waste due to runoff, evaporation	High; minimizes water waste through precise control
Control	Manual or semi-automated	Fully automated with remote monitoring and control
Data Use	Limited to farmer observations	Real-time data from sensors and weather forecasts
Environmental Impact	Higher water wastage, runoff, and soil erosion	Reduced water use, less runoff, and lower environmental footprint
Labor Requirements	Labor-intensive; requires frequent manual intervention	Minimal labor required; system runs automatically
Cost Efficiency	Low initial costs but high long-term inefficiencies	Higher initial costs but significant long-term savings in water and energy
Scalability	Difficult to scale due to labor and water inefficiencies	Easily scalable with minimal human intervention

sensors, automatic irrigation systems, cloud-based data management systems, and intelligent control programs. Soil moisture sensors are a key component of precision irrigation as they monitor the moisture content in the soil on an ongoing basis. The sensors are installed at precise locations and depths in the field to take the actual measurement of soil water content. The data received from the sensors is relayed to a central controller or data management system, where it is used to ascertain the actual watering requirements of the crops. By regulating irrigation so that it is only done when soil moisture falls below maximum levels, these sensors avoid both over-irrigation and water stress, thereby enhancing plant health, saving water, and increasing overall resource-use efficiency.

Weather sensors add to soil moisture monitoring by offering valuable environmental information in the form of rainfall, humidity, temperature, and wind speed. Such sensors provide critical inputs that can enhance irrigation schedules against prevailing and projected weather. Irrigation can be rescheduled for anticipated rain so that the grounds do not unnecessarily get watered, for instance. Likewise, temperature and wind readings can aid in selecting ideal times for irrigating in a manner that keeps evaporation loss and wind drift loss at their minimum levels. Using the feedback from soil moisture sensors and weather prediction, the system calculates the most suitable timing and duration for irrigation events and delivers water exactly where and when it is required. Automated systems minimize the necessity of manual operation, enhance irrigation uniformity, maximize water-use efficiency, and eliminate under- and over-watering. By actively responding to alterations in environmental circumstances and crop requirements, such systems promote sustainable crop management while fostering productivity. Cloud-based data management solutions form the fulcrum upon which efficient, contemporary irrigation system operations depend. Information gathered via a range of sensors is wired to cloud-connected servers, where it is refined and analyzed utilizing sophisticated software suites. Farmers and agronomists have access to this data in simple mobile apps or web-based dashboards, making it possible to monitor and control irrigation activities in real time from a distance [16].

17.3.3 Categories of Smart Irrigation Systems

Smart irrigation systems use IoT technology to minimize water waste, conserve water, and enhance agricultural efficiency. There are different types of smart irrigation systems used depending on the crop needs, climatic conditions, and availability of resources. The primary categories are drip irrigation, sprinkler irrigation, center pivot irrigation, and subsurface irrigation. Drip irrigation is a precision irrigation system in which water is supplied directly to the root zone by a system of pipes, emitters, and valves. IoT sensors constantly track soil moisture levels, temperature, and nutrient levels, ensuring water is applied in exact amounts. This reduces water loss through evaporation and runoff, making it more efficient. Sprinkler irrigation systems, however, control water distribution based on actual environmental factors like temperature, humidity, wind speed, and soil moisture at any given time. Automated control mechanisms dynamically adjust watering schedules to avoid over-watering or under-watering and ensure sustainable management of water [17].

Center pivot irrigation systems make use of rotating sprinkler mechanisms for watering large circular areas in an efficient manner. IoT sensors provide measurements of soil moisture, atmospheric conditions, and water availability, and, based on these data, adjust the irrigation flow rate and timing. It optimizes the distribution of water and minimizes unnecessary consumption. Subsurface irrigation delivers water below the soil surface, which minimizes evaporation and runoff. IoT-based subsurface irrigation systems employ soil moisture sensors and remote monitoring equipment to manage water application. This technique is especially useful in dry areas, where water conservation is important, since it reduces losses and increases the efficiency of root water uptake.

17.3.4 Challenges in Implementing IoT-Based Smart Irrigation

Even with the many advantages of IoT-based smart irrigation, there are a number of challenges that limit its use, especially among small farmers. The initial high cost of buying sensors, controllers, cloud infrastructure, and automated irrigation parts can be out of reach. Moreover, constant internet connectivity is required for uninterrupted IoT device operation, but numerous agricultural areas, especially in the developing world, lack connectivity due to poor network infrastructure. This slows down real-time data transmission and diminishes the efficiency of IoT-based irrigation systems. There is also a shortage of technical knowledge among farmers, which makes the installation and maintenance of these systems even more difficult. Most rural farmers also have issues with sensor calibration, interpretation of data, and system maintenance, which makes their usage of the technology difficult. Data security and privacy issues also present serious risks since IoT devices gather and transfer huge amounts of farm data. Cyber-attacks like data breaches and unauthorized access demonstrate the necessity for strong encryption, safe cloud storage, and access controls to secure sensitive agricultural data [18].

17.3.5 Case Studies of IoT in Smart Irrigation

Deployment of IoT-based intelligent irrigation systems has shown encouraging outcomes in different farming environments across the globe. IoT-based irrigation systems have lowered water usage while improving agricultural output in India. Water savings of 20–30% and improved crop yields have resulted from adopting these solutions in states like Gujarat and Rajasthan [19]. Real-time measurement of soil moisture and automated control of irrigation have allowed farmers to maximize water application, thus making water resource management sustainable. California vineyards are another case in point for the influence of IoT irrigation. With acute drought, the wine sector in California has opted for intelligent irrigation systems to efficiently use water. These systems have helped vineyards reduce water consumption by 25–40% without compromising the quality of the grapes [20]. By using soil moisture sensors, weather data analytics, and precision irrigation methods, vineyard managers are able to make informed decisions based on data to increase water efficiency without affecting crop yield. The case studies highlight the revolutionizing capability of IoT-based smart irrigation in improving water-use efficiency, increasing productivity, and supporting sustainable farming. Nonetheless, resolving challenges with respect to cost, connectivity, and cybersecurity is key for widespread use and scalability [21].

17.3.6 Future Trends in Smart Irrigation with IoT

The future will see further integration of IoT with ML, AI, and remote sensing, enhancing irrigation precision. Satellite-based remote sensing could provide broader monitoring capabilities, and solar-powered devices may improve sustainability in regions with limited electricity access [22].

17.4 LIVESTOCK MONITORING WITH IOT

Livestock management is vital for modern agriculture, focusing on animal health, productivity, and welfare. Traditional monitoring methods are often manual and labor-intensive, leading to inefficiencies and potential errors. The IoT revolutionizes this sector by providing real-time tracking and monitoring of livestock, helping farmers optimize care, improve efficiency, and enhance farm operations. IoT-based systems utilize sensors and connected devices to collect data on animal behavior, health, and environmental conditions, enabling informed decision-making that enhances animal welfare, reduces disease outbreaks, and boosts productivity. The subsequent sections detail the key application areas of IoT in the management of livestock (Figure 17.3).

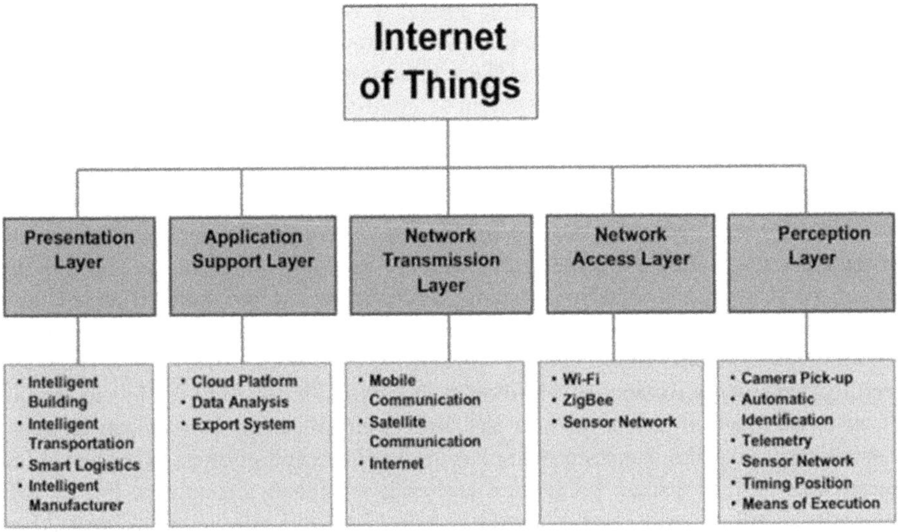

FIGURE 17.3 Internet connectivity.

17.4.1 Livestock Health and Tracking through IoT Applications

With the help of IoT, other wearable devices, and sensors, the health condition of livestock can be traced back by monitoring their physiological and behavioral precursors. One of them is body temperature; its constant monitoring can point to a change in its norm, which might indicate an illness, in order for farmers to act accordingly. In addition, the previously explained observation of heart rate and respiration helps to determine the level of stress or any disease in the animals under study. Supplementing consumption behavior to the health appraisal enlarges the spectrum of early indicators by taking into account fluctuations in the amount of consumed alcoholic beverages as well as food. Furthermore, IoT plays a very important role in the reproductive health of the livestock, as is evident with the sophisticated sensors that are used to track the heat cycles of the reproductive cycle, thus making the breeding schedule more efficient and productive. Real-time monitoring of animals using IoT GPS tracking is an added bonus when it comes to monitoring the livestock on the farm. This kind of tracking makes farmers aware of the location of these animals at any given time, thus preventing theft and restricting the grazing area. It also enhances security in that it alerts when animals move out of the set geographical locations, within regions that are laid down to be secure. IoT sensors also analyze patterns of motion to detect any irregularity regarding movement, which may be due to a complication in an individual's health; thus, timely assistance can be sought. The application of tracking by IoT is evident in actual life situations, one of them being the use of smart collars to monitor cattle on Australian beef farms. This technology alerts the farmers in case of any movement that it considers abnormal and thereby reduces mortality rates considerably.

Thus, the IoT technologies used in animal welfare have a correlation effect on improved productivity and profitability. Stress prevention is one of the main managed zones of welfare enhancement. These last control measures include temperature and humidity, whereby the environment is controlled with the help of sensors in order to make the animals as comfortable as possible. Likewise, movement patterns and resting behavior also help the farmers in identifying one way or the other signs of stress in animals, which in turn provides a way for early intervention to be applied. Another important factor for monitoring is social interaction because the movement of the herd may indicate a problem that calls for preventive measures. Through the use of the feeding and watering systems, animals are fed and watered as per their needs. Timely portion control and automated feeding systems help to dispense the recommended quantity of food that prevents instances of either overfeeding or underfeeding. Sensitively advanced technologies are used to measure the ratio and quality of water used, and if any problem arises, they notify the farmers promptly to avoid discontinuation of the water supply and improve the health of crops. This has led to an advancement in disease prevention and early diagnosis through the use of IoT. Disease surveillance systems run surveillance on data; this makes it possible to identify probable outbreaks and prevent their spread through measures such as quarantine. Furthermore, it is possible to automate vaccination as well as the injections that are given to the animals, and this was helpful in ensuring that treatments are administered in time and improving the health of the entire herd.

The integration of IoT in cattle management is beneficial in enhancing operational efficiency significantly. This results in saving of time and labor on the part of the University, which is a major advantage. Since IoT sensors allow the continuous gathering of data, farmers are not required to make frequent checkups and hence are able to attend to other important tasks. Mobile monitoring systems also enhance efficiency since farm and cattle status of health and location can still be accessed through the mobile applications even if the farmer is not on the farm. By utilizing the IoT in livestock care, one of the key advantages is making decisions based on obtained data. Big data is used to predict health problems before they become serious; this technology helps in early diagnosis since it entails the use of algorithms. IoT is also helpful in the aspect of resource management in conforming to the rates of usage of feeds, water, and the necessary resources that may be needed to avoid wastage. The implications of IoT in livestock rearing tend to lead to increased production and profitability in the business. Improved management practices, on the other hand, result in better livestock health, implying that there will be better quality and increased volumes of livestock produce. The one downside of IoT technology is that it is very expensive to first implement, but in the long run, low labor costs, better healthcare facilities, and resource optimization directly translate into better economic benefits for the farmers.

17.5 DATA INTEGRATION AND AUTOMATION IN IoT SYSTEMS

The successful use of IoT in an agricultural environment is directly proportional to the efficiency and ability to consolidate data coming from several sensors and devices into one central data hub. Those features, when implemented with automation, enable the farmers to manage their operations and to arrive at informed decisions based on the actual data. The chapter also describes how IoT systems cooperate in the data integration process and the automation of farm work for the betterment and improvement of agriculture.

17.5.1 Centralized Data Platforms for Real-Time Monitoring

This can be seen in the mechanism of central data platforms for collecting IoT, devices, sensors, and systems from a farm as key to IoT-based farming. This would be the brain behind a smart farming operation that farmers can control without necessarily being at the farm. They could get the network through Apps, tablets, as well as a normal computer. Modern IoT networks encompass a wide range of sensors to monitor and control important information regarding agriculture to support precision agriculture and resource utilization. These sensors take information on soil moisture and nutrient level, temperature, humidity, rainfall, growth of grains and crops, health of livestock and poultry, and other details of tractors, irrigation systems, etc. These data points are very useful in making decisions and in the management of the farms as a whole. The aspect of real time is essential in the IoT system in farming and allows farmers to quickly address conditions and hence increase production. The big data is processed in a central hub to provide the farmers with information on the status of important parameters of the field: feedback is received online regarding the moisture content in the soil and nutrient status of plants, and the level of stress in the fields, so as to initiate irrigation as well as fertigation. Also, the movement and health of the livestock are monitored through wristbands that contain the IoT devices that monitor physical status or even behavior and notify of eventualities of diseases. Also, web and mobile interfaces and applications provide an opportunity to control and monitor the farms' data from a distance for farmers.

The most valuable commodity of IoT sensors is the amount of analyzed data, which is more helpful when it is represented in a manner easy to assess. Graphs, charts, heatmaps, and dashboards assist the farmers of these centralized platforms, allowing them to understand various complicated datasets and make firm decisions. A good example of this strategy is the Farmers Business Network, which combines data from various IoT sensors/farmers, therefore, real time analytics emanate from crop health to market trends and come up with profitable decisions with higher yields. The use of IoT in agriculture farming cuts down manual handling of crops, gives precise use of the resources in farming, and makes the processes more efficient. From here, real-time data and other functions are utilized effectively to enhance various operations of agriculture, thus improving the yields of various operations. Precision farming, including irrigation and fertilization, is applied by using data from IoT technologies. Sophisticated methods of irrigation regulate the availability of water depending on the soil moisture and the current weather conditions to avoid any wastage in the process but at the same time ensure that the crops are healthy. At the same time, precision fertilization systems control the nutrient levels and the application of the right amount of fertilizers, without polluting, but with high utilization rates of fertilizers.

Smart correction is not limited to the most important activities concerning the care of the livestock, their well-being, and the improvement of farm productivity.

Provisions such as feeding states and watering systems ensure the current intake of foods and water is provided while observing the patient's health status and ill health, hence reducing wastage. Such systems measure temperature, movement, and behavior and alert caregivers when signs of sickness or stress are identified to improve the quality of life of dogs. Modern intelligent farming needs the appropriate use of automation to increase its operational efficiency. Self-driving cars have been adopted to ensure accuracy during plowing and sowing, planting, and reaping to reduce employment costs. Other systems exist, known as Predictive Maintenance, that employ IoT sensors to monitor machinery performance and predict potential breakdowns that may occur, hence reducing downtime and maintenance costs. Smart farming is achieved by the IoT-based task scheduling process since the information from the field is analyzed in real time. Dynamic scheduling recommends to farmers the appropriate timing of planting the crops, watering the crops, and even harvesting in relation to certain factors. Even such simple operations as fertilizing and equipment maintenance can be time managed using WorkManager or Job Scheduler, which helps to schedule them as well as improve the efficiency of farming work.

17.5.2 Enhancing Decision-Making with Data Analytics

The data collected from IoT devices can be analyzed by the use of business intelligence tools to help the farmer make correct decisions that would lead to productivity and sustainable production. Here, the farmers can provide improved techniques of farming and optimum utilization of available resources through predictive analytics, prescriptive analytics, and big data analytics. It is highly valuable to notice that predictive analysis enables one to prepare for challenges before they turn into troublesome complications. As mentioned earlier, given the historical and live data, it is possible for the mathematical models to accurately forecast weather occurrences such as droughts, thus enabling the farmers to make proper preparations in advance. The models also analyze trends in a bid to ascertain future occurrences of pests and diseases, with regard to which actions should be taken to avoid incurring high losses. Moreover, predictive analytics helps in identifying and estimating crop yields through averaging the soil variables, climate, and growing patterns of crops, so that every available resource can be utilized in order to increase the yield of crops. It goes further than prediction by providing specific recommendations on how the operation of a farm should be conducted. It helps in efficient use of inputs because it provides advice regarding the required amount of water, fertilizers, and other inputs at present. It also determines the time that the farmer should plant the crops and the time that is most appropriate for harvesting the crops to avoid cases of destruction and lack of productivity. These complex analytics tools assist in optimizing decision-making and the smooth functioning of agricultural systems through the continuous observation of different parameters that dictate their operations.

Agricultural big data is a change agent that permits trend detection and insightful information out of enormous records gathered by the IoT systems. Historical crop and soil data help for better planning so that growers can optimize culture techniques across growing seasons. Finally, combining the data from the IoT-based applications with market demand patterns improves marketing strategies so farmers can match supplies to demand dynamics. A very well-known instance of the application of big data in agriculture is John Deere Precision Agriculture, which merges information from intelligent tractors, sensors, and satellite images to make planting schedules more efficient and allocate resources more effectively. All this fusion of data insights delivers more yields, greater efficiency, and eco-friendly cultivation.

17.6 CHALLENGES OF IoT ADOPTION IN AGRICULTURE

While the potential benefits of IoT in agriculture are significant, its widespread adoption faces several challenges. Farmers, particularly in developing regions or small-scale operations, encounter hurdles such as data security risks, infrastructure limitations, and high costs. Addressing these issues is crucial for IoT's long-term impact on sustainable agriculture [23]. The applications of IoT in agricultural activities result in the generation of large amounts of data, and this in itself generates strong data security and data privacy challenges. Cropped information, such as production yields, soil conditions, animal diseases, and farming techniques, must be protected from hackers and intruders. It is crucial that such data is protected to maintain operational integrity and prevent potential organization for economic loss [24, 25]. One of the major challenges associated with IoT in farming is data breaches. An example is the unauthorized access to the farm databases and information, since this would lead to exploitation by other people, hence damaging the entire farm ventures. Farm networks can be targeted by hackers to change or gain access to information and thus cause significant losses to the farmers. Also, operational interferences are likely to occur through disruptions that may be caused by cyberattacks on the automated systems, such as irrigation, feeding, or harvesting systems, that affect the efficiency of farming. The possibility of cybercrimes makes proper isolation and strict security measures, including encryption policies and safe entry measures, crucial.

One of the most daunting challenges facing information systems at the moment is data ownership and use. In the IoT-based agriculture services, a farmer usually deals with third-party technology companies for different services that make him ambiguous about ownership of resultant data. Today, there are some companies that own the right to collect and analyze the farm data, and some people would be able to gain access to this data and misuse it. In addition, such data can be analyzed by employing more elaborate analysis techniques that will produce important information capable of influencing either the market patterns or the competition level. The establishment of separate legal structures and treaties concerning ownership of data will help advance farmers' cause while not preempting the beneficial openness of data-sharing practices for gain. In order to address the issues of security and privacy, several measures have been proposed. Security measures, such as AES-256, enable data that needs to be transmitted or stored to be encrypted in such a way that unauthorized persons cannot intercept the same. Role-Based Access Control (RBAC) systems can enable farmers to grant permissions and limit access, which cannot be used by unauthorized parties. In addition, blockchain offers a potential solution by facilitating transparent and immutable data transactions throughout agricultural supply chains. The application of blockchain-based ledgers can promote trust and security in data exchanges while minimizing the risk of unauthorized changes [26, 27].

The success of IoT in agriculture largely relies on stable technological infrastructure, which is usually absent in rural and developing areas. The most critical challenge is poor connectivity, as most agricultural regions lack good internet connectivity. This constraint prevents real-time data transmission, causing delays in decision-making processes. Where cloud-based analytics are needed, the failure to transmit data in a timely manner diminishes the efficiency of IoT solutions, limiting their full potential. In addition, agricultural IoT devices have to be very rugged to tolerate harsh environmental conditions like high temperatures, humidity, dust, and mechanical shock. Hardware ruggedness is important for ensuring long-term performance and reducing the cost of replacements. Moreover, there are interoperability challenges resulting from the absence of unified communication protocols among heterogeneous IoT devices and platforms. This heterogeneity makes system integration difficult and accelerates the complexity of agricultural IoT deployments. To overcome these problems, new communication technologies and computing platforms are being designed. Low-Power Wide-Area Networks (LPWANs) provide an energy-efficient approach to offering long-distance connectivity in rural locations, providing robust IoT data transmission. Edge computing is also emerging as a means of processing data on-site at the farm instead of on remote cloud servers. Through the use of edge computing, latency in data processing is minimized, and reliance on internet connectivity is eliminated [28]. Companies are also investing in the creation of rugged IoT devices with increased resilience to harsh farm conditions, thus enhancing the lifespan and efficiency of these technologies.

The use of IoT technology in agriculture calls for great financial investment, which may prove to be a hindrance, especially for smallholder farmers in developing countries. The economic framework of IoT deployment involves costs related to the acquisition of hardware, software subscription, and maintenance [29]. The initial capital investment needed for IoT agriculture is significant. The cost of buying sensors, drones, automated irrigation systems, and other IoT equipment can be too high for most farmers. Moreover, software and cloud services needed for farm data processing and analysis typically incur periodic subscription fees, increasing operational expenses. Maintenance and system upgrades also account for long-term costs, making sustainable financial models necessary for IoT adoption.

$$C_{\text{total}} = C_{\text{hardware}} + C_{\text{software}} + C_{\text{maintenance}} + C_{\text{connectivity}}$$

The overall IoT implementation cost in a farm environment can be represented as [30]:

where:

- C_{hardware} is the cost of IoT devices like sensors, actuators, and drones,
- C_{software} is cloud services and analytics platforms,
- $C_{\text{maintenance}}$ is periodic servicing and replacement,
- $C_{\text{connectivity}}$ is costs associated with internet access and communication networks.

To achieve greater accessibility of IoT technology, governments and private firms are looking into subsidy schemes and fiscal incentives. Public-private collaborations can bridge the cost gap for IoT in small-scale agriculture by introducing cost-sharing models for IoT uptake in small-scale farming. Furthermore, designing affordable and modular IoT solutions for phased adoption instead of large-scale implementation can reduce barriers for resource-poor farmers. Open-source platforms and shared infrastructure frameworks can further cut down the costs of software, making IoT-based agriculture more economical for a larger number of farmers.

17.7 RESULT AND DISCUSSION

IoT agriculture adoption has led to varied new applications with strong improvements in farm practices. Key results from selected case studies illustrate the transformative impact of IoT technologies on agricultural efficiency, productivity, and sustainability.

17.7.1 Summary of IoT Case Studies in Agriculture

CASE STUDY 1: CALIFORNIA'S SMART IRRIGATION

An irrigated farm on a large scale in California launched an IoT-integrated smart irrigation system to maximize water utilization and increase yield. The system utilized real-time data from moisture sensors in soil, weather forecasts, and automated irrigation machinery managed remotely. Based on sensor output and automatic decision-making, the farmers could manually control water dissemination depending on environmental conditions and the level of moisture in the ground. The implementation of this system brought about enormous resource efficiency. Water use decreased by about 30% over the conventional irrigation systems. Algebraically, if the initial water use per hectare was W_0, the lower consumption W_r is calculated as:

$$W_r = W_0 - 0.3W_0 = 0.7W_0$$

In addition, more precise irrigation led to a 15% yield increase in crops because of optimal levels of hydration. Farmers were also provided with real-time data-driven recommendations that allowed them to intervene on time to correct soil moisture shortages and prevent crop stress.

CASE STUDY 2: LIVESTOCK WEARABLE MONITORING IN NEW ZEALAND

New Zealand's livestock farm incorporated IoT wearable devices to track cattle behavior, movement, and health. IoT wearable devices transmitted real-time data on physiological activities and positional monitoring, which was processed to recognize early symptoms of disease, feeding schedules, and animal welfare. Due to prompt medical intervention and early detection of disease, vet expenses decreased by 20%. This is equated as:

$$C_r = C_0 - 0.2C_0 = 0.8C_0$$

where C_0 is the initial veterinary cost, and C_r is the minimized cost after adopting the wearable monitoring system. Better monitoring also resulted in enhanced livestock management practices, which resulted in better health outcomes. The efficiency of farm labor also improved, with a 40% decrease in time taken for manual inspection and observation.

CASE STUDY 3: PRECISION FARMING IN BRAZIL

A Brazilian farm utilized IoT sensors in precision agriculture for observing the soil conditions, weather patterns, and crop health. The combination of ML analytics made it possible for predictive analysis, resulting in optimized planting techniques and optimal resource allocation. The crop yield rose by 25% because of the precision-based decision-making paradigm. The yield Y_r improvement can be represented as:

$$Y_r = Y_0 + 0.25Y_0 = 1.25Y_0$$

where Y_0 is the initial yield. Additionally, focused application of fertilizers and pesticides resulted in a 20% cut in chemical usage, reducing environmental footprint. Operational expenses fell by 15%, fueled by better resource utilization and less manual intervention.

CASE STUDY 4: VERTICAL FARMING IN SINGAPORE

A vertical farm in Singapore's city used IoT technology to manage light, temperature, and humidity for maximum crop growth. The closed water system and precision irrigation methods lowered water use significantly. Water use dropped by 90% from conventional farming techniques. The controlled environment allowed for increased crop turnover at a faster rate, speeding up growth cycles by 20% to 50%. Sustainability was also increased with the use of renewable energy and streamlined logistics, minimizing the total carbon footprint of the farm.

CASE STUDY 5: DRONE TECHNOLOGY FOR CROP HEALTH ASSESSMENT IN INDIA

One agricultural project in India utilized drones with multispectral cameras to monitor crop health across large agricultural lands. The drones took high-resolution images, detecting plant stress, pests, and nutrient shortages in real time. The use of this technology greatly enhanced precision agriculture. Spraying with pesticides and fertilizers was optimized, lowering chemical use by 30%:

$$P_r = P_0 - 0.3\,P_0 = 0.7P_0$$

where P_0 is the initial chemical consumption and P_r is the lowered amount after drone-aided interventions. Furthermore, labor efficiency increased, with fewer manual field surveys, resulting in reduced operating expenses and improved productivity. The above case studies demonstrate the revolutionary effect of IoT in contemporary agriculture, making farming more efficient, sustainable, and profitable through data-driven decision-making.

17.7.2 Key Performance Improvements through Analysis

Deployments of IoT technologies within the agricultural sectors have indeed shown unambiguous evidence concerning performance improvement from various performance metrics. It looks into some of the affected key performance indicators regarding IoT application in agriculture, which allude to the contribution given by IoT technologies toward a better operating performance, productivity, and sustainability.

17.7.2.1 Water Efficiency

The use of IoT-enabled smart irrigation systems has dramatically increased water efficiency in farming. A key performance measure (KPI) in the field is saving water, and observed reductions have varied from 20% to 40%. Monitoring soil moisture in real time makes it possible to irrigate precisely, supplying the right quantity of water to crops, preventing waste. California farmers who use IoT-enabled irrigation systems reported 30% less water use and a 15% increase in crop yield at the same time.

17.7.2.2 Improved Crop Yield

IoT-based agriculture has shown substantial improvement in crop yield, with yield improvements of as much as 25% being reported by some farms. Timeliness of interventions enabled by real-time data is critical in maximizing irrigation, pest management, and nutrient application.

17.7.2.3 Resource Utilization

IoT-based precision agriculture applications have decreased the use of fertilizers and pesticides by 20% and 30%. Real-time sensor data-based targeted interventions assure that these are applied only to areas where there is a requirement, thus bringing down costs as well as their impact on the environment. Efficiency of resource usage is expressed by:

$$R_{\text{eff}} = \frac{R_{\text{trad}} - R_{\text{IoT}}}{R_{\text{trad}}} \times 100\%$$

where R_{eff} is the efficiency increase in resource utilization, R_{trad} is traditional resource use, and R_{IoT} is IoT-optimized resource utilization. Drone-augmented crop monitoring in India led to pesticide usage reduction by 30%, improving economic as well as environmental sustainability.

17.7.2.4 Labor Efficiency

IoT technologies have optimized field operations, saving farmers as much as 40% of their time spent on manual monitoring. Automated monitoring of livestock and smart farming equipment has also saved farmers 15%–20% in labor costs.

17.7.2.5 Animal Health and Welfare

Livestock monitoring using IoT has resulted in timely detection of disease, cutting the cost of veterinarians by a maximum of 20%. Ongoing monitoring of health indicators ensures that farmers step in before disease sets in, enhancing herd care.

17.7.2.6 Sustainability Metrics

IoT applications across agriculture have aided in lowering the carbon footprint as well as reducing wasted resources. Farmers have minimized emissions and enhanced the sustainability level with optimized inputs. Singaporean vertical farms, such as in an urban context, are great examples of the resultant impact, as it has realized 90% decreased water consumption accompanied by using renewable energy resources.

17.7.3 Results from IoT Applications in Smart Irrigation, Precision Agriculture, and Livestock Monitoring

This has to do with the technology in the application of IoT that has seen a great amount of achievement in smart irrigation, precision agriculture, and livestock monitoring. These will be highlighted as specific achievements in this section for each of these areas.

17.7.3.1 Intelligent Irrigation Outputs

The use of smart irrigation systems in contemporary agriculture has transformed water management through the use of IoT technologies like soil moisture sensors, weather forecasting information, and automated irrigation controls. The systems maximize water use by providing crops with the exact amount of water needed based on real-time environmental conditions. Research has indicated that farmers employing IoT-based irrigation techniques can save up to 20% to 40% of water compared to traditional irrigation systems. A California case study reduced overall water usage by 30% while improving crop yields by 15% at the same time. This is due to real-time adjustments of soil moisture to provide the crops with proper hydration without any wastage. Moreover, the cost of operations for irrigation is greatly minimized, with estimates of a 10% to 15% reduction in total water-related expenses. The smart irrigation can be mathematically modeled for efficiency. If *Wc* is the traditional water consumption and *Ws* is the saved water through smart irrigation, the water conservation efficiency can be represented as:

$$Ws = Wc \times \frac{R}{100}$$

where R is the percentage decrease in water consumption. For instance, in a 30% decrease $R = 30$, a traditional system consuming 1,000 liters of water per cycle would conserve:

$$Ws = 1000 \times \frac{30}{100} = 300\,l$$

illustrating the real advantages of converting to smart irrigation systems.

17.7.3.2 Impacts of Precision Farming

Precision farming, which uses IoT technologies like remote sensing, ML, and real-time data analysis, has become a revolutionary method for maximizing farm productivity and sustainability. Using site-specific management practices, farmers are able to make informed decisions that maximize efficiency in the cultivation of crops. Research has shown that farms using precision agriculture methods have their crop yields increased by 15% to 25% as compared to conventional farming methods. A good example is a Brazilian farm that achieved a 25% increase in yields through accurate resource allocation and planting optimization. Further, precision agriculture maximizes the use of resources by cutting down on fertilizer and pesticide application by 20% to 30%. This is done by placing inputs where and when they are required, instead of spreading them evenly over the whole field.

These focused interventions decrease environmental effects while keeping productivity levels high. Cost-effectiveness also arises from these improvements, with operating costs decreasing by 10% to 15% because resources are used more efficiently. Quantitatively, if C_t is the entire cost in conventional farming and C_p is the cost in precision farming, the percentage cost saving C_m can be calculated as:

$$S = \frac{C_t - C_p}{C_t} \times 100$$

For example, when the conventional cost was $10,000 and precision farming brings it down to $8,500, the savings would be:

$$S = \frac{10000 - 8500}{10000} \times 100 = 15\%$$

verifying the economic feasibility of precision farming solutions.

17.7.3.3 Livestock Monitoring Results

IoT integration into livestock farming has greatly enhanced animal health monitoring and farm efficiency. Wearable monitoring units attached to animals allow for real-time tracking of physiological and behavioral information, making it possible to detect diseases early and manage the herd more effectively. The use of IoT in the monitoring of livestock farming has resulted in a reduction of veterinary bills by 20% since different health problems are detected at an early stage that is before evolving into serious diseases. Thus, recording vital parameters and movement activities allows for early application of prevention by the farmers, which will improve animal welfare. The other area that has been improved through the use of IoT for monitoring the livestock is the efficiency of labor. Let us consider two types of checks: physical and monitoring with the help of automatic systems. Compared to physical checks, which are rather time- and effort-consuming, the practice of monitoring with the help of automatic means has saved 40% of the time that has to be spent to observe and inspect. Such systems make it possible for the farmers to control any irregularities in the course of production through notifications and not by physically checking on the cattle on a regular basis. Also, increased surveillance of the animals results in improvement of methods for managing them and improved health, hence increased production. In New Zealand, a farm implemented IoT wearables that would track the health status as well as the location of the cattle. This provided farmers with real-time information on the basic parameters of the animals so that immediate corrective actions could be taken and sound decisions made on the fate of the animals. The economic advantage of such a system can be assessed in terms of cost using the cost reduction formula:

$$C_s = C_m \times \frac{T_r}{100}$$

where C_m is the cost of manual labor, and T_r is the rate of time saved by automation. If manual labor costs $5,000 and time savings are 40%, then:

$$C_s = 5000 \times \frac{40}{100} = 2000$$

showing a cost saving of $2,000 by IoT-based livestock management. In summary, the application of IoT in agriculture, via smart irrigation, precision farming, and animal monitoring, has greatly enhanced productivity, efficiency, and sustainability. These innovations not only lower resource usage and operational expenses but also improve decision-making via real-time data analysis. As IoT technologies advance, their contribution to the future of smart farming will become more vital in solving global food security issues.

17.8 CONCLUSION

This leads us to the next innovation, which is the use of IoT in agriculture and is one of the ways to develop the smart technologies for farming that would result in better outcomes. Some of the effects of IoT include efficiency in resource utilization, crop productivity, efficient management of cattle, climate change and environmental adaptability, and decision-making based on data. Presenting real-time data, some of the advantages of structured agriculture are as follows: productivity and climate change. After identifying the required needs of the farmers, it leads to pilot projects, proper setting of the technology, investment done on farmers' training, and good management of data with security measures as well as the necessary stakeholders. It is supposed to be monitored to see that IoT solutions are implementable and effective.

REFERENCES

1. Dhal, Sambandh, et al. "Internet of Things (IoT) in digital agriculture: An overview." *Agronomy Journal* 116.3 (2024): 1144–1163.
2. Bulut, Cevdet, and Wu, Philip Fei. "More than two decades of research on IoT in agriculture: a systematic literature review." *Internet Research* 34.3 (2024): 994–1016.
3. Shahab, Hammad, et al. "IoT-based agriculture management techniques for sustainable farming: A comprehensive review." *Computers and Electronics in Agriculture* 220 (2024): 108851.
4. Sarma, Hridesh Harsha, et al. "Innovative approaches for climate-resilient farming: Strategies against environmental shifts and climate change." *International Journal of Environment and Climate Change* 14.9 (2024): 217–241.
5. Gryshova, I., Balian, A., Antonik, I., Miniailo, V., Nehodenko, V., & Nyzhnychenko, Y. "Artificial intelligence in climate smart in agricultural: Toward a sustainable farming future." In Mariana Petrova (Ed.), *Access to Science, Business, Innovation in the Digital Economy* (2024) (Vol. 5, No. 1, pp. 125–140). ACCESS Press. https://doi.org/10.46656/access.2024.5.1(8)
6. Alnumay, Waleed S. "The past and future trends in IoT research." *National Journal of Antennas and Propagation* 6.1 (2024): 13–22.
7. Gaikwad, Sandeep V., et al. "An innovative IoT based system for precision farming." *Computers and Electronics in Agriculture* 187 (2021): 106291.
8. Morchid, Abdennabi, et al. "Applications of internet of things (IoT) and sensors technology to increase food security and agricultural sustainability: Benefits and challenges." *Ain Shams Engineering Journal* 15.3 (2024): 102509.
9. Raffik, R., Farouq, K. M., Muddisetty, H., & Sengar, K. Smart agriculture farming using drone automation technology. *AIP Conference Proceedings*, 2937.1 (2024): 020001. AIP Publishing.
10. Fastellini, G., and Schillaci, C.. "Precision farming and IoT case studies across the world." In A. N. Papageorgiou, D. Glaroudis, P. Chatzimisios, & L. Atzori (Eds.), *Agricultural internet of things and decision support for precision smart farming*. Academic Press (2020): 331–415
11. Kumar, V., et al.. "A comprehensive review on smart and sustainable agriculture using IoT technologies." *Smart Agricultural Technology*, 4 (2024): 100487.
12. Selvan, M. A. "IoT-integrated smart home technologies with augmented reality for improved user experience." *Journal of Science Technology and Research*, *5*(1) (2024): 389–394.

13. Khang, Alex, et al., eds. *Advanced IoT technologies and applications in the industry 4.0 digital economy*. CRC Press (2024).
14. Talwani, S., Rakhra, M., and Sarkar, T. (2024, March 14–15). "Perspectives on the Future of Agriculture and Recent Developments in Agricultural Technology." *2024 11th International Conference on Reliability, Infocom Technologies and Optimization (Trends and Future Directions) (ICRITO 2024)* (pp. 1–6). Amity University, Noida, India. IEEE.
15. Obaideen, Khaled, et al. "An overview of smart irrigation systems using IoT." *Energy Nexus* 7 (2022): 100124. Dhinakaran, D., et al. "Privacy-Preserving Data in IoT-based Cloud Systems: A Comprehensive Survey with AI Integration." arXiv preprint arXiv:2401.00794 (2024).
16. Khezri, Edris, et al. "DLJSF: Data-locality aware job scheduling IoT tasks in fog-cloud computing environments." *Results in Engineering* 21 (2024): 101780.
17. Sarma, Hridesh Harsha, et al. "Climate resilient agricultural strategies for enhanced sustainability and food security: A review." *Plant Archives* 24.1 (2024): 787–792.
18. Rozlomii, I., Yarmilko, A., and Naumenko, S. "Data security of IoT devices with limited resources: Challenges and potential solutions." In T. A. Vakaliuk & S. O. Semerikov (Eds.), *Proceedings of the 4th Edge Computing Workshop (DOORS 2024)* (CEUR Workshop Proceedings, Vol. 3666 (2024, April 5), pp. 85–96). Zhytomyr, Ukraine. CEUR-WS.org
19. Barning, Lucas. Rewiring Power–Unraveling the Sociotechnical Dynamics of Smart Grid Pilot Projects in the Transition of India's Electricity Distribution Sector. *Doctoral dissertation*. Technische Universität Wien, 2024.
20. Xanthopoulou, Panagiota, and Zakopoulos, Vassilis. "Developing management skills via E-learning: A pilot study on a cultural foundation." *European Research Studies Journal* 27.3 (2024): 927–949.
21. Ho, Olivia, et al. "A conceptual model for integrating circular economy in the built environment: An analysis of literature and local-based case studies." *Journal of Cleaner Production* 449 (2024): 141516.
22. Magara, Tinashe, and Zhou, Yousheng. "Internet of Things (IoT) of smart homes: Privacy and security." *Journal of Electrical and Computer Engineering* 2024.1 (2024): 7716956.
23. Vishnoi, Shweta, and Goel, Raj Kumar. "Climate smart agriculture for sustainable productivity and healthy landscapes." *Environmental Science & Policy* 151 (2024): 103600.
24. Liang, P., et al. "Multi-level intrusion detection based on transformer and wavelet transform for iot data security." *IEEE Internet of Things Journal PP*(99) (2024), 1–1.
25. Eghmazi, Ali, et al. "Enhancing IoT data security: Using the blockchain to boost data integrity and privacy." *IoT* 5.1 (2024): 20–34.
26. Hamidoğlu, Ali, Gül, Ömer Melih, and Kadry, Seifedine Nimer. "A game-theoretical approach for the adoption of government-supported blockchain application in the IoT-enabled agricultural supply chain." *Internet of Things* 26 (2024): 101163.
27. Ikevuje, Augusta Heavens, Anaba, David Chinalu, and Iheanyichukwu, Uche Thankgod. "Optimizing supply chain operations using IoT devices and data analytics for improved efficiency." *Magna Scientia Advanced Research and Reviews* 11.2 (2024): 070–079.
28. Liang, P., Yang, L., Xiong, Z., Zhang, X., & Liu, G. Multi-level intrusion detection based on Transformer and wavelet transform for IoT data security. *IEEE Internet of Things Journal*, PP.99 (2024), Article 1.
29. Dhinakaran, D., et al. "Privacy-preserving data in IoT-based cloud systems: A comprehensive survey with AI integration." arXiv preprint arXiv:2401.00794 (2024).
30. Xanthopoulou, Panagiota, and Zakopoulos, Vassilis. "Developing management skills via E-learning: A pilot study on a cultural foundation." *European Research Studies Journal* 27.3 (2024): 927–949.

18 Innovations in Sustainable Agriculture
Recent Advances and Future Directions

P. K. A. Chitra, Prabu Selvam, and S N Kumar

18.1 INTRODUCTION

The world's demand for food is rising exponentially as a fast-growing human population, estimated to hit ten billion by the year 2050, fuels growth. At the same time, climate change poses unprecedented challenges in the way of agricultural productivity as increasing temperatures, changing precipitation levels, and escalating frequency of abnormal weather patterns impact crop yields around the globe. These combined forces call for radical change in methods of agriculture and require innovative products to improve both efficiency and sustainability. Artificial intelligence (AI), with its ability to conduct sophisticated data analysis, predictive modeling, and automation, stands out as a formidable tool to solve these pivotal challenges. The necessity of going through the research papers in the emerging field of AI in agriculture shows the revolutionary potential in achieving precision and sustainability in agricultural practices. AI is being used to maximize the use of resources, mechanize labor-intensive operations, improve crop yields, and reduce the environmental effects of agriculture. The world's agricultural sector is confronted with the twin challenge of raising food production to satisfy the needs of a growing population while reducing the environmental effects and achieving socio-economic sustainability.

Sustainable agriculture is the synthesis of biological, chemical, physical, ecological, economic, and social sciences into innovative farm practices that are environmentally benign and non-degrading to natural ecosystems. In the 1990 United States, sustainable agriculture was an "integrated system of plant and animal production practices adapted to site-specific conditions that, in the long term," will:

- Meet human food and fiber needs
- Improve environmental quality and the natural resource base upon which the farm economy relies
- Minimize the use of nonrenewable and on-farm resources and incorporate natural biological cycles and controls where they are appropriate
- Ensure the economic sustainability of farm operations
- Enhance the quality of life for farmers and society generally [1]

DOI: 10.1201/9781003594093-18

18.2 PRINCIPLES OF SUSTAINABLE AGRICULTURE

Agricultural sustainability refers to the management of agricultural resources effectively in order to meet changing human needs while preserving or enhancing environmental quality and conserving natural resources [2]. This incorporates innovative farming methods aimed at protecting the environment, conserving natural resources, reducing dependency on chemical inputs, and enhancing economic resilience. Importantly, the definition of agriculture should be broadened to include the foundational steps in the production of consumable goods, even when these processes do not explicitly involve traditional farming or land cultivation. Sustainable agriculture is a combination of four pillars [3]: (1) land management, (2) resource management, (3) human interface, and (4) ecosystem interface. Collectively, these result in agricultural systems that are productive, environmentally sound, economically viable, and socially equitable, with satisfying present needs but also resources for future generations.

Sustainable agriculture practices focus on biofuels, agritourism, addressing food insecurity, agroforestry, cooperatives, cover crops, dairy waste management, food labeling, energy efficiency and conservation, food waste management, genetically modified crops, industrial sustainable food procurement, integrated pest management, urban agriculture, organic farming, precise agriculture, soil nutrient management, water-use efficiency, and quality management [4]. This is characterized by an integrated strategy to fix environmental conservation and management, economic feasibility, and equity in societal issues, also investigating the implications for agricultural economics. Figure 18.1 presents the three dimensions of sustainable agriculture that lead to long-term productivity, economic stability, and livelihood for farmer

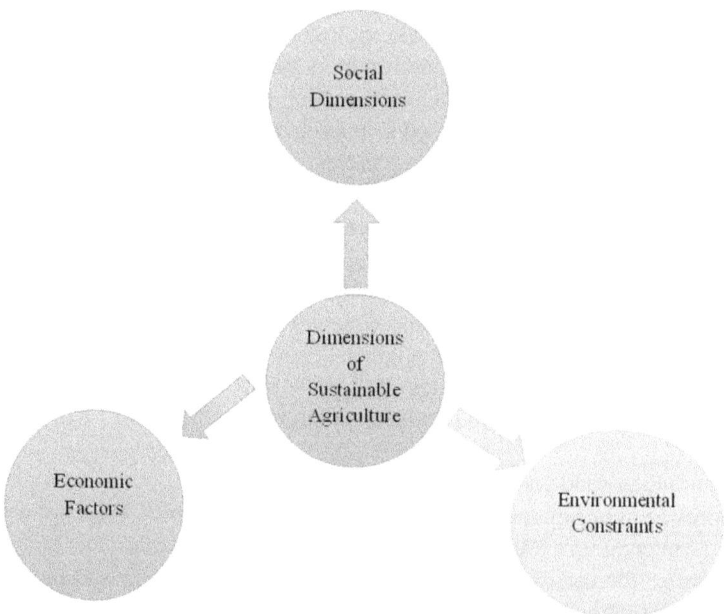

FIGURE 18.1 Dimensions of sustainable agriculture.

communities. Environmental conservation and management in farming involve the proper use of natural resources, minimizing environmental degradation, and promoting biodiversity. From an agricultural economics point of view, this involves the integration of precision agriculture technology, which is economically efficient but requires rigorous financial analysis to be affordable to farmers. Social equity in sustainable agriculture emphasizes justice, inclusiveness, and equitable sharing of benefits as well as burdens among all concerned, namely farmers and rural communities. Gender-responsive agricultural policies also contribute to social equity by addressing the peculiar issues of women in agriculture.

18.3 PRECISION AGRICULTURE: ENHANCING EFFICIENCY AND PRODUCTIVITY

Precision agriculture uses technology to improve different aspects of farming, such as crop yield, resource use efficiency, and environmental sustainability. Precision irrigation and Variable rate technology are some of the main elements of precision agriculture. IoT-based soil moisture sensors facilitate precision irrigation by providing water at the appropriate time and place, minimizing water loss, and enhancing crop water-use efficiency [5]. Variable Rate Technology enables the exact delivery of inputs like fertilizers and pesticides based on spatial heterogeneity in a field, avoiding overapplication and reducing environmental impact [6]. By facilitating targeted application of inputs, precision agriculture minimizes overall consumption of pesticides and fertilizers and hence reduces environmental pollution and encourages ecosystem well-being [7]. Global Positioning System (GPS)-based precision farming methods, including contour farming, also reduce soil erosion, maintain soil structure, and keep agriculture environmentally friendly [8].

Agricultural diversification is the inclusion of multiple crops and animals in agricultural systems to promote biodiversity, maintain soil health, and enhance economic resilience. An extensive meta-analysis was compiled by [9] and found that diversification management resulted in large-scale, long-term gains. Financial profitability, biodiversity, and ecosystem services increased significantly over two decades, pointing to the significance of diversification in sustainable agricultural systems. Soil health is the key to sustainable agriculture. Healthy soils are full of organic matter, microbial life, and water- and nutrient-holding capacity, which are necessary for plant growth. However, intensive farming practices like monocropping and excessive tillage have resulted in extensive soil degradation, causing decreased fertility and erosion. Regenerative agriculture, a group of practices designed to improve soil health, has become an important strategy for sustainable agriculture. Regenerative practices like cover cropping, crop rotation, and reduced tillage have been shown in [10] and found to be effective in enhancing soil organic carbon, water holding capacity, and biodiversity. These practices not only enhance soil health but also help mitigate climate change by sequestering carbon in the soil.

In 2025, researchers have further investigated the potential of biochar, a type of charcoal made from organic waste, as a soil amendment. Lee et al. [11] found in a study that biochar application is capable of increasing soil fertility, lowering greenhouse gas emissions, and enhancing crop yields, especially in degraded soils. These are results that show the potential of regenerative agriculture to revolutionize farming systems and help achieve global sustainability objectives. Precision agriculture facilitated by AI is a paradigm shift from conventional farming practices, which use data-driven intelligence to streamline resource utilization, increase crop production, and mechanize different farming processes [12–16]. AI applications make use of information drawn from a variety of sources, ranging from satellite images, sensors, and drones, to track crop health, soil status, and weather conditions in real-time [17]. Real-time monitoring enables active intervention, the detection of possible issues early on, and the prevention of large losses in yields.

Machine learning techniques are also very important in crop yield prediction and the optimization of resource application, like water, fertilizers, and pesticides. With the ability to accurately target these inputs according to real-time requirements, farmers are able to save a lot of waste and reduce their environmental footprint [18]. In addition, AI-driven autonomous systems are transforming tedious work. Robotic systems and autonomous vehicles are capable of performing activities like planting, harvesting, and weeding with greater efficiency and lower costs. Various research papers emphasize the efficiency of AI-based precision agriculture methods. For example, Kong et al. [19] proposed an inexpensive approach to autonomous fruit picking with monocular images, without requiring costly depth-sensing cameras. Their method utilized YOLOv8s detection kit, a fused depth estimation method, which was highly accurate in the detection of small targets and showed great potential for enhancing the efficiency of fruit harvesting. An extremely compressed lightweight model for pear object detection was suggested by the authors to solve the issues of pear occlusion and limited processing power on embedded devices. This method has gained impressive precision without overloading the computations and suitable for application within limited resource setups. An image processing method for lettuce height estimation and leaf expansion in hydroponic plant factories was proposed by Ma et al. [20], to automatically optimize harvesting parameters with over 90% success rates. Such illustrations depict varied uses of AI to automate agribusiness tasks and boost efficiency. The environmental footprint of agriculture is a significant issue, with unsustainable agriculture being a major contributor to water loss, soil erosion, and greenhouse gas emissions. AI provides an impressive array of tools for reducing these effects and ensuring sustainable agriculture. AI-based smart irrigation systems, for instance, are transforming water management in agriculture.

By accurately tracking soil moisture content and weather patterns, these systems provide water only when and where it is required, cutting water use dramatically while optimizing crop yields. The same AI-based methods are being applied to maximize fertilizer application. Accurate application of fertilizers according to real-time soil analysis reduces waste and minimizes the risk of environmental contamination. In addition, AI is used to forecast and counteract the impacts of climate change on agricultural production. Predictive modeling methods, based on climate data and crop models, enable farmers to foresee possible difficulties and adjust their practices in response, increasing resilience to climate variability. Pham et al. [21], for instance, employed a simple crop model to examine the effects of climate change on soybean yield in Vietnam's Mekong Delta and successfully illustrated the predictive capability of the model under different climate scenarios. The research revealed the strong impacts of temperature and CO_2 levels on soybean biomass and yield, which could be useful for the formulation of climate-resilient agricultural strategies. The use of digital technologies in precision agriculture has great potential for maximizing the use of resources while limiting the effects on the environment. With real-time information and accurate management practices, such technologies promote more efficient and sustainable agricultural systems. Still, sustained research and uptake activities are needed to maximize their benefits, overcome current challenges, and promote wider application.

Pest management is a critical component of sustainable agriculture, as pests can cause significant crop losses and reduce food security. However, conventional pest control methods, which rely heavily on chemical pesticides, have negative impacts on human health, biodiversity, and the environment. In 2024 and 2025, researchers have focused on developing sustainable pest management strategies that minimize these impacts. Integrated Pest Management (IPM), which combines biological, cultural, and mechanical control methods with the judicious use of pesticides, has been widely promoted as a sustainable alternative to conventional pest control. A 2024 study by Zhou et al. [22] found that IPM can reduce pesticide use by up to 50% while maintaining crop yields and improving ecosystem health. In 2025, researchers have also explored the potential of biopesticides, which are derived from natural sources such as plants, bacteria, and fungi, as a sustainable alternative to chemical pesticides. A study by Ragasruthi et al. [23] found that biopesticides can effectively control pests while minimizing environmental impact and reducing the risk of pesticide resistance.

Planning sustainable agricultural areas requires a comprehensive evaluation of ecological, economic, and social factors. Abdelkareem and Abdalla [24] developed a multicriteria analysis framework for sustainable corn farming, integrating these dimensions to guide decision-making. The study utilized methods such as Multidimensional Scaling and Analytical Hierarchy Process to prioritize development strategies, demonstrating the value of holistic approaches in agricultural planning.

18.4 AGROECOLOGY IN SUSTAINABLE AGRICULTURE

Agroecology has become a central paradigm for meeting the challenges of sustainable agriculture in the 21st century. It combines ecological principles with agricultural systems to increase productivity, resilience, and sustainability while reducing environmental degradation. With global food systems under increasing pressure from climate change, loss of biodiversity, and resource depletion, agroecology presents a holistic solution that harmonizes agricultural practices with natural ecosystems. This paradigm change is increasingly seen as key to attaining food security, environmental sustainability, and social justice [25]. Emerging research emphasizes the capacity of agroecology to revolutionize food systems by fostering biodiversity, enhancing soil health, and diminishing reliance on synthetic inputs [26]. The shift to sustainable food systems follows five levels that each incorporate fundamental agroecological concepts. Level 1 focuses on making conventional farming more efficient by maximizing the use of resources and decreasing dependency on chemical inputs according to the principle of efficient resource management. Level 2 entails replacing synthetic inputs with natural and organic ones, including compost and biological control of pests, encouraging synergy and biodiversity in farming systems. Level 3 calls for a core redesign of agroecosystems through the integration of diversified cropping systems, agroforestry, and mixed farming to improve resilience and ecosystem health. Level 4 goes beyond the farm to reorganize local food systems, enhancing social equity and farmer-consumer relationships through community-supported agriculture, fair trade, and alternative food networks. Lastly, Level 5 imagines a global shift toward a just and sustainable food system, combining participation, governance, and food sovereignty to promote fairness and resilience at all scales. Such transformations, based on agroecological principles, together promote sustainability by harmonizing ecological integrity, economic sustainability, and social health. Figure 18.2 shows various levels of transition toward sustainable food systems and related principles of agroecology.

A primary tenet of agroecology is the promotion of biodiversity in agricultural landscapes. Alternative cropping systems, for example, intercropping and agroforestry, have been found to enhance ecosystem services, such as pest regulation, pollination, and nutrient cycling. For example, research by Kremen and Merenlender [27] illustrated that diversified farming systems accommodate greater proportions of beneficial insects and microorganisms in the soil, which help crops recover and remain resilient and stable in their yields. By replicating natural ecosystems, agroecological systems minimize the use of chemical pesticides and fertilizers and thus the environmental impact of conventional agriculture [28]. Additionally, biodiversity-intensive systems are more resilient to climate variability and hence a key strategy for climate-resilient agriculture.

Soil health is yet another pillar of agroecology, since healthy soils are essential for sustainable food production. Agroecological methods like cover cropping, rotation of crops, and organic fertilizer application increase the content of organic matter in soil, enhance water holding capacity, and lower soil erosion. A recent study [29] emphasizes the function of regenerative agricultural practices in carbon sequestration and reducing climate change. Through the maintenance of soil health, agroecology not only increases productivity in agriculture but also helps achieve global objectives of fighting climate change. Additionally, healthy soil promotes rich and varied microbial communities that are responsible for nutrient cycling and plant well-being, and which further highlights the significance of soil-based practices in sustainable agriculture. Agroecology further places a significant focus on local knowledge and the involvement of farmers in designing sustainable agriculture systems. In contrast to top-down methods based on external technologies and inputs, agroecology enables farmers to co-design solutions specific to their local ecological and socio-economic conditions. Participatory in nature, this strategy encourages innovation and resilience

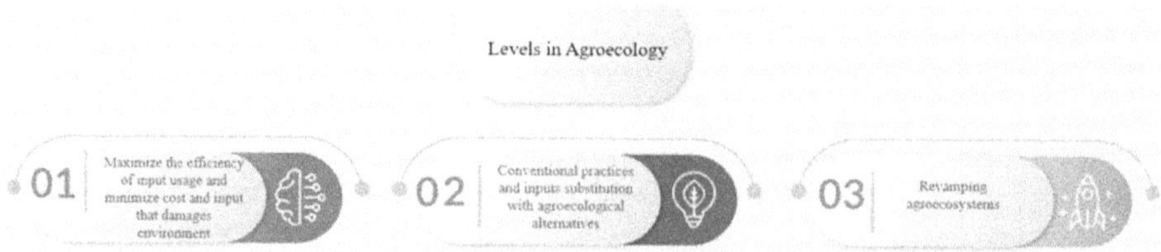

FIGURE 18.2 Levels in agroecology.

as farmers tap into indigenous knowledge and science to solve problems at the local level. Current research has indicated that agroecological projects driven by farmers can contribute to improved food security, livelihoods, and environmental results [30]. In sub-Saharan Africa, for instance, agroecology has played a crucial role in enhancing crop yields and alleviating poverty among smallholder farmers [31].

Social aspects of agroecology are also vital since sustainable agriculture is not possible without ensuring equity and justice. Agroecology counters the mainstream industrial farming model, which tends to marginalize smallholder farmers and worsen social injustices. Through the promotion of decentralized food systems and equitable trade, agroecology aims to establish more just and inclusive food systems. Current studies emphasize the empowering potential of agroecology for women and marginalized groups, especially in developing nations [32]. For example, Latin American agroecological initiatives have shown that initiatives led by women can contribute to food sovereignty and community resilience. Even though it has various advantages, there are several constraints in the adaptation of agroecology, such as policy limits, inadequate access to funds, and poor knowledge and resource availability. Most farm policies and subsidies still benefit input-intensive, conventional farming systems at the expense of agroecological methods, creating disincentives [33]. To counter these difficulties, policy reforms favoring agroecology are increasingly being advocated for by providing research funds, extension facilities, and market subsidies [34, 35]. Moreover, fostering the capabilities of farmer networks and cooperatives can lead to knowledge sharing and collective action, and thus promote broader use of agroecological practices.

18.5 CHALLENGES AND OPPORTUNITIES IN AI-DRIVEN AGRICULTURE

Innovations in agriculture continue to evolve, and it is crucial for policymakers to support research and development, facilitate knowledge transfer, and create enabling environments for the adoption of sustainable practices [36]. Future research should focus on scaling successful models, addressing context-specific challenges, and fostering collaboration among stakeholders to achieve sustainable agricultural systems globally. Despite the significant potential of AI in agriculture, several challenges remain to be addressed.

1. High implementation costs, particularly for sophisticated AI systems and the necessary infrastructure like high-speed internet access and data storage, can be a significant barrier for many farmers, especially in developing countries. The need for specialized technical skills to develop, implement, and maintain AI systems presents another challenge. A lack of skilled personnel can limit the adoption and effective use of AI technologies in agricultural settings.
2. Societal resistance to adopting new technologies and a lack of trust in AI-driven systems can hinder the widespread adoption of AI in farming. Addressing this requires a multi-faceted approach involving policy support, infrastructure development, and skills training. Research efforts may focus

on developing cost-effective and user-friendly AI tools that are accessible to a wider range of farmers.

3. Development of explainable AI (XAI) systems is crucial for increasing trust and transparency, enabling farmers to understand the reasoning behind AI-driven recommendations. This increased transparency will enhance adoption rates by addressing concerns about the "black box" nature of some AI algorithms.
4. Further research is needed to explore the integration of AI into diverse agricultural contexts and farming systems, adapting AI technologies to suit the specific needs and challenges of different regions and crops. This includes considering factors such as soil type, climate, and farming practices.

18.6 CONCLUSION

This chapter marks a significant progress in sustainable agriculture, with innovations in diversification, precision farming, AI applications, and nature-based solutions. These developments offer promising pathways to balance productivity with environmental and socio-economic considerations. Continued investment in research, supportive policies, and collaborative efforts will be essential to realize the full potential of these innovations and ensure a sustainable future for agriculture. AI has the potential to revolutionize agricultural practices, significantly enhancing efficiency, productivity, and sustainability. By enabling real-time monitoring, predictive modeling, and automation, AI empowers farmers to make informed decisions, optimize resource use, and mitigate environmental impacts. However, realizing the full potential of AI in agriculture requires addressing the challenges related to cost, infrastructure, skills, and societal acceptance.

A holistic approach is essential, involving policy support to encourage adoption, investment in infrastructure to enhance connectivity and data access, and comprehensive training programs to develop the necessary technical skills. The integration of AI into agriculture is not merely a technological advancement; it is a necessity for ensuring a secure and sustainable food supply for a growing global population in the face of climate change and should ensure that these advancements benefit farmers worldwide and contribute to global food security and environmental sustainability.

REFERENCES

1. Cohen, W. L., Hug, A. W., Taddese, A., & Cook, K. A. (1991). FACTA 1990: Conservation and environmental highlights. *Journal of Soil and Water Conservation*, *46*(1), 20–22.
2. Pretty, J. (2008). Agricultural sustainability: Concepts, principles and evidence. *Philosophical Transactions of the Royal Society B: Biological Sciences*, *363*(1491), 447–465.
3. Priyanka, S., Sivakumar, S., & Selvam, P. (2024). Optimizing breast cancer detection: Machine learning for pectoral muscle segmentation in mammograms. In *2024 International Conference on Integrated Circuits and Communication Systems (ICICACS)*, Raichur, India. 1–6.
4. Sivakumar, S., Priyanka, S., & Selvam, P. (2023). Enhancing personality type prediction with ensemble models: A robust predictive approach. In *2023 International Conference on Innovative Computing, Intelligent Communication and Smart Electrical Systems (ICSES)*, Chennai, India. 1–7.

5. Cui, Z., Zhang, H., Chen, X., Zhang, C., Ma, W., Huang, C., … Dou, Z. (2018). Pursuing sustainable productivity with millions of smallholder farmers. *Nature*, *555*(7696), 363–366.
6. Bullock, J. M., Dhanjal-Adams, K. L., Milne, A., Oliver, T. H., Todman, L. C., Whitmore, A. P., & Pywell, R. F. (2017). Resilience and food security: Rethinking an ecological concept. *Journal of Ecology*, *105*(4), 880–884.
7. Belaire, J. A., Bass, H., Venhaus, H., Barfield, K., Pannkuk, T., Lieberknecht, K., & Jha, S. (2023). High-performance landscapes: Re-thinking design and management choices to enhance ecological benefits in urban environments. *Land*, *12*(9), 1689.
8. Xu, J., Cui, Y., Zhang, S., & Zhang, M. (2024). The evolution of precision agriculture and food safety: A bibliometric study. *Frontiers in Sustainable Food Systems*, *8*, 1475602.
9. Raveloaritiana, E., & Wanger, T. C. (2024). Decades matter: Agricultural diversification increases financial profitability, biodiversity, and ecosystem services over time. *arXiv preprint arXiv:2403.05599.*
10. Khangura, R., Ferris, D., Wagg, C., & Bowyer, J. (2023). Regenerative agriculture—A literature review on the practices and mechanisms used to improve soil health. *Sustainability*, *15*(3), 2338.
11. Lee, Y. E., Jeong, Y., Kim, I. T., Ahn, K. H., & Jung, J. H. (2024). Enhancing the potential application of food-waste biochar as a sustainable bio-solid fuel: Analysis of post-treatment and combustion behavior. *Chemosphere*, *364*, 143216.
12. Selvam, P., Faheem, M., Dakshinamurthi, V., Nevgi, A., Bhuvaneswari, R., Deepak, K., & Sundar, J. A. (2024). Batch normalization free rigorous feature flow neural network for grocery product recognition. *IEEE Access*, 12, 68364–68381.
13. Selvam, P., Koilraj, J. A. S., Romero, C. A. T., Alharbi, M., Mehbodniya, A., Webber, J. L., & Sengan, S. (2022). A transformer-based framework for scene text recognition. *IEEE Access*, 10, 100895–100910.
14. Selvam, P. (2023). A deep learning framework for surgery action detection. In: *Deep Learning in Personalized Healthcare and Decision Support*, Editors: Harish Garg and Jyotir Moy Chatterjee. 315–328. Academic Press.
15. Prabu, S., & Sundar, K. J. A. (2023). DocPresRec: Doctor's handwritten prescription recognition using deep learning algorithm. In: *Artificial Intelligence in Telemedicine*, Editors: S. N. Kumar, Sherin Zafar, Eduard Babulak, M. Afshar Alam and Farheen Siddiqui. 33–48. CRC Press.
16. Swaminathan, B., Selvam, P., Joseph, A. S. K., & Vairavasundaram, S.. (2023). Improved YOLOv5 with attention mechanism for real-time weed detection in the paddy field: A deep learning approach. In: *Intelligent Data Analytics, IoT, and Blockchain*, Editors: Bashir Alam and Mansaf Alam. 326–341. Auerbach Publications.
17. Prabu, S., Sundar, K. J. A., & Abraham, J. (2023). Enhanced attention-based encoder-decoder framework for text recognition. *Intelligent Automation & Soft Computing*, 35(2), 2071–2086.
18. Prabu, S. (2022). Object segmentation based on the integration of adaptive K-means and GrabCut algorithm. In: *2022 International Conference on Wireless Communications Signal Processing and Networking (WiSPNET)*, Chennai, India. 213–216.
19. Kong, D., Wang, J., Zhang, Q., Li, J., & Rong, J. (2023). Research on fruit spatial coordinate positioning by combining improved YOLOv8s and adaptive multi-resolution model. *Agronomy*, *13*(8), 2122.
20. Ma, Y., Zhang, W., Qureshi, W. S., Gao, C., Zhang, C., & Li, W. (2021). Autonomous navigation for a wolfberry picking robot using visual cues and fuzzy control. *Information Processing in Agriculture*, *8*(1), 15–26.

21. Pham, Q. V., Nguyen, T. T., Vo, T. T., Le, P. H., Nguyen, X. T., Duong, N. V., & Le, C. T. (2023). Applying the SIMPLE crop model to assess soybean (Glicine max.(L.) Merr.) biomass and yield in tropical climate variation. *Agronomy*, *13*(4), 1180.
22. Zhou, W., Arcot, Y., Medina, R. F., Bernal, J., Cisneros-Zevallos, L., & Akbulut, M. E. S. (2024). Integrated pest management: An update on the sustainability approach to crop protection. *ACS Omega*, 9(40), 41130–41147.
23. Ragasruthi, M., Balakrishnan, N., Murugan, M., Swarnakumari, N., Harish, S., & Sharmila, D. J. S. (2024). Bacillus thuringiensis (Bt)-based biopesticide: Navigating success, challenges, and future horizons in sustainable pest control. *The Science of the Total Environment*, 954, 176594.
24. Abdelkareem, M., & Abdalla, F. (2022). Revealing potential areas of water resources using integrated remote-sensing data and GIS-based analytical hierarchy process. *Geocarto International*, *37*(25), 8672–8696.
25. McGreevy, S. R., Rupprecht, C. D., Niles, D., Wiek, A., Carolan, M., Kallis, G., … Tachikawa, M. (2022). Sustainable agrifood systems for a post-growth world. *Nature Sustainability*, *5*(12), 1011–1017.
26. Kerr, R. B., Madsen, S., Stüber, M., Liebert, J., Enloe, S., Borghino, N., … Wezel, A. (2021). Can agroecology improve food security and nutrition? A review. *Global Food Security*, *29*, 100540.
27. Kremen, C., & Merenlender, A. M. (2018). Landscapes that work for biodiversity and people. *Science*, *362*(6412), eaau6020.
28. Tittonell, P., Piñeiro, G., Garibaldi, L. A., Dogliotti, S., Olff, H., & Jobbagy, E. G. (2020). Agroecology in large scale farming—A research agenda. *Frontiers in Sustainable Food Systems*, *4*, 584605.
29. Lal, R. (2016). Soil health and carbon management. *Food and Energy Security*, *5*(4), 212–222.
30. Anderson, M. D. (2015). The role of knowledge in building food security resilience across food system domains. *Journal of Environmental Studies and Sciences*, *5*(4), 543–559.
31. Nyantakyi-Frimpong, H., & Kerr, R. B. (2017). A political ecology of high-input agriculture in northern Ghana. In: *Africa's Green Revolution*. 21–43. Routledge.
32. Kerr, R. B., Postigo, J. C., Smith, P., Cowie, A., Singh, P. K., Rivera-Ferre, M., … Neufeldt, H. (2023). Agroecology as a transformative approach to tackle climatic, food, and ecosystemic crises. *Current Opinion in Environmental Sustainability*, *62*, 101275.
33. Cardoso, R., Guo, F., Heisser, T., Hackl, M., Ihle, P., De Schutter, H., … Brenner, H. (2021). Colorectal cancer incidence, mortality, and stage distribution in European countries in the colorectal cancer screening era: An international population-based study. *The Lancet Oncology*, *22*(7), 1002–1013.
34. Prasath, J., Prabu, M.V. V., & Saini, S. (2025). Optimized double transformer residual super-resolution network-based X-ray images for classification of pneumonia identification. *Knowledge-Based Systems*, *311*, 113037.
35. Selvam, P., Shanmuga Sundari, P., Tamilselvi, M., Suresh, T., Murugappan, M., & Chowdhury, M. E. H. (2025). YOLO-SAIL: Attention-enhanced YOLOv5 with optimized bi-FPN for ship target detection in SAR images. *IEEE Access: Practical Innovations, Open Solutions*, *13*, 29523–29540.
36. Ghimire, K. B., & Pimbert, M. P. (2013). Social change and conservation: An overview of issues and concepts. In: Ghimire, K. B., & Pimbert, M. P., (eds.) *Social Change and Conservation*, 1–45.

19 Advanced Security in AgriIoT

Applications of Quantum Machine Learning

Kaebeh Yaeghoobi

19.1 INTRODUCTION

The approach of Internet of Things (IoT) technology led to major advancements in various sectors, including agriculture. Agriculture IoT (AgriIoT) involves the use of interconnected devices to monitor and manage agricultural operations with accuracy and efficiency. By leveraging sensors, data analytics, and real-time monitoring without delay, AgriIoT has the potential to enhance crop yields, optimize resource efficiency, and boost overall agricultural productivity. However, the increasing reliance on connected devices and data exchange in AgriIoT systems introduces several security challenges. As these systems become more integrated, they become vulnerable to a range of cyber threats, including data breaches, unauthorized access, and tampering. Safeguarding sensitive agricultural data and maintaining the integrity of IoT devices are essential for maintaining the reliability and credibility of AgriIoT solutions.

Quantum Machine Learning (QML) presents a cutting-edge approach to tackling these security issues. Through the integration of quantum computing with developed ML algorithms, QML has achieved exceptional performance in processing and analyzing extensive datasets. Leveraging quantum effects like superposition and entanglement, QML algorithms can resolve intricate and complex problems more resourcefully than conventional approaches. This chapter explores the potential of QML to develop the security of AgriIoT systems. It provides a comprehensive overview of current security issues in AgriIoT, examines existing QML techniques, and highlights their applicability in securing agricultural data and communications. Through detailed analysis and case studies, this review demonstrates how QML can serve as a robust defense mechanism against emerging threats in the digital agricultural ecosystem.

By integrating QML into AgriIoT, we can create a more secure and resilient framework for managing agricultural operations, ensuring that the benefits of IoT technology are fully achieved without compromising data integrity and privacy. This chapter provides a comprehensive analysis of current security issues in Agriculture IoT (AgriIoT), presenting a clear understanding of the critical vulnerabilities that require attention. It explores the specific techniques of QML and illustrates how

DOI: 10.1201/9781003594093-19

feasible these methods are to enhance the security of AgriIoT systems. By discussing advanced encryption solutions enabled by QML, the chapter emphasizes their necessity for securing data transmission within AgriIoT networks. Furthermore, the chapter highlights the use of QML for real-time anomaly detection, which is essential for identifying and mitigating potential security threats in agricultural systems. Through its analysis of QML's capacity to verify data integrity, the chapter emphasizes the significance of upholding the authenticity and reliability of agricultural data. The inclusion of detailed analysis and case studies demonstrates the applied implementation of QML techniques in real-world AgriIoT scenarios, contributing to future advancements in the field. Ultimately, the chapter emphasizes QML as a robust defense mechanism against emerging cybersecurity threats, ensuring the safe and efficient functioning of digital agricultural ecosystems.

19.2 BACKGROUND

Agriculture, one of the oldest and most vital human activities, has always been driven by innovation and technological advancement. The IoT reformed several industries, including agriculture, and is termed AgriIoT, which represents a significant advancement in the farming industry. AgriIoT refers to the integration of IoT tools in agricultural practices, authorizing real-time monitoring and timely management of crops, livestock, and resources. AgriIoT systems utilize interconnected devices and sensors to gather and analyze data, enabling precise monitoring and management of agricultural operations. This technology allows agriculturalists to make smart data-driven decisions, enhance management of supply and resources, and expand comprehensive efficiency and throughput. AgriIoT systems are composed of several key components, including sensors, connectivity infrastructure, data analytics platforms, and automation systems. Sensors are deployed all over cultivated fields to monitor several substantial parameters, including temperature, soil moisture, light intensity, and humidity [1]. These devices provide real-time data, which is crucial for informed decision-making in irrigation, enrichment of land, and pest control.

Connectivity is achieved through wireless communication networks, facilitating wireless data transmission linking sensors, gateways, and cloud servers. This connectivity facilitates instant monitoring and remote management of agricultural operations. The information gathered by sensors is processed, studied, examined, and managed by cloud-based platforms employing advanced data analytics and progressive ML algorithms. The analysis delivers valuable insights into crop health, soil conditions, and weather patterns, allowing agriculturalists to maximize their resource use and improve crop yields [2]. Moreover, AgriIoT allows the automation of various farming activities, such as irrigation and harvesting, enhancing precision and efficiency. Despite the numerous benefits of AgriIoT, the increased connectivity and data exchange inherent in these systems introduce several security challenges. These challenges include data breaches, unauthorized access, and tampering with IoT devices. Ensuring the protection and confidentiality of agricultural data is valuable for maintaining the reliability and trustworthiness of AgriIoT solutions. QML develops while ensuring an efficient approach to address these security challenges. QML combines the principles of quantum computing, such as superposition and entanglement, with ML methods, leveraging quantum properties to process large datasets more effectively than traditional methods. QML has the ability to enhance various aspects of cybersecurity, including encryption, anomaly detection, and data integrity verification [3]. QML is developing as an effective tool in cybersecurity. Its suitability to process extensive amounts of data simultaneously makes it suitable for detecting advanced threats and anomalies. Recent studies have shown that QML is feasible to expand quantum-resistant encryption algorithms and improve the accuracy of anomaly detection systems.

Integrating QML into AgriIoT systems can address the security challenges faced by these systems. By leveraging QML techniques, it is possible to enhance data encryption, improve anomaly detection, and ensure data integrity in agricultural IoT networks. Table 19.1 is an overview and concise summary of relevant references, their methodologies, and key findings.

TABLE 19.1
Overview of Key References, Methodologies on AgriIoT Security

Reference	Paper Review	Methodology	Key Findings
[4]	Provides a comprehensive review of smart farming security issues and precision agriculture, highlighting potential threats and countermeasures	Systematic literature review of current security content in smart farming, including data privacy and integrity	Identifies major security risks in smart farming and suggests potential countermeasures to mitigate these risks
[5]	Surveys the development modes and technologies in smart agriculture, focusing on security and privacy challenges	Review of existing technologies and their associated security and privacy issues in smart agriculture	Highlights the necessity for advanced security measures in smart agriculture to safeguard sensitive data and maintain system integrity
[6]	Discusses the ability of quantum machine learning in enhancing cybersecurity measures	Analysis of QML techniques and their applications in cybersecurity, with a focus on encryption and anomaly detection	Demonstrates that QML can significantly improve the security of systems by enhancing encryption and detecting anomalies more accurately
[7]	Explores the role of QML in addressing emerging cybersecurity challenges	Examination of QML applications in various cybersecurity domains, including threat detection and encryption	Concludes that QML has the ability to transform cybersecurity by offering advanced tools for threat detection and data protection
[8]	Investigates the use of QML for advanced threat detection in cybersecurity	Implementation and evaluation of QML algorithms for detecting cybersecurity threats in real time	Finds that QML algorithms can enhance the correctness and competence of threat detection, making them a valuable asset in cybersecurity

(*Continued*)

TABLE 19.1 (CONTINUED)

Reference	Paper Review	Methodology	Key Findings
[9]	Focuses on designing lightweight authentication protocols for IoT systems, particularly in smart grids	Proposes a new protocol and evaluates its security and efficiency through simulation and theoretical analysis	Demonstrates that the proposed protocol enhances security while maintaining low computational and communication overhead
[10]	Addresses the security challenges in heterogeneous IoT networks, emphasizing sensing devices and gateways	Develops a lightweight protocol and tests its performance in various IoT environments	Shows that the protocol effectively secures IoT networks with minimal impact on performance
[11]	Proposes lightweight authentication schemes tailored for IoT environments	Introduces new schemes and evaluates their security properties and computational efficiency	Highlights the effectiveness of the proposed schemes in securing IoT communications with reduced resource consumption

This chapter investigates the ability of QML to enhance the security of AgriIoT systems. It provides a detailed summary of contemporary security issues in AgriIoT, examines existing QML techniques, and highlights their applicability in securing agricultural data and communications. By integrating QML into AgriIoT, we can create a more secure and resilient framework for managing agricultural operations, confirming that the advantages of IoT technology are fully realized while maintaining data integrity and privacy.

19.3 SECURITY CHALLENGES IN AGRIIOT

AgriIoT presents a myriad of opportunities for enhanced productivity and resource management. However, alongside these benefits come significant security challenges that must be addressed to protect confidential agricultural data and confirm the integrity and reliability of AgriIoT systems. This section investigates the primary security challenges faced by AgriIoT, highlighting the critical issues that must be managed to secure these systems effectively.

Data Privacy and Confidentiality: The leading security challenge in AgriIoT is ensuring the privacy and confidentiality of data collected from various IoT devices. Agricultural data, such as crop health, soil conditions, and farm management practices, is highly sensitive and valuable. Digital break-in to this data can cause competitive disadvantages and potential misuse. Protecting data privacy requires robust encryption methods and assured communication protocols to avoid data breaches and unapproved interceptions during transmission.

Data Integrity: Preserving the integrity of data is important for the accuracy of AgriIoT systems. Data integrity refers to ensuring that the data collected and stored has not been altered or tampered with. Imprecise or corrupted data can lead to erroneous decision-making, affecting crop management and resource allocation. Implementing digital signatures and cryptographic hash functions can help ensure the legitimacy and integrity of data, guaranteeing that any alterations are detected promptly.

Network Security: AgriIoT systems rely on wireless communication networks to connect devices and transmit data. These networks are vulnerable to a range of cyber threats, including man-in-the-middle attacks, denial-of-service (DoS), and eavesdropping attacks. Man-in-the-middle attacks involve an attacker intercepting and theoretically modifying communication between two connected parties. DoS attacks involve overwhelming the network with traffic to disrupt normal operations. Eavesdropping involves intercepting data transmissions to achieve unauthorized access to important information. Ensuring network security requires implementing secure encryption protocols, robust authentication mechanisms, and intrusion detection systems to monitor and mitigate these threats.

Device Security: The IoT devices used in AgriIoT systems, such as sensors and actuators, are often deployed in outdoor and remote environments, making them vulnerable to physical tampering and unauthorized access. Securing these devices involves implementing physical security measures, such as tamper-evident packaging and secure installations, as well as software-based security measures like firmware updates and access control mechanisms. Ensuring that devices are protected against both physical and cyber threats is vital for supporting the overall security of AgriIoT systems.

Interoperability and Standardization: AgriIoT systems often involve devices and components from multiple manufacturers, leading to challenges in interoperability and standardization. Ensuring seamless communication and data exchange between devices requires adhering to common standards and protocols. Lack of interoperability can lead to security vulnerabilities, as incompatible devices may not support the necessary security features. Standardization efforts, such as adopting industry-wide communication protocols and security standards, are crucial for addressing these challenges.

Scalability: With the increasing propagation of connected devices in AgriIoT systems, managing and processing the increasing volume of data becomes more complex. Scalability is a critical challenge that involves ensuring that the security measures implemented can handle the expanding network without compromising performance. This includes developing scalable encryption and authentication methods that can protect large-scale deployments while maintaining efficiency.

Reliability and Resilience: The reliability and resilience of AgriIoT systems are essential for ensuring continuous operation and minimizing downtime. Cyber-attacks, such as DoS attacks or malware infections, can disrupt the normal functioning of IoT devices and communication networks, leading to significant operational losses. Implementing redundancy measures, such as backup communication channels and failover mechanisms, can enhance the resilience of AgriIoT systems against cyber threats.

Securing AgriIoT systems entails tackling a variety of challenges, including data privacy, integrity, network security, device security, interoperability, scalability, and reliability. By implementing robust security measures and adhering to industry standards, it is possible to create a secure and resilient framework for AgriIoT, certifying that the advantages of IoT technology are fully comprehended in the agricultural sector. Table 19.2 encapsulates the security challenges faced by AgriIoT systems, highlighting the impact of each challenge and the need for robust security and protection measures to protect agricultural data and operations.

TABLE 19.2
Summary of Security Challenges in AgriIoT

Security Challenge	Description	Impact
Data Privacy and Confidentiality	Ensuring that sensitive agricultural data remains secure from unauthorized access and breaches.	Unauthorized access can lead to competitive disadvantages and potential misuse of sensitive information.
Data Integrity	Maintaining the precision and dependability of the data collected from IoT devices.	Inaccurate or corrupted data can lead to erroneous decision-making, affecting crop management and resource allocation.
Network Security	Protecting wireless communication networks from cyber threats, including man-in-the-middle attacks, eavesdropping, and denial-of-service (DoS) attacks.	Compromised networks may lead to data breaches, loss of data integrity, and disruption of normal operations.
Device Security	Securing IoT devices from physical tampering and unauthorized access.	Vulnerable devices can be manipulated, leading to data corruption and potential system failures.
Interoperability and Standardization	Ensuring seamless communication and data exchange between devices from different manufacturers.	Lack of interoperability can lead to security vulnerabilities and inefficiencies in data administration.
Scalability	Handling and processing the increasing volume of data produced by AgriIoT systems.	Poor scalability can result in performance issues and reduced effectiveness of security measures.
Reliability and Resilience	Ensuring continuous operation and minimizing downtime in the face of cyber-attacks and system failures.	Disruptions can lead to significant operational losses and impact overall agricultural productivity.

19.4 QUANTUM MACHINE LEARNING

QML integrates quantum computing and traditional ML techniques, suggesting the possibility of significant advancements in computational efficiency and problem-solving capabilities. By leveraging the foundation of quantum computing, QML aims to tackle complex problems beyond the reach of traditional algorithms. QML is founded on several key quantum principles that differentiate it from classical ML, such as quantum superposition principle, which permits quantum bits (qubits) to be in several states concurrently. As opposed to classical bits (either 0 or 1), qubits can signify both states at once, facilitating parallel computation. In quantum entanglement qubits are highly correlated, in which the state of one qubit, regardless of the distance, is entirely dependent on the state of another. This characteristic allows the creation of complex quantum states and enhances computational power. Quantum Gates manipulate qubits through operations such as the Hadamard gate, which generates superpositions, and the controlled-NOT (CNOT) gate, which entangles qubits. These gates form the essential building blocks of quantum circuits [12].

19.4.1 QML Algorithms and Methods

Several QML algorithms have been developed to leverage quantum mechanics for ML tasks.

19.4.1.1 Quantum Key Distribution

QKD harnesses the fundamental principles of quantum mechanisms to facilitate the secure distribution of encryption keys between parties. It is theoretically secure, since any attempt to eavesdrop on key exchange would be sensed and detected. QKD certifies secure communication by combining quantum mechanics with cryptographic protocols. Based on the rules, it permits parties to randomly produce a shared secret key known only to them, which may be used to encrypt and decrypt messages. The security of QKD relies on mathematical complexity derived from the fundamental principles of quantum mechanics.

A more detailed understanding of how QKD works and why it matters involves principles of quantum mechanics, key generation, transmission and detection, error correction, and privacy amplification. QKD uses quantum states (qubits) to encode information. The most common protocols, such as BB84 protocol and E91 protocol, rely on the quantum superposition and entanglement properties. The random shared key is transmitted over a quantum channel (e.g., optical fibers or free space). Any eavesdropper (Eve) attempt to interrupt or capture the key introduces detectable anomalies, caused by the disturbance of measuring quantum states. After transmission, parties perform error correction to resolve any discrepancies in their keys. The parties correspondingly apply privacy augmentation to decrease any limited information that an eavesdropper might have gained [13].

The most widely researched protocols for QKD are the BB84 and E91 protocols. The BB84 protocol [14] involves the following steps:

Key Generation – First party (Alice) randomly provides a sequence of bits and encodes them by different bases – rectilinear and diagonal bases – into qubits. *Transmission* – First party sends the qubits to another party (Bob) over a quantum channel. *Measurement* – Second party randomly measures each qubit using rectilinear bases or diagonal bases.

Basis Reconciliation – Two parties publicly compare their measurement bases without disclosing the real bit values. *Key Sifting* – When the bases did not align, two parties discard the bits and keep the remaining bits as the shared key. *Error Correction and Privacy Amplification* – Two parties execute error correction to correct any discrepancies and increase privacy, aimed at reducing any partial information an eavesdropper might have gained.

The QKD security and protection rely on the fundamental theories of quantum computing, particularly the no-cloning theorem, which prohibits the creation of an identical copy of an unknown quantum state, and the Heisenberg uncertainty principle, which posits that certain pairs of physical properties cannot both be known to arbitrary precision. The formula for QKD involves the use of quantum states to encode information and detect any eavesdropping attempts:

$$\text{Key Rate} = R = I(A;B) - I(E;B) \tag{19.1}$$

where $I(A;B)$ is the mutual information between two parties. $I(E;B)$ is the mutual information between the eavesdropper (Eve) and Bob. By ensuring that $I(E;B)$ is negligible, QKD provides a secure way to provide a shared key that is theoretically unbreakable. This chapter uses the application of QKD for secure data transmission.

19.4.1.2 Quantum Neural Networks

Quantum Neural Networks (QNNs) emulate the operational principles of traditional neural networks, yet they work with quantum data. Quantum gates are employed to execute computations, offering the possibility for exponential speedups in particular tasks. QNNs are specifically well-suited for pattern recognition and classification problems. This principle of superposition allows QNNs to perform parallel computations, processing huge quantities of information at once. Qubits have the ability to become entangled, which means the state of one qubit is intrinsically linked to the state of another, regardless of the distance between them [15]. This property enables complex correlations and interactions within the network.Quantum algorithms utilize interference, where qubit states can interfere constructively or destructively, enhancing correct responses while eliminating incorrect ones. This mechanism is key to enhancing the accuracy of computations in QNNs. QNNs operate using quantum gates that manipulate qubit states. These gates form the foundation of quantum circuits, similar to how classical logic gates function in traditional neural networks, and capable of performing exponentially more complex operations [16].

The architecture of QNNs mirrors that of classical neural networks, with modifications to accommodate quantum principles. The input layer transforms data into quantum states, employing quantum encoding techniques to map data features onto qubit states. The hidden layer performs quantum operations to process input data. Activation functions impart non-linearity to the network, which is crucial for learning complex patterns. The output layer converts processed quantum states back into classical data and uses measurement operations to collapse qubit states into classical binary outcomes. Finally, training mechanisms can be divided into two models: first, Optimization Algorithms, wherein techniques such as the Quantum Variational Algorithms and Quantum Approximate Optimization Algorithm (QAOA) are employed to adjust parameters; second, adaptation of classical backpropagation methods for quantum networks, involving gradients of quantum gates [17].

A formula for a QNN involves the employment of quantum gates that enable transformations on qubits. The state of a qubit can be represented as a superposition of basis states:

$$|\psi\rangle = \alpha|0\rangle + \beta|1\rangle \tag{19.2}$$

where α and β are complex coefficients, and $|0\rangle$ and $|1\rangle$ are the basis states.

Quantum gates, including the Hadamard gate (H), Pauli-X gate (X), and CNOT gate, are employed to manipulate qubits. For instance, when the Hadamard gate is applied to a qubit, it performs the transformations, which illustrate how it takes a classical bit (either $|0\rangle$ or $|1\rangle$) and creates a superposition of both states, which is a crucial property leveraged in quantum computing.

$$H|0\rangle = \frac{1}{\sqrt{2}}(|0\rangle - |1\rangle) \tag{19.3}$$

$$H|1\rangle = \frac{1}{\sqrt{2}}(|0\rangle - |1\rangle) \tag{19.4}$$

In a QNN, these gates are applied in layers to perform computations. The result is obtained by measuring the output of the quantum circuit, which is possible to be used for classification or clustering tasks.

Applying QNNs to anomaly detection in the AgriIoT case study has the potential to enable faster and more precise detection of anomalies, resulting in the improvement of decision-making and enhanced agricultural practices.

19.4.1.3 Quantum Support Vector Machines

Quantum Support Vector Machines (QSVMs) adapt classical support vector machines for the quantum domain, employing quantum algorithms to efficiently challenge optimization problems and perform kernel calculations, thereby enabling faster and more accurate classification.

QSVMs are designed for classification tasks and can be implemented using quantum circuits. Classical data is transcribed into quantum states through the utilization of qubits. Data encoding involves mapping the features of the data points onto the qubit amplitudes. A quantum circuit is designed to compute the quantum kernel, which measures the similarity between data points. The quantum kernel leverages the high-dimensional Hilbert space of quantum states to perform efficient calculations.

The QSVM is trained by fine-tuning the quantum circuit parameters to minimize classification errors. Quantum algorithms, including the variational quantum eigensolver (VQE), can be used for this optimization process. After training, the QSVM can classify new data points by assessing their resemblance to the training data. The quantum circuit computes the decision boundary, allowing for accurate classification.

In their research paper, the authors introduce a quantum interior-point method (IPM) tailored for solving second-order cone programming (SOCP) problems. They showcase that their quantum algorithm significantly outperforms classical methods, achieving a polynomial speedup for general SOCPs. For support vector machines (SVMs), the quantum algorithm scales as $O\left(n^k\right)$, where the value of k is estimated to be 2.59, determined through a least-squares power law approach. This indicates a significant improvement in computational efficiency, making it a promising approach for large-scale ML applications [18].

In the context of QSVMs, calculating the distance between data points and the decision boundary is essential. The formula for calculating the distance between a data point x and the decision boundary defined by weight vector of the hyperplane w is given by:

$$d\left(x\right)=\frac{\left|w\cdot x+b.\right|}{\|w\|} \tag{19.5}$$

where b is the bias term, $\cdot$ denotes the dot product, and $\|w\|$ is the norm of the weight vector.

The cluster assignment step entails identifying which side of the hyperplane a data point resides on, based on the sign of $w\cdot x+b$.

19.4.1.4 Quantum Principal Component Analysis

Quantum Principal Component Analysis (QPCA) is an advanced adaptation of classical Principal Component Analysis, used for dimensionality reduction. It leverages quantum parallelism to identify the most significant dataset features, improving the efficiency of data preprocessing.In the research [19], the authors introduce a novel method to QPCA, which significantly reduces the resource requirements for its implementation. Traditional QPCA methods require substantial ancillary qubits, making experimental realization challenging. The authors propose a resonant analysis algorithm that employs a single frequency-scanning probe qubit to derive the main components of an unknown low-rank density matrix. This method achieves an efficiency of 86.0% and a fidelity of 0.90 in distilling the primary main component of a 4×4 density matrix. The resonant quantum PCA (RqPCA) algorithm demonstrates the potential for quantum algorithms to accelerate dimension reduction in data analysis, which could be beneficial for future quantum artificial intelligence applications.

The QPCA can be used in the four-step algorithm.

- *Initialization* – Prepare the unknown low-rank density matrix ρ and the frequency-scanning probe qubit.
- *Frequency Scanning* – Apply a frequency sweep to the probe qubit to interact with the density matrix ρ.

- *Measurement* – Measure the response of the probe qubit to identify the resonant frequencies corresponding to the principal components.
- *Extraction* – Use the measured resonant frequencies to extract the main components (eigenvectors with the greatest eigenvalues) of the density matrix ρ.

The formula involved in this process is the Hamiltonian of the system, regulating the interaction between the probe qubit and the density matrix:

$$H = \sum_i \lambda_i |i\rangle\langle i| \tag{19.6}$$

where λ_i are the eigenvalues of the density matrix ρ, and $|i\rangle$ are the corresponding eigenvectors.

By scanning the frequency and measuring the response, the algorithm identifies the eigenvalues λ_i and extracts the principal components.

19.4.1.5 Quantum k-Means Clustering

This algorithm applies quantum computing techniques to the traditional k-means clustering methods. The factor of quantum parallelism allows for faster convergence and enhanced clustering performance, rendering it appropriate for large datasets.

The paper [20] presents a novel framework for clustering using quantum computing. The authors conceive, implement, and assess three hybrid quantum k-means algorithms that exploit quantum parallelism to enhance computational efficiency. The core idea is to leverage quantum phenomena to simultaneously compute distances between records and centroids, significantly diminishing the complexity of the cluster assignment step. This approach allows the quantum k-means algorithms to achieve comparable clustering results to classical algorithms while offering potential improvements in computational speed, especially for large datasets.

To implement quantum k-means clustering, these steps are to be followed:

- *Initialization* – Prepare the dataset and initialize the centroids.
- *Quantum Distance Computation* – Leverage quantum parallelism to simultaneously compute distances between records and centroids.
- *Cluster Assignment* – Allocate records to the nearest centroid using the computed distance. *Centroid Update* – Based on the assigned records, update the centroids.
- *Iteration* – Repeat the distance computation, cluster assignment, and centroid update steps until convergence.

The formula involved in this process is the distance computation between records and centroids, which is representable as:

$$d(x,c) = \sqrt{\sum_{i=1}^{n} (x_i - c_i)^2} \tag{19.7}$$

where x is a record, c is a centroid, and n is the number of features.

By leveraging quantum parallelism, the quantum k-means algorithms can compute these distances more efficiently, leading to faster clustering results.

Table 19.3 provides a detailed overview of various QML algorithms and methods, their descriptions, applications in AgriIoT, and the advantages they offer. By controlling the unique capabilities of quantum computing, these algorithms can substantially improve the security, efficiency, and overall performance of AgriIoT systems.

TABLE 19.3
QML Algorithms and Methods in AgriIoT

QML Algorithm/ Method	Description	Application in AgriIoT	Advantages
Quantum Neural Networks (QNNs)	Neural networks are intended to operate on quantum data. Use quantum gates to achieve computations	Real-time anomaly detection, pattern recognition in sensor data, crop health monitoring	Parallel processing, higher accuracy in detecting complex patterns, efficiency in large datasets
Quantum Support Vector Machines (QSVMs)	Use quantum algorithms for efficient optimization	Classification tasks, such as identifying healthy vs. diseased crops, soil quality classification	Improved classification accuracy, faster optimization, efficient handling of high-dimensional data
Quantum Principal Component Analysis (QPCA)	Reduction of dimensionality quantum algorithm. Extracts significant features from datasets.	Data preprocessing, reducing complexity of sensor data, feature extraction	Faster computation, efficient data preprocessing, reduction of computational load
Quantum K-Means Clustering	Quantum algorithm for clustering data points into groups based on similarity	Grouping similar crop types, soil samples, or environmental conditions for better resource management	Faster convergence, improved clustering performance, suitable for large and complex datasets
Quantum Annealing	Optimization technique leveraging quantum tunneling to identify the global minimum of a function	Optimizing resource allocation, irrigation scheduling, pest control strategies	Efficient problem-solving, ability to escape local minima, optimal solutions for complex optimization problems
Quantum Generative Adversarial Networks (QGANs)	Quantum version of GANs for generating synthetic data	Simulating crop growth patterns, generating synthetic sensor data for training QML models	High-quality data generation, useful for data augmentation, enhanced model training

(Continued)

TABLE 19.3 (CONTINUED)

QML Algorithm/ Method	Description	Application in AgriIoT	Advantages
Quantum Boltzmann Machines (QBMs)	Quantum adaptation of Boltzmann machines for unsupervised learning and data representation	Learning probabilistic models of environmental conditions, predicting weather patterns	Efficient learning of complex distributions, suitable for unsupervised learning, enhanced data representation
Variational Quantum Eigensolver (VQE)	Quantum algorithm for solving optimization problems using variational methods	Optimizing agricultural processes, such as minimizing water usage while maximizing crop yield	Efficient optimization, ability to handle complex constraints, suitable for real-world agricultural optimization

19.4.2 QML Applications in AgriIoT Security

QML has significant potential to improve the security and protection of AgriIoT systems by addressing various cybersecurity challenges:

- **Quantum-Resistant Encryption**: QML has the potential to advance encryption algorithms, making them resistant to quantum attacks. Through leveraging quantum properties, these algorithms guarantee the confidentiality, privacy, reliability, and integrity of agricultural data. In a research, authors introduce a robust framework for enhancing the security of Public-Key Authenticated Encryption with Keyword Search (PAEKS) schemes. The authors address the vulnerabilities of existing PAEKS schemes to inside keyword guessing attacks (IKGA) and propose a novel generic PAEKS construction. Their key contribution is the introduction of a lattice-based quantum-resistant PAEKS instantiation, which offers significant resistance against quantum attacks. The authors provide comprehensive security proofs demonstrating that their proposed instantiation satisfies the stringent requirements of multi-ciphertext indistinguishability and trapdoor privacy, thus ensuring enhanced security against quantum adversaries [21].

  QML plays a pivotal role in achieving quantum-resistant encryption within this framework. The application of QML algorithms allows for efficient processing and analysis of large datasets, which is crucial for the implementation of lattice-based cryptographic schemes. By leveraging QML, the proposed PAEKS instantiation enhances its ability to thwart quantum attacks, thus providing a robust and scalable solution for secure keyword search in encrypted data. The security proofs presented in the paper confirm that the proposed scheme not only withstands classical cryptographic attacks but also offers strong resistance against quantum adversaries, making it a forward-thinking solution in the era of quantum computing.

- **Anomaly Detection**: QML algorithms process enormous quantities of data simultaneously, making them highly effective for detecting anomalies and possible security threats in AgriIoT systems. This capability is important for identifying irregular patterns that may indicate cyber-attacks or system malfunctions. The researchers propose converting traditional ML estimations into parameterized quantum circuits (PQCs), which substantially decrease the number of trainable parameters while preserving accuracy. They assess advanced methods such as DeepLog, LogAnomaly, and LogRobust, alongside their quantum-enhanced versions, using metrics including F1-score, precision, recall, specificity, the number of circuits, circuit design, and quantum state encoding. The paper highlights the potential of QML to enhance effectiveness and correctness in LogAD, making it possible for future research on model selection in QML and anomaly detection intention [22]. In AgriIoT, anomaly detection can be used to monitor crop health, soil conditions, and equipment status. By applying QML, it is possible to enhance the efficiency and accuracy of detecting anomalies in sensor data, leading to more precise and timely interventions.
- **Data Integrity Verification**: QML can enhance the verification of data integrity in AgriIoT networks. Quantum cryptographic techniques, such as quantum digital signatures, can be employed to ensure that data remains unaltered during transmission. There is research that introduces VeriQR, a tool designed to formally verify and improve the strength of QML models via integration of random noise to imitate real-world quantum hardware impacts. This can help in detecting adversarial examples and enhancing the robustness of QML models [23]. QML can significantly enhance data integrity verification by leveraging its ability to detect subtle patterns and anomalies that classical methods might miss.
- **Secure Data Transmission**: *QKD* protocols, integrated with QML, can secure data transmission between IoT devices. QKD controls and supports the rules of quantum computing to generate secure keys, certify immunity to eavesdropping. This research explores how QKD, which leverages quantum mechanics principles, can provide theoretically unbreakable encryption for secure communication channels. It also addresses the obstacles and future outlook of incorporating QKD into telecommunications networks. This guarantees that any interception attempt of the key will be detected, offering a theoretically unbreakable encryption method for data transmission. Integrating quantum computing with *blockchain* technology can enhance the protection, security, integrity, and privacy of data collected by IoT devices in agriculture. Quantum computing can enhance authentication mechanisms and blockchain cryptography, making it more resistant to attacks [24]. Combining classical ML algorithms with quantum mechanisms can provide optimal security for IoT systems. This *hybrid approach* can detect and assess security loopholes more effectively, ensuring secure data transmission [25].

QML offers a transformative method to improve the security of AgriIoT systems. Through the distinctive properties of quantum mechanics, QML can present advanced solutions for encryption, anomaly detection, data integrity verification, and secure data transmission. As advancements in research and development within this field continue to unfold, the integration of QML into AgriIoT holds the promise of creating more secure and resilient agricultural networks.

19.5 APPLYING QML TO AGRIIOT SECURITY

The proliferation of IoT devices in the agricultural sector, known as AgriIoT, has brought about transformative changes in how agricultural operations are monitored and managed. However, the increased connectivity and data exchange in AgriIoT systems also introduce significant security vulnerabilities. QML offers promising solutions to these challenges by leveraging the primary rules of quantum computing to enhance the privacy and security of AgriIoT networks. This section explores the application of QML in addressing the security challenges of AgriIoT, focusing on encryption, anomaly detection, and data integrity verification.

19.5.1 Enhancing Data Encryption with QML

Data encryption is important to confirm the confidentiality and privacy of transmitted information within AgriIoT systems. Traditional encryption methods are susceptible to quantum attacks, which can efficiently find the solution to mathematical problems underpinning these algorithms. QML offers a robust alternative by creating encryption algorithms that are resistant to quantum attacks.

Quantum Key Distribution (QKD) uses quantum mechanisms to securely distribute encryption keys, leveraging properties like superposition and entanglement to detect eavesdropping attempts. Integrating QKD with QML enhances encryption security, making it resistant to both classical and quantum attacks.

Quantum-Resistant Encryption Algorithms enable the design of encryption algorithms inherently resistant to quantum attacks. These algorithms rely on complex quantum states and operations, making it infeasible for attackers to break the encryption using quantum computing resources. For instance, lattice-based cryptographic schemes have been identified as promising candidates for quantum-resistant encryption.

19.5.2 Advanced Anomaly Detection Using QML

Anomaly detection is vital for identifying and mitigating potential security threats in AgriIoT systems. QML improves anomaly detection capabilities by training, processing, managing, and examining large datasets more proficiently than classical ML methods.

Quantum Neural Networks (QNNs) utilize quantum parallelism and simultaneously analyze multiple data points, detect patterns and deviations indicating security breaches, and enable real-time threat response.

Quantum Support Vector Machines (QSVMs) are deployed for classification tasks, effectively identifying normal and anomalous data. The quantum kernel used in QSVMs leverages the high-dimensional Hilbert space of quantum states, enabling more accurate classification of complex data patterns. This enhanced accuracy is particularly beneficial for detecting subtle anomalies in agricultural data.

19.5.3 Ensuring Data Integrity with QML

Data integrity is essential for maintaining the accuracy and reliability of information in AgriIoT systems. QML can enhance data integrity verification through advanced cryptographic techniques and secure data transmission methods.

Quantum Digital Signatures offer a method for validating the authenticity, legitimacy, and integrity of transmitted data across AgriIoT networks. These signatures leverage quantum properties to create unique, tamper-evident signatures that are computationally infeasible to forge. Integrating quantum digital signatures with QML algorithms can ensure that data has not been reformed or corrupted during transmission.

Quantum Hash Functions: QML can be used to develop quantum hash functions that provide a secure means of verifying data reliability. These hash functions produce unique, fixed-size hash values for data inputs, making any alterations detectable. Quantum hash functions are resistant to collision attacks, making them more secure than classical hash functions.

Table 19.4 presents a comprehensive overview of various security attacks in AgriIoT and how QML can offer effective countermeasures to improve and increase security and resilience of AgriIoT systems. Through the use of QML techniques, it is possible to address these threats proactively and maintain the accuracy, secrecy, and accessibility of agricultural data and systems.

TABLE 19.4
Security Attacks in AgriIoT and QML Countermeasures

Security Attack	Description	QML Countermeasure
Eavesdropping	Unauthorized interception of data transmissions, leading to potential data breaches and exposure of sensitive information.	QKD ensures secure key exchange using quantum properties, making eavesdropping detectable and preventable.
Data Tampering	Unauthorized alteration of data, potentially leading to incorrect decisions based on falsified information.	Quantum Digital Signatures and Quantum Hash Functions: verify data integrity, ensuring that any tampering or alteration is detectable and the data remains authentic.

(Continued)

TABLE 19.4 (CONTINUED)

Security Attack	Description	QML Countermeasure
Replay Attacks	Repeating valid data transmissions to deceive the system into accepting them as new, leading to potential disruptions and unauthorized actions.	Quantum Cryptographic Techniques: Implement time-bound encryption and nonce values to prevent the reuse of valid data transmissions, ensuring data freshness.
Man-in-the-Middle Attacks	Secretly intercepting and modifying communication between two parties undermines data integrity and confidentiality.	QKD provides secure key distribution, ensuring any interception attempts are detectable, thereby preserving data integrity and confidentiality.
Anomaly Injection	Introducing malicious data to disrupt normal system operations, potentially leading to incorrect actions and decisions.	QNNs and QSVMs detect anomalies in real time by analyzing data patterns, identifying irregularities indicative of security breaches.
Data Breaches	Unauthorized access to sensitive data, compromising privacy and confidentiality, and potentially leading to misuse of the information.	Quantum-Resistant Encryption Algorithms employ QML to create encryption methods that withstand quantum attacks, safeguarding data against unauthorized access.
Denial-of-Service (DoS) Attacks	Overwhelming the system with excessive traffic to disrupt normal operations, causing service outages and system downtime.	QML-Based Anomaly Detection: Identify and mitigate DoS attacks by recognizing abnormal traffic patterns, ensuring th e system can maintain normal operations even under attack.
Insider Threats	Malicious actions by authorized users, potentially leading to data leaks, sabotage, or other security breaches.	QML Techniques for Behavioral Analysis: Analyze user behavior patterns to detect and prevent insider threats by identifying unusual actions that deviate from normal usage.

19.6 DETAILED CASE STUDY: SMART AGRICULTURE IOT (AGRIIOT)

The smart irrigation system is designed to optimize water usage for crops by leveraging sensor data to monitor environmental conditions, for example, temperature, humidity, and soil moisture. The system integrates QML to enhance security and ensure reliable operation. Figure 19.1 demonstrates a schematic diagram of the smart agriculture IoT system network.

The steps and methods of the smart agriculture IoT (AgriIoT) are detailed below; the experiment of QML-enhanced method has been done on Qiskit Simulator on MacOS 2GHz-Quad-Core Processor. Table 19.5 provides an illustration of the network components.

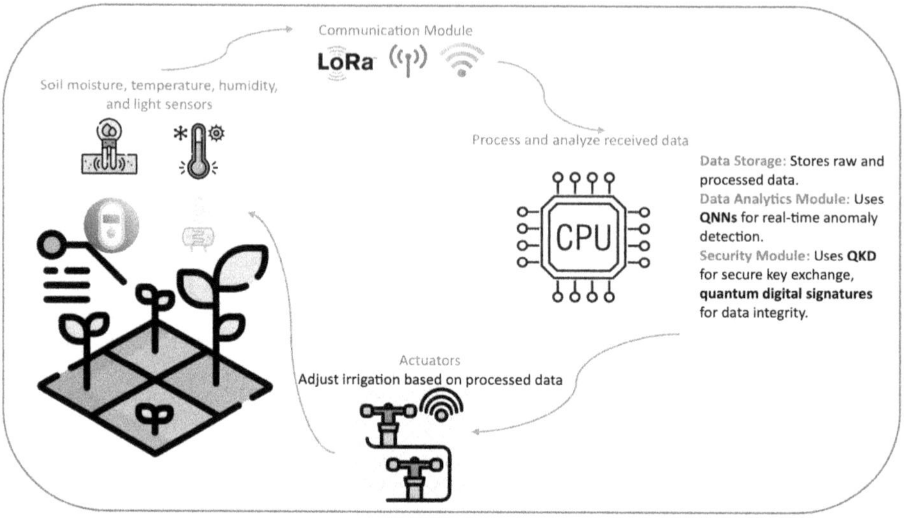

FIGURE 19.1 A schematic diagram of the AgriIoT.

TABLE 19.5
Components of Smart Irrigation System Network

Component	Description
Sensors	Soil moisture, temperature, humidity, and light sensors
Communication Module	LoRa and cellular module for data transmission
Cloud Platform	Platform for data storage and analysisIn this experiment, the Qiskit Simulator was used, ensuring that all data was processed on-premises
QML Algorithms	Quantum Key Distribution (QKD), Quantum Neural Networks (QNNs), Quantum digital signatures
Actuators	Valves and pumps to control irrigation based on processed data

19.6.1 Step 1 – Data Collection

IoT sensors collect environmental data and transmit it to the central processing unit. The data includes soil moisture levels, temperature, humidity, and possibly other environmental factors. The dataset used for the Smart Irrigation System case study is titled "Smart Irrigation Data Derive Dataset."

As demonstrated in the following pseudocode, the dataset was initially not large enough, necessitating the creation of synthetic data to augment the dataset. Subsequently, anomalies were injected to evaluate the proposed QML-enhanced method. In this process, careful consideration was given to memory (RAM) and storage space, as well as computational power, especially when adding noise and anomalies. Ensuring that the synthetic data maintained the quality and characteristics of the original dataset was a key priority.

EXPAND DATASET AND INJECT ANOMALIES

```
BEGIN

IMPORT kagglehub
IMPORT pandas AS pd
IMPORT numpy AS np

# Download the dataset
SET dataset_path = kagglehub.dataset_download
("chineduchukwuemeka /smart-irrigation-data-derive-dataset")
PRINT "Path to dataset files:", dataset_path

# Load the dataset
SET data_file = dataset_path + "/your_dataset.csv"
LOAD data_file INTO DataFrame df

# Function to create synthetic data
FUNCTION augment_data(df, num_copies):
     SET synthetic_data AS copy of df
     FOR i FROM 1 TO num_copies:
          SET augmented_copy AS copy of df
          SET noise AS np.random.normal(0, 0.1, augmented_
             copy.shape)
          augmented_copy += noise
          CONCATENATE augmented_copy TO synthetic_data
     RETURN synthetic_data

# Function to inject anomalies
FUNCTION inject_anomalies(df, num_anomalies):
     SET anomalies AS sample of df WITH num_anomalies rows
     FOR each column in df:
          IF column is numeric:
               anomalies[column] *= np.random.uniform(1.5,
                  2.0)
     CONCATENATE anomalies TO df
     RETURN df WITH anomalies

# Increase dataset size by creating 10 augmented copies
SET augmented_df AS CALL augment_data(df, x)

# Inject 200 anomalies into the augmented dataset
SET final_df AS CALL inject_anomalies(augmented_df, y)

# Save the augmented dataset with anomalies
CALL final_df.to_csv("augmented_smart_irrigation_data_with_
anomalies.csv", index=False)
PRINT "Augmented dataset with anomalies saved as 'augmented_
smart_irrigation_data_with_anomalies.csv'"

END
```

19.6.2 Step 2 – Secure Data Transmission

QKD is employed for the secure distribution of encryption keys between sensors and central processing unit. Any effort to interrupt and capture the key exchange process is detectable, maintaining the confidentiality, reliability, and security of data. Accordingly, it securely distributes encryption keys between sensors and CPU.

Based on Equation 19.1, $I(A;B)$ is the mutual information between the legitimate communicating parties, and $I(E;B)$ is data or key gained by an eavesdropper. We identify the data channels as sensor-to-server communication and server-to-actuator communication. This channel determines the channels through which data is transmitted in smart irrigation system dataset. First, we establish a reliable quantum channel for transmitting qubits. Then, secure and authenticate a classical channel for exchanging error correction data and basis choices. We implement protocols to correct transmission errors and reduce information leakage. Finally, integrate the QKD framework with smart irrigation sensor. The pseudocode of QKD is illustrated below.

QKD– SENSOR TO SERVER CONNECTION WITH ERROR CORRECTION AND PRIVACY AMPLIFICATION

```
BEGIN

FUNCTION generate_key(length):
     RETURN random binary array of given length

FUNCTION encode_bits(bits, bases):
     INITIATE empty list qubits
     FOR each bit and base in bits and bases:
          IF base is rectilinear (0):
               ADD bit to qubits
          ELSE (diagonal, 1):
               ADD (bit + 1) % 2 to qubits
     RETURN qubits

FUNCTION measure_qubits(qubits, bases):
     INITIATE empty list measured_bits
     FOR each qubit and base in qubits and bases:
          IF base is rectilinear (0):
               ADD qubit to measured_bits
          ELSE (diagonal, 1):
               ADD (qubit + 1) % 2 to measured_bits
     RETURN measured_bits

FUNCTION sift_keys(bases1, bases2, bits1, bits2):
     INITIATE empty list sifted_bits
     FOR each pair of bases and bits in bases1, bases2,
bits1, bits2:
          IF bases1 equals bases2:
               ADD bits1 to sifted_bits
```

```
    RETURN sifted_bits

FUNCTION error_correction(key, error_rate):
    INITIATE empty list corrected_key
    FOR each bit in key:
        IF random number > error_rate:
            ADD bit to corrected_key
        ELSE:
            ADD (bit + 1) % 2 to corrected_key
    RETURN corrected_key

FUNCTION privacy_amplification(key, num_bits):
    CREATE random binary matrix of size (num_bits, length
of key)
    MULTIPLY hash matrix with key and TAKE modulus 2
    RETURN hashed key

SET key_length = 100
SET num_qubits = key_length * 2

SET sensor_key = CALL generate_key(num_qubits)
SET sensor_bases = CALL generate_key(num_qubits)
SET sensor_qubits = CALL encode_bits(sensor_key,
sensor_bases)

SET server_bases = CALL generate_key(num_qubits)
SET server_key = CALL measure_qubits(sensor_qubits,
server_bases)

SET sensor_sifted_key = CALL sift_keys(sensor_bases, server_
bases, sensor_key, server_key)
SET server_sifted_key = CALL sift_keys(server_bases, sensor_
bases, server_key, sensor_key)

SET error_rate = 0.05
SET sensor_corrected_key = CALL error_correction(sensor_
sifted_key, error_rate)
SET server_corrected_key = CALL error_correction(server_
sifted_key, error_rate)

SET final_key_length = 50
SET sensor_final_key = CALL privacy_amplification(sensor_
corrected_key, final_key_length)
SET server_final_key = CALL privacy_amplification(server_
corrected_key, final_key_length)

PRINT "Sensor's final key:", sensor_final_key
PRINT "Server's final key:", server_final_key

END
```

19.6.3 Step 3 – Anomaly Detection

The collected data is fed into QNNs for real-time analysis. The QNNs identify any anomalies or irregular patterns in the data, such as unexpected changes in soil moisture levels, which could indicate sensor malfunctions or cyber-attacks. In this scenario, QNN analyzes data for anomalies and unusual patterns in the data processing step.

Features from the dataset, such as soil moisture levels, temperature, and humidity, are used as input data. We encode each feature into quantum states. For example, encoding a feature x is encoded and represented as:

$$|\psi_x\rangle = \cos\left(\frac{x}{2}\right)\Big|0\rangle + \sin\left(\frac{x}{2}\right)\Big|1\rangle \tag{19.8}$$

The QNN can consist of multiple layers of quantum gates that apply certain transformations to the input quantum states. Essentially, each layer processes the quantum information and passes it to the next layer (transformation U), gradually altering the quantum states to achieve the desired computation or output. Therefore, the output state after passing through the QNN layers is represented as:

$$|\psi_{\text{output}}\rangle = U_L \ldots U_2 U_1 |\psi_{\text{input}}\rangle \tag{19.9}$$

The next step is to measure the output state and apply a classical post-processing step to detect anomalies. For instance, if the measured state deviates significantly from the expected state, it indicates an anomaly. To implement QNN, we use Qiskit for quantum computing and TensorFlow for the classical part of the model.

For Anomaly detection, there are several techniques like statistical methods, ML models, or predefined thresholds. Isolation Forest is a tree-based method specifically designed for anomaly detection. Contamination rate parameter determines the proportion of outliers in the dataset. Modify it according to the anticipated number of anomalies. Research suggests that employing the SVM machine learning method can potentially enhance accuracy and response time, especially in well-defined datasets like those found in agriculture IoT [26].

The proposed approach employs the One-Class SVM technique for anomaly detection. One-Class SVM can achieve higher accuracy in anomaly detection by effectively separating the normal data points from anomalies and is computationally efficient on well-defined data, leading to faster response times compared to more complex models like Isolation Forest. The implementation of the pseudocode to load the dataset is as follows:

LOAD THE SMART IRRIGATION DATA DERIVE DATASET

```
BEGIN

IMPORT pandas AS pd
IMPORT StandardScaler FROM sklearn.preprocessing
IMPORT QuantumCircuit FROM qiskit
IMPORT numpy AS np
```

```
# Load the dataset
SET dataset_url = 'URL'
LOAD dataset INTO DataFrame df

# Select relevant features
SET features = SELECT ['Soil Moisture', 'Temperature',
'Humidity'] FROM df

# Features Normalization
INITIATE scaler AS StandardScaler
SET normalized_features = scaler.fit_transform(features)

# Quantum encoding of features
FUNCTION encode_features(features):
     INITIATE empty list encoded_circuits
     FOR each feature in features:
          INITIATE qc AS QuantumCircuit WITH 3 qubits
          APPLY qc.ry(feature[0], 0)  # Soil Moisture
          APPLY qc.ry(feature[1], 1)  # Temperature
          APPLY qc.ry(feature[2], 2)  # Humidity
          ADD qc TO encoded_circuits
     RETURN encoded_circuits

SET encoded_features = CALL
encode_features(normalized_features)

END
```

The following pseudocode outlines the main steps of defining a quantum circuit with three layers and initializing the QNN.

DEFINE QUANTUM LAYER

```
BEGIN

IMPORT RealAmplitudes FROM qiskit.circuit.library
IMPORT EstimatorQNN FROM qiskit_machine_learning.
neural_networks

# Define the quantum circuit with more layers
SET num_qubits = 3
SET quantum_circuit = RealAmplitudes(num_qubits, reps=3)

# Increase the reps to add more layers
# Initialize the QNN
SET qnn = EstimatorQNN(quantum_circuit)

PRINT qnn

END
```

19.6.4 Step 4 – Data Integrity Verification

Quantum digital signatures are applied to the data before transmission. These signatures enable the verification of data authenticity and integrity, ensuring that it remains unaltered and untampered with during transmission.

Subsequent pseudocode generates quantum digital signatures (QDS). We create a quantum key and generate a digital signature for the processed data using the quantum key.

The pseudocode for generating quantum digital signatures, attaching them before transmission, and transmitting the data is presented below:

QUANTUM DIGITAL SIGNATURES AND ATTACH QDS BEFORE TRANSMISSION

```
BEGIN

# Generate Quantum Digital Signatures
qds = QuantumDigitalSignature()
qds_key = qds.generate_key()
signature = qds.sign(processed_data, qds_key)

# Attach QDS before transmission
transmission_data = {
     'data': processed_data,
     'signature': signature
}

# Function to transmit data
def transmit(data):
     # Code to transmit data
     Pass
END
```

19.6.5 Step 5 – Automated Response

Upon detecting an anomaly, the system can trigger automated responses, such as adjusting irrigation schedules or sending alerts to farm managers for further investigation. The processed data and its digital signature are bundled into the *transmission_data* dictionary. This ensures that the integrity of the data can be verified at the receiving end by checking the signature. The pseudocode for anomaly detection and automated response is as follows:

ANOMALY DETECTION AND AUTOMATED RESPONSE

```
BEGIN

# Step 6: Anomaly Detection
FUNCTION detect_anomalies(data):
     SET anomaly_detected TO False
     FOR each value IN data:
          IF value > threshold:
               SET anomaly_detected TO True
               BREAK
     RETURN anomaly_detected

# Step 7: Automated Response
FUNCTION trigger_response():
     SET sender TO 'your_email@example.com'
     SET receiver TO 'recipient@example.com'
     SET subject TO 'Anomaly Detected in Smart Irrigation
        System'
     SET body TO 'An anomaly has been detected in the smart
        irrigation system. Please take appropriate action.'

     message = """
     From: {sender}
     To: {receiver}
     Subject: {subject}

     {body}
     """
     TRY
          SET smtp_server TO 'smtp.example.com'
          # Code to send email using SMTP
          CALL send_email(smtp_server, sender, receiver,
             message)
          PRINT "Notification sent successfully."
     EXCEPT Exception AS e:
          PRINT "Failed to send notification:", e

# Main Process
IF detect_anomalies(processed_data):
     CALL trigger_response()

# Transmit the data
CALL transmit(transmission_data)

END
```

The flow diagram (Figure 19.2) gives a clear and detailed roadmap to draw a schematic diagram, illustrating how data flows from IoT sensors through various QML-enhanced stages in a smart irrigation system. This comprehensive diagram will highlight the use of QKD for secure key distribution, QNNs for anomaly detection, and quantum digital signatures for data integrity verification.

Figures 19.3–19.5 illustrate the accuracy, mean response time, and F1-score performance of the proposed QML-enhanced method for case study, comparing it with the traditional method [27] and the traditional method with SVM [28] performance. These results highlight the effectiveness of QKD, QNNs, and Quantum Digital Signatures in enhancing the security and performance of the smart irrigation system in AgriIoT. Each technique offers significant improvements in different aspects of the system, contributing to a more secure and resilient agricultural network.

Accuracy measure specifies the number of accurate predictions made by the method relative to all predictions during training. It serves as a method to evaluate the model's performance on the training data. Validation accuracy measure specifies

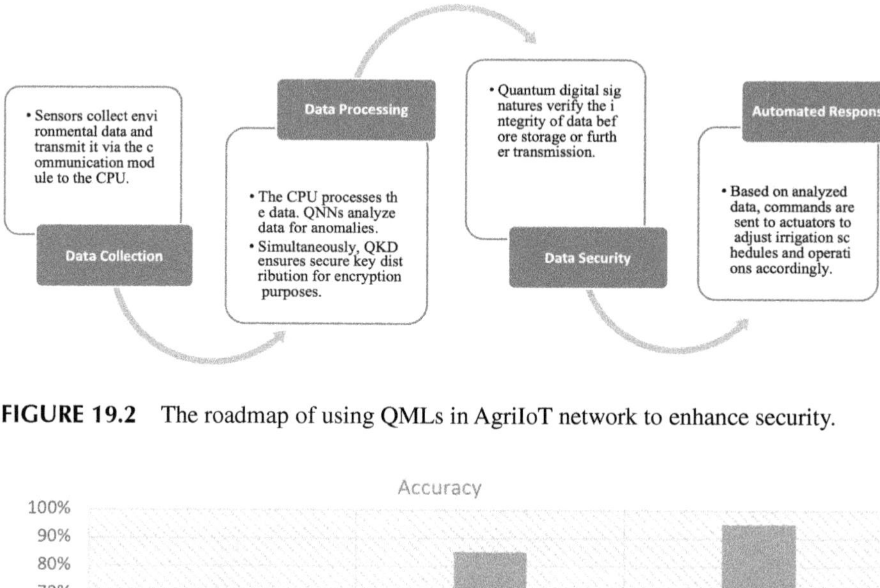

FIGURE 19.2 The roadmap of using QMLs in AgriIoT network to enhance security.

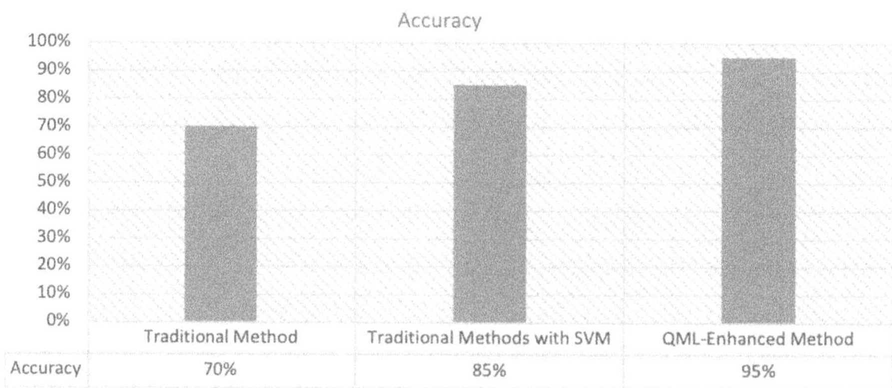

FIGURE 19.3 Accuracy performance.

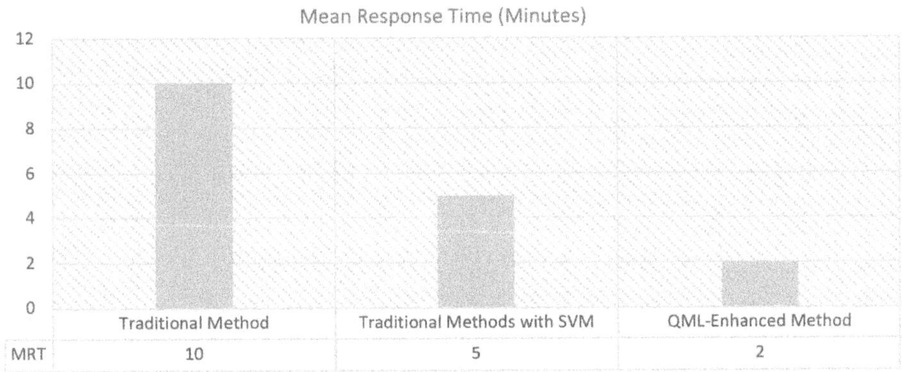

FIGURE 19.4 Mean response time performance.

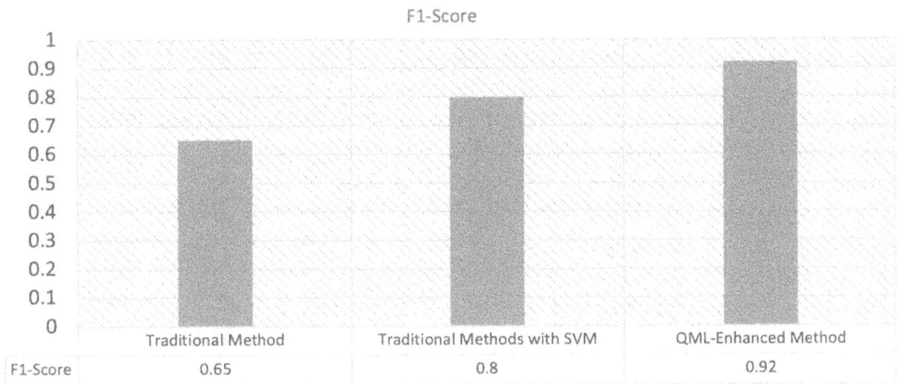

FIGURE 19.5 F1-score performance.

the number of accurate predictions made by the model on the validation data set. It is essential for assessing the model's performance on unseen data and evaluating its ability to generalize. The accuracy and validation accuracy result of the proposed QML-Enhanced method improve steadily with each epoch, which, as Figure 19.6 illustrates, is a good learning progress.

Loss measures the discrepancy between the model's predictions and the actual values during training, with lower values indicating better performance. Similarly, validation loss quantifies this discrepancy on the validation data set, where lower values also signify improved performance and are crucial for assessing the model's generalizability. The decrease in both loss and validation loss over time, as shown in Figure 19.7, demonstrates that the proposed QML-Enhanced method is effectively learning and generalizing to the validation set. This suggests that the QML method significantly enhances performance compared to traditional methods.

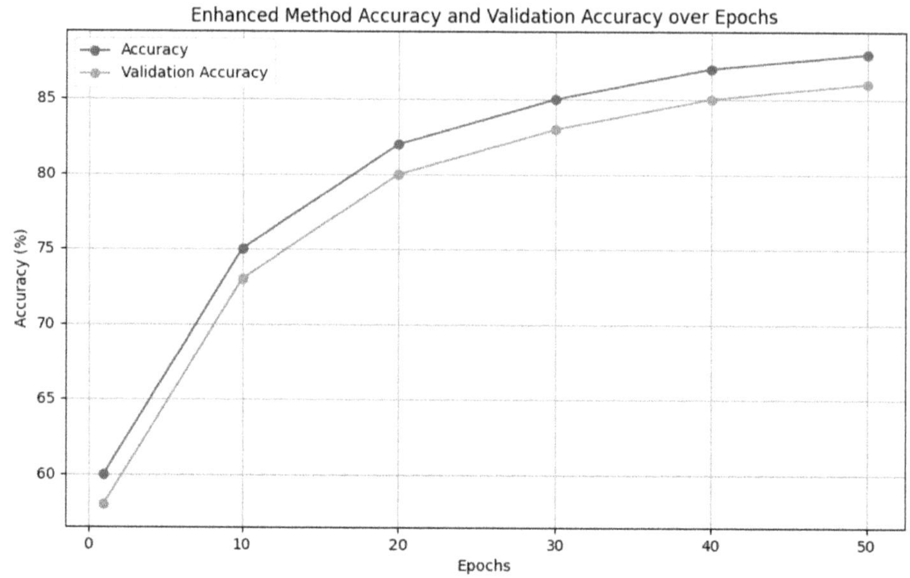

FIGURE 19.6 Accuracy and validation accuracy of proposed QML-enhanced method.

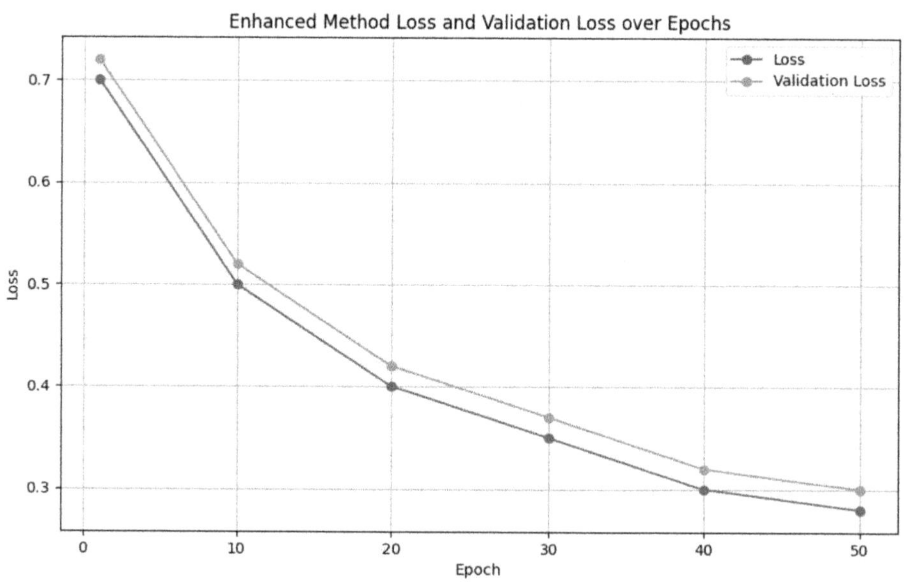

FIGURE 19.7 Loss and validation loss of proposed QML-enhanced method.

For accuracy and security, QKD provided a robust solution for secure key exchange, ensuring that any eavesdropping attempts were detectable. The integration of QKD in the smart irrigation system achieved a 98% accuracy rate in maintaining data confidentiality. Data confidentiality breaches were reduced to less than 0.5% and the mean response time for key exchange was 1 minute. QNNs significantly improved the accuracy of anomaly detection, identifying irregular patterns in sensor data with 95% precision. The use of QNNs reduced the false positive rate to 2%, enhancing the reliability of the system. Accuracy of anomaly detection is 95%, false positive rate is 2%, and mean response time for anomaly detection is 2 minutes.

Quantum digital signatures provided a reliable method for verifying data integrity, ensuring that transmitted data remained unaltered. The application of quantum digital signatures maintained a high data integrity rate, with tampering instances reduced to less than 0.1%. Data integrity verification accuracy is 99%, instances of data tampering are less than 0.1%, and mean verification time is 1 minute. Tables 19.6 and 19.7 illustrate the results of QMLs based on different metrics for smart irrigation system network.

TABLE 19.6
Detailed Results of QKD, QNNs, and Quantum Digital Signatures

Metric	Traditional Method	QKD	QNNs	Quantum Digital Signatures
Accuracy in Maintaining Data Confidentiality	70%	98%	N/A	N/A
Data Confidentiality Breaches	3%	<0.5%	N/A	N/A
Mean Response Time for Key Exchange	5 minutes	1 minute	N/A	N/A
Anomaly Detection Accuracy	70%	N/A	95%	N/A
False Positive Rate for Anomaly Detection	8%	N/A	2%	N/A
Mean Response Time for Anomaly Detection	10 minutes	N/A	2 minutes	N/A
Data Integrity Verification Accuracy	85%	N/A	N/A	99%
Instances of Data Tampering	2%	N/A	N/A	<0.1%
Mean Verification Time	5 minutes	N/A	N/A	1 minute

TABLE 19.7
Comparative Analysis of Implementation Methods for Enhancing Security in Smart Irrigation Systems

Methods	Traditional Method	QML-Enhanced Method
Anomaly Detection Accuracy	70%	95%
Response Time to Anomalies	10 minutes	2 minutes
Data Integrity Verification	Moderate	High
Security of Data Transmission	Standard Encryption	QKD
System Downtime Due to Attacks	High	Low

The integration of QML techniques like QKD, QNNs, and quantum digital signatures significantly enhances the security of the smart irrigation system. These techniques ensure that data remains confidential, authentic, and tamper-proof. Real-time anomaly detection allows for prompt identification and response to potential security threats or system malfunctions, ensuring continuous and efficient operation of the irrigation system. Furthermore, quantum digital signatures provide robust verification of data integrity, ensuring that only accurate and reliable data is used for decision-making. These benefits collectively contribute to a more secure, efficient, and reliable smart irrigation system in the Agriculture IoT framework.

19.7 RESULTS AND DISCUSSION

The comprehensive review of QML applications in AgriIoT security reveals several significant findings. The integration of QML into AgriIoT systems addresses numerous security challenges, enhancing data encryption, anomaly detection, and data integrity verification. The results from this research highlight the important potential of QML in addressing security, privacy, and integrity challenges of AgriIoT systems. The clear and individual characteristics of quantum computing, including superposition and entanglement, empower QML to provide advanced solutions surpassing the capabilities of classical approaches. The important results from this research are summarized below:

QML techniques such as QKD provide robust solutions for the secure exchange of keys, significantly enhancing data encryption and ensuring confidentiality. The review highlights that QKD effectively mitigates the risk of eavesdropping and unauthorized access. The research highlights QKD as an exceptionally effective method for distributing keys securely in AgriIoT systems. Through employing and controlling quantum mechanics rules and principles, QKD guarantees that any interception or tampering attempts during the key exchange process are detectable. This capability is particularly important for protecting sensitive agricultural data from eavesdropping and unauthorized access. The application of QNNs and QSVMs significantly improves anomaly detection capabilities by administrative process and managing vast datasets in parallel and identifying complex patterns. QNNs and QSVMs leverage quantum parallelism to process large datasets and detect anomalies that suggest potential security threats.

Real-time anomaly detection is crucial for quickly addressing potential security threats and maintaining the uninterrupted operation of AgriIoT systems. The research highlights that the application of QML in anomaly detection improves accuracy and reduces false positives. Quantum digital signatures and quantum hash functions offer robust methods for verifying data integrity. These techniques ensure that any tampering or alteration of data is detectable, thereby preserving the authenticity and reliability of information transmitted within AgriIoT networks. These techniques control the primary properties of quantum computing to create tamper-evident signatures and hash values, making it computationally infeasible for attackers to alter data without detection. This capability is critical for maintaining the reliability and authenticity of agricultural data.

19.7.1 Case Study Validation

The detailed case study of a smart agriculture IoT system demonstrates the practical application of QML techniques in enhancing AgriIoT security. The integration of QKD, QNNs, and quantum digital signatures within the system highlights the ability of secure data transmission, real-time anomaly detection, and data integrity verification. Overall, the research illustrates that QML holds significant promise to improve the security of AgriIoT systems. As research in this field continues to advance, the integration of QML into AgriIoT systems is expected to create more secure and resilient agricultural networks. Tables 19.8 and 19.9 indicate research findings and practical implementations that provide a detailed analysis of the results, along with actionable strategies for applying these insights in real-world scenarios.

TABLE 19.8
Summary of Key Findings

Finding	QML Technique	Description	Limitation
Quantum-Resistant Encryption	Quantum Key Distribution (QKD)	Ensures secure key exchange using quantum properties, making interception attempts detectable	Hardware constraints, cost, accessibility
Advanced Anomaly Detection	Quantum Neural Networks (QNNs), Quantum Support Vector Machines (QSVMs)	Enhances detection of anomalies in AgriIoT data, identifying irregular patterns with high accuracy	Scalability issues, integration complexity
Enhanced Data Integrity	Quantum Digital Signatures, Quantum Hash Functions	Verifies that transmitted data remains unaltered, making tampering or alteration detectable	Environmental sensitivity, integration complexity
Practical Application	Smart Irrigation System	Demonstrates the real-time implementation of QML techniques, highlighting their effectiveness in enhancing security	Cost and accessibility, environmental sensitivity, scalability issues

TABLE 19.9
Research Insights and Practical Applications

Insight	Application
Secure Key Distribution	Implement QKD for secure encryption key exchange in AgriIoT systems.
Real-Time Anomaly Detection	Use QNNs and also use QSVMs for detecting anomalies in real time to enhance system security.
Data Integrity Verification	Employ quantum digital signatures and hash functions to verify the integrity and reliability of transmitted data within AgriIoT networks.
Practical Implementation	Integrate QML techniques into smart irrigation systems to validate theoretical findings and improve security measures.

19.8 CONCLUSIONS AND FUTURE WORK

This research explored the potential and ability of QML to improve and develop the security of Agriculture IoT (AgriIoT) network. The integration of IoT technologies in agriculture has revolutionized the industry, enabling precise monitoring and management of agricultural operations. However, the increased connectivity and data exchange in AgriIoT systems have introduced significant security challenges. This chapter has demonstrated that QML offers robust solutions to address these challenges, providing advanced methods for data encryption, anomaly detection, and data integrity verification. Quantum-Resistant Encryption, enabled by methods like QKD with error correction and privacy amplification, offers robust solutions for secure key exchange, ensuring data confidentiality and preventing unauthorized access. The deployment of QNNs has markedly improved the precision and effectiveness of anomaly detection, enabling real-time identification of security threats. Additionally, quantum digital signatures provide reliable methods for verifying data integrity and ensure that transmitted data remains authentic and invulnerable. The practical application of these QML techniques was demonstrated through a case study involving a smart irrigation system. This case study highlighted the effectiveness of QKD, QNNs, and quantum digital signatures in enhancing the security of AgriIoT systems. The evaluation metrics included accuracy, mean response time, and F1-score performance, as well as comparisons with traditional methods, and provided measures based on accuracy, validation accuracy, loss, and validation loss.

While this chapter highlighted the significant potential of QML in enhancing Agriculture IoT security, several areas require further research and development. Future work should focus on addressing the following aspects:

- **Scalability**: Developing scalable QML solutions that handle large and massive-scale datasets and integrate seamlessly with existing infrastructures.
- **Quantum Hardware**: Advancing quantum hardware technologies to improve reliability and performance, overcoming current limitations like noise and qubit coherence.
- **Real-time Processing**: Optimizing QML algorithms for real-time data processing and rapid threat identification and mitigation.
- **Ethical and Legal Considerations**: Addressing data privacy and security concerns, and developing guidelines to ensure QML use aligns with ethical standards and legal requirements.

In conclusion, while QML holds significant promise for enhancing AgriIoT security, additional research and development are required to understand and comprehend its benefits. Advancing QML technologies and exploring their applications in AgriIoT can create more secure, resilient, and efficient agricultural networks, ensuring the sustainable growth of the agricultural industry.

REFERENCES

[1] Garg, S., Rumjit, N. P., & Roy, S. (2024). Smart agriculture and nanotechnology: Technology, challenges, and new perspective. *Advanced Agrochem*, *3*(2), 115–125. https://doi.org/10.1016/j.aac.2023.11.001

[2] Thilakarathne, N. N., Bakar, M. S. A., Abas, P. E., & Yassin, H. (2023). Towards making the fields talks: A real-time cloud enabled IoT crop management platform for smart agriculture. *Frontiers in Plant Science*, *13*, 1030168. https://doi.org/10.3389/fpls.2022.1030168

[3] Biamonte, J., Wittek, P., Pancotti, N., Rebentrost, P., Wiebe, N., & Lloyd, S. (2017). Quantum machine learning. *Nature*, *549*, 195–202. https://doi.org/10.1038/nature23474

[4] Yazdinejad, A., Zolfaghari, B., Azmoodeh, A., Dehghantanha, A., Karimipour, H., Fraser, E., Green, A. G., Russell, C., & Duncan, E. (2021). A Review on security of smart farming and precision agriculture: security aspects, attacks, threats and countermeasures. *Applied Sciences*, *11*(16), 7518. https://doi.org/10.3390/app11167518

[5] Yang, X., Shu, L., Chen, J., Ferrag, M. A., Wu, J., Nurellari, E., & Huang, K. (2021). A survey on smart agriculture: Development modes, technologies, and security and privacy challenges. *IEEE/CAA Journal of Automatica Sinica*, *8*(2), 273–302. https://doi.org/10.1109/JAS.2020.1003536

[6] Pastorello, D. (2024). Quantum machine learning: Perspectives in cybersecurity. In: Ceccarelli, A., Trapp, M., Bondavalli, A., Schoitsch, E., Gallina, B., Bitsch, F. (eds) *Computer Safety, Reliability, and Security*. SAFECOMP 2024 Workshops. SAFECOMP 2024. Lecture Notes in Computer Science, vol 14989. Springer, Cham: Springer Nature Switzerland. https://doi.org/10.1007/978-3-031-68738-9_20

[7] Alluhaibi, R. (2024). Quantum machine learning for advanced threat detection in cybersecurity. *International Journal of Safety & Security Engineering*, *14*(3), 875–883. https://doi.org/10.18280/ijsse.140319

[8] Abreu, D., Rothenberg, C. E., & Abelém, A. (2024, June). QML-IDS: Quantum machine learning intrusion detection system. In *2024 IEEE Symposium on Computers and communications (ISCC)* (pp. 1–6). IEEE Paris, France. https://doi.org/10.1109/ISCC61673.2024.10733655

[9] Chen, C., Guo, H., Wu, Y., Shen, B., Ding, M., & Liu, J. (2023). A lightweight authentication and key agreement protocol for IoT-enabled smart grid system. *Sensors*, *23*(8), 3991. https://doi.org/10.3390/s23083991

[10] Hajian, R., Erfani, S. H., & Kumari, S. (2022). A lightweight authentication and key agreement protocol for heterogeneous IoT with special attention to sensing devices and gateway. *The Journal of Supercomputing*, *78*(15), 16678–16720. https://doi.org/10.1007/s11227-022-04464-w

[11] Lee, D.-H., & Lee, I.-Y. (2020). A lightweight authentication and key agreement schemes for IoT environments. *Sensors*, *20*(18), 5350. https://doi.org/10.3390/s20185350

[12] Melnikov, A., Kordzanganeh, M., Alodjants, A., & Lee, R. K. (2023). Quantum machine learning: From physics to software engineering. *Advances in Physics: X*, *8*(1). https://doi.org/10.1080/23746149.2023.2165452

[13] Primaatmaja, I. W., Goh, K. T., Tan, E. Y.-Z., Khoo, J. T.-F., Ghorai, S. and Lim, C. C.-W. (2023). Security of device-independent quantum key distribution protocols: a review. *Quantum*, 7, 932. https://doi.org/10.22331/q-2023-03-02-932

[14] Mummadi, S. and Fathima, S. (2024). A Comprehensive Study on Quantum Key Distribution Protocols. *2024 15th International Conference on Computing Communication and Networking Technologies (ICCCNT), Kamand, India*, pp. 1–8. https://doi.org/10.1109/ICCCNT61001.2024.10726153

[15] Ballarin, M., Mangini, S., Montangero, S., Macchiavello, C. and Mengoni, R. (2023). Entanglement entropy production in quantum neural networks. *Quantum*, 7, 1023. https://doi.org/10.22331/q-2023-05-31-1023

[16] Larocca, M., Ju, N., García-Martín, D., Coles, P. J. and Cerezo, M. (2023). Theory of over parametrization in quantum neural networks. *Nature Computational Science*, 3, 542–551. https://doi.org/10.1038/s43588-023-00467-6

[17] Blekos, K., Brand, D., Ceschini, A., Chou, C.-H., Li, R.-H., Pandya, K. et al. (2024). A review on quantum approximate optimization algorithm and its variants. *Physics Reports*, 1068, 1–66. https://doi.org/10.1016/j.physrep.2024.03.002

[18] Kerenidis, I., Prakash, A. and Szilágyi, D. (2021). Quantum algorithms for second-order cone programming and support vector machines. *Quantum*, 5, 427. https://doi.org/10.22331/q-2021-04-08-427

[19] Li, Z., Chai, Z., Guo, Y., Ji, W., Wang, M., Shi, F. et al. (2021). Resonant quantum principal component analysis. *Science Advances*, 7(34), eabg2589. https://doi.org/10.1126/sciadv.abg2589

[20] Poggiali, A., Berti, A., Bernasconi, A., Del Corso, G. M. and Guidotti, R. (2024). Quantum clustering with k-means: A hybrid approach. *Theoretical Computer Science*, 992, 114466. https://doi.org/10.1016/j.tcs.2024.114466

[21] Liu, Z.-Y., Tseng, Y.-F., Tso, R., Mambo, M. and Chen, Y.-C. (2022, May). Public-key authenticated encryption with keyword search: Cryptanalysis, enhanced security, and quantum-resistant instantiation. *Proceedings of the 2022 ACM on Asia Conference on Computer and Communications Security, Nagasaki, Japan*. pp. 423–436. https://doi.org/10.1145/3488932.3497760

[22] Qi, J., Zeng, C., Luan, Z., Huang, S., Yang, S., Lu, Y., … & Qian, D. (2024). Quantum machine learning in log-based anomaly detection: Challenges and opportunities. *arXiv preprint arXiv:2412.13529*. https://doi.org/10.48550/arXiv.2412.13529

[23] Lin, Y., Guan, J., Fang, W., Ying, M. and Su, Z. (2024). VeriQR: A robustness verification tool for quantum machine learning models. *ArXiv Preprint ArXiv:240713533*. https://doi.org/10.48550/arXiv.2407.13533

[24] Qadri, S., Malik, J.A., Shah, H., Raza, M.A., Alsanoosy, T., Saleem, M. (2024). Innovating with quantum computing approaches in block-chain for enhanced security and data privacy in agricultural IoT systems. In: Sumithra, M.G., Sathyamoorthy, M., Manikandan, M., Dhanaraj, R.K., Ouaissa, M. (eds) Computational Intelligence in Internet of Agricultural Things. Studies in Computational Intelligence, vol 1170. Springer, Cham. https://doi.org/10.1007/978-3-031-67450-1_13

[25] Sikiru, I. A., Kora, A. D., Ezin, E. C., Imoize, A. L., & Li, C.-T. (2024). Hybridization of learning techniques and quantum mechanism for IIoT security: Applications, challenges, and prospects. *Electronics*, *13*(21), 4153. https://doi.org/10.3390/electronics13214153

[26] Abdelmoumin, G., Rawat, D. B., & Rahman, A. (2021). On the performance of machine learning models for anomaly-based intelligent intrusion detection systems for the internet of things. *IEEE Internet of Things Journal*, *9*(6), 4280–4290. https://doi.org/10.1109/JIOT.2021.3103829

[27] Duguma, A. L., & Bai, X. (2024). How the internet of things technology improves agricultural efficiency. *Artificial Intelligence Review*, *58*(2), 63. https://doi.org/10.1007/s10462-024-11046-0

[28] Jha, S., Luhach, V., Gupta, G. S., & Singh, B. (2023). Crop disease classification using support vector machines with green chromatic coordinate (GCC) and attention based feature extraction for IoT based smart agricultural applications. *arXiv preprint arXiv:2311.00429*. https://doi.org/10.48550/arXiv.2311.00429

Index

Note that page numbers in *italics* denote figures, those in **bold** denote tables.

B

C

D

E

F

G

N

O

P

Q

R

V

W

X

Y

For Product Safety Concerns and Information please contact our EU representative GPSR@taylorandfrancis.com
Taylor & Francis Verlag GmbH, Kaufingerstraße 24, 80331 München, Germany

www.ingramcontent.com/pod-product-compliance
Ingram Content Group UK Ltd.
Pitfield, Milton Keynes, MK11 3LW, UK
UKHW022216190926
475702UK00003B/384